AF540238

Sustainable Resource Development

Policy, Problem and Prescription

Udai Prakash Sinha

CONCEPT PUBLISHING COMPANY PVT. LTD.
NEW DELHI-110 059

ISBN-13: 978-81-8069-746-3

First Published 2011

Published and Printed by

Concept Publishing Company Pvt. Ltd.
Regd. Office:
A/15-16, Commercial Block, Mohan Garden
New Delhi-110059 (India)
Phones : 25351460, 25351794, *Fax* : 091-11-25357109
Email : publishing@conceptpub.com,
Website : www.conceptpub.com

Editorial Office:
H-13, Bali Nagar, New Delhi-110 015, India.

Cataloging in Publication Data--*Courtesy:* D.K. Agencies (P) Ltd. <docinfo@dkagencies.com>

Sinha, Udai Prakash, 1956-
Sustainable resource development : policy, problem and prescription / Udai Prakash Sinha.
p. cm.
Includes bibliographical references (p.) and index.
ISBN 9788180697463

1. Natural resources--India--Munger (District)--Management. 2. Sustainable development--India--Munger (District) 3. Sustainable development--Government policy -India--Munger (District) 4. Quality of life--India--Munger (District) I. Title.

DDC 333.70954 22

Prof. (Dr.) K. N. Dubey
M.Sc., Ph. D.
VICE-CHANCELLOR

T.M. BHAGALPUR UNIVERSITY
BHAGALPUR - 812007
Phone No. : + 91 - 641 -2620100 (O)
2620600 (R)
Fax No. : + 91 - 641 - 2620240
Mob. No. : 09430872759

Ref. No. :...............

Dated 05.09.2010

Foreword

Among various initiatives for promoting sustainable development, it has been realized that effective management of natural resources, assumes prime importance for improving the standards of living of the people and at the same time, the environmental protection and conservation, can also be ensured. It has been widely recognised that people should manage their own resources by means of their active participation, so that our developmental programmes can be need-based, pro-people and more dynamic in nature.

Economic performance in the 1980s was erratic and varied widely among countries and continents. In general, economic growth was much slower in the 1980s than in the 1970s. In the developing countries, external debt has escalated, prices for raw commodities have fallen, and adjustment policies have not functioned as per expectation. One-third of the population of the developing countries—some 1.2 billion people live under the poverty line of $370/person/year. For these poor, structural adjustment has done virtually nothing, or has even increased their plight even in the 1990s many Governments in the developing countries remained preoccupied with short-term economic and political crises and could assign only a very low priority to environmental and natural resource management and conservation .

In this thought provoking research anthology, Dr. Udai Prakash Sinha, Professor, Dept. of Economics, T. M. Bhagalpur University, Bhagalpur shows how the best use of resources is possible in an economic sense in the heartland of North Indian agricultural sector. The findings are interesting and it will help the policy-makers, students intending for higher degrees, regional planners, economists, geographers as well as the resource assessors of various disciplines. It has been tried to avoid the sophisticated statistical techniques but the author has used simple statistics and cartographic illustrations, so that it may help common readers as well. The author also exhibits considerable facility for bringing into focus issues of interdisciplinary nature and making the subject more stimulating for further research. This book is valuable as the resource inventory, highlights the problems and prospects of resource development in a backward region where the pressure of population is high and the cultivable land is scarce due to intermingling nature of economy in between plains and plateaus.

I hope this book will certainly be useful especially for economists, geographers, environmentalists and will find a respectable place in the growing academic community of the world.

K.N. Dubey
Vice-chancellor

Preface

In Indian Planning, rural development has taken long strides today since immemorial past, due to increased attention of scientists, administrators along with the enhanced local participation especially in the field of microlevel planning. The integrated development in the Munger division has been taken here as the problem of this study. The author also tries to search out the present distribution and future growth potential of resources with a view to highlight the effect of planning process which is going on with the assistance of Central and the State Government. This study throws light on the problems associated with the multiple growth of resources for social justice and economic prosperity of the society under differential environmental conditions. The method of study is synoptic-regional analysis and further it is supported by integrated growth of various resources available in the Munger division in varying proportions during different historical periods.

The Munger division is an area of chequered economic growth. Geographically, it is a region of varied landforms from the flood plains of the Ganga in the north up to the hilly, forested and rugged terrain of Chotanagpur Scrap Belt to the South. Economically, the regional imbalance is created by the availability of minerals, like mica and bauxite in the extreme south, growth of urban centres on the bank of the Ganga and agriculturally fertile plains in the north. In such a situation this study is highly needed in order to plan the resources of the region scientifically, and in a co-ordinated manner. Integrated resource development planning is the sole aim of this project which has been compiled after a collection of huge data from fieldwork, statistical analysis and scientific appraisal of the available resources for the welfare of the society.

The systematic planning of resources is not only imperative but needs immediate attention, as the pressure of population is increasing rapidly rather the development of other resources. This has led to numerous problems of malnutrition, deficiency of utility services, unemployment situation, unbalanced regional development etc. which should be reformed on planned basis. The present work is a modest attempt to highlight the problems and prospects of resource assessment and the need for their development in the Munger division. An attempt has also been made to suggest feasible and practical plan with the help of integrating different resources of the region so that a viable economic system may evolve in the area under study. There are twelve chapters in this work and the base of resource planning has been adopted through available resources of the region only which are crucial for regional economic development. So far as the author is aware, no study has yet been made in this direction with special reference to the Munger division, except laudable works of the National Institute of Rural Development, Hyderabad, which has attempted to some extent on Indian level study.

The extensive use of maps, figures and diagrams will help illustrate the problems that have exercised the minds of scholars and researchers in this wide-ranging subject.

Udai Prakash Sinha

Contents

List of the Tables

List of the Figures

1

Introduction

The Quest for Sustainable Development of Resources

The term 'Sustainable Development' was first brought into use in the year 1987 by the World Commission on Environment and Development (the Bruntland Commission). The Commission defined it to means "......... Development which meets the need of the current generation without jeopardizing the needs of future generations."

According to another definition of sustainable development : "Sustainable development is the management and conservation of the natural resource base and the orientation of technological and institutional changes in such a manner as to ensure the attainment and continued satisfaction of number needs for present and future generation. Such sustainable development in the agriculture, forestry and fishery sector, conserves land, water and animals genetic resources in environmentally degradable, technologically appropriable, economically viable and socially Burntland acceptable environs".

It is, however, rather difficult to precisely articulate the concept of 'sustainability' as it is not possible to argue that all natural resource be preserved as they are, from generation to generation. There now is an argument that all natural resource be preserved in some aggregate sense, with losses in one area being compensated elsewhere. In other words, sustainable development means the development that lasts (Fig. 1.1). (Bruntland Commission, F.A.O., U.N., Rome).

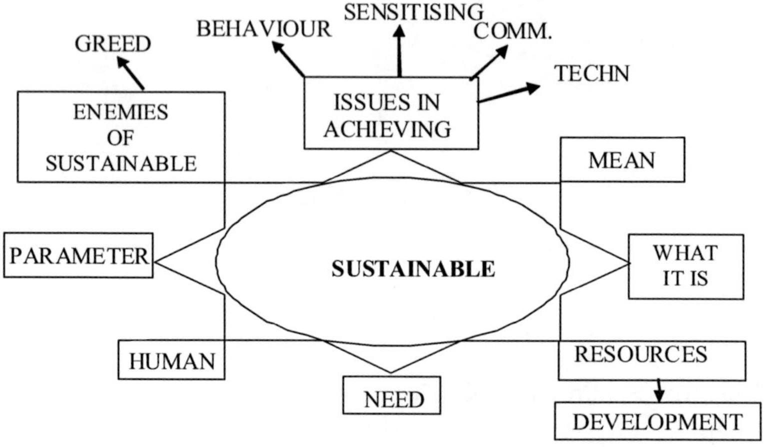

Fig. 1.1 : Sustainable Concept and Command.

The natural world provides raw materials and energy inputs without which production and consumption would be impossible. Some of these inputs have economic restrictions in their use because these are 'owned' and traded in the market. But then, there are other biotic or a-biotic resources that do not come within the ambit of the market economy and therefore do not have any economic restriction in their use. Again, the production and consumption processes create waste or residual that cause damage to the environment. This, however, is hardly captured in the economics of cost analysis because these remain external to the economic agents that create the waste or residual that causes the damage. As a result, there is a strong possibility that the process of production and consumption might overuse some 'non-priced' resources and create waste or residual to such an extent that the development process would become unsustainable. A large and diverse literature has emerged in recent years that concentrate on this issue. The present discourse attempts to present a brief review of the literature on this subject.

The Concept of Sustainability

The concept of sustainability has a long historical lineage. The Greek had the notion of *'Gala'*, the goddess of the Earth, the mother figure of natural replenishment. The idea was that the goddess of the Earth creates environmental balance (Donald Huges, 1983, p. 55). But this works up to a limit, after which 'Monis', the goddess, of justice punishes those who treat the earth ill.

The idea that there exist ecological bounds to the scale of economic activity had been there in the works of Malthus, Ricardo and J. S. Mill. The conventional economic wisdom that views increased production and consumption as end in itself had been rejected by Sismondi (1773-1842) and John Ruskin (1819-1900). A. Pigou in his *The Economics of Welfare* first analysed the environmental damage caused by pollution as 'social cost' (1920) and L. Gray developed the theory of optimal rate of depletion of an exhaustible resource in his article *Rent Under the Assumption of Exhaustibility* (1914) and this was formalised by H. Hotelling (1931). The theory of optimal use of a renewable resource was developed by H.S. Gordon (1954). [For an overview of this history, see Pearce, in R.K. Turner (ed.), 1993].

In the area of economic literature, the recent discussion on growth and sustainability has two distinct phases. The first phase (late 60s and early 70s) is characterised by the debate about environmental quality versus economic growth. The most famous publication in this phase is *Limits to Growth*. Rationale for preserving environment, as discussed in this phase was that 'even a stationary state level of economic activity could not be preserved unless the environment was conserved' [Pearce, in R.K. Turner (ed.), 1993, p. 70]. Limits to growth, as discussed in this phase, are set by the exhaustible and not by renewable resources.

In the 1980s, the emphasis shifted from exhaustible to renewable resources. It was admitted that the technical development may provide economy as regards the use of the exhaustible resource and thereby the pace of growth may be maintained without damaging the environmental quality. The problem, as it was realised, was from the other side i.e., from the side of the renewable resources. It becomes clear, from the development of scientific knowledge on ecosystem that ecosystem functions, in general, behave as renewable resources. Thus, the carbon cycle behaves as a balanced system, in which carbon is emitted from the living things and is absorbed by carbon 'sink', notably oceans and forests. It was also pointed out that 'renewability' of the renewable resources is not automatically assured. They may be overexploited and thereby get exhausted or create hazardous waste that turns the system unsustainable. The theoretical literature developed in this context has drawn attention to the conditions, which are likely to give rise to exhaustion of a renewable resource. These are (1) high discounting of the future, (2) a high ratio of resource price to cost of harvesting, and (3) the extent to which the management regime limits access to resources. The technological development can arrest the process by reducing the cost of harvesting in some cases, but then, the constraint may operate from the waste receiving capacity of the environment.

It has therefore been observed that a development process that threatens the ecosystem by overexploiting the renewable resources may become unsustainable.

Intergenerational Equity

Burntland Commission (World Commission, 1987 on Environment and Development) did put the problem of sustainable development more firmly on the international political agenda. The Commission defined sustainable development as 'development that meets the needs of the present without compromising the ability of the future generation to meet their own needs' (WCED Report, *Our Common Future*, p. 43). The emphasis is on intergenerational equity. The technological option that the society exercises today has a trade-off with the environmental quality that the future generation can obtain. This can be described in terms of

Production Possibility Curve (PPC) for the present and the future generations. In Fig.1.2, we describe the situation under the assumption that the PPC shifts towards left under the ongoing technology so that the future generation can get the same level of market goods only at the cost of degrading the environmental quality; or else, it can maintain the same level of environmental quality (expressed in terms of index of environmental quality derived from the data on different dimensions of ambient environment; for example, airborne SO_2 (Sulphur-di-Oxide) concentrations, water quality data, etc.) but only with a reduced level of marketed output (e_3); some level of environmental quality may even become unattainable (e_1).

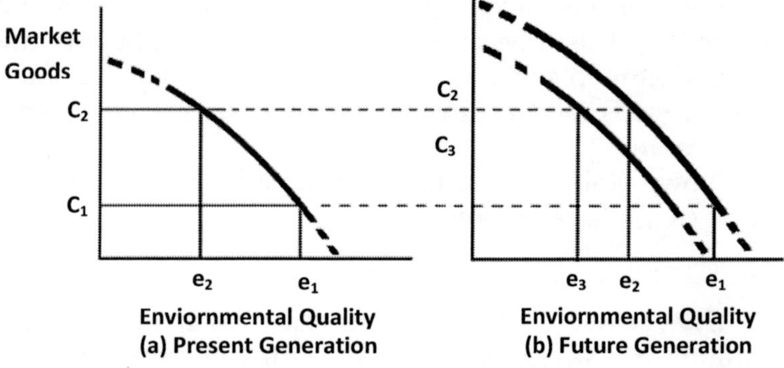

Fig. 1.2 : Production Possibility Curves for Present and Future Generations.

The essence of the discussion on sustainability is that the present generation should not exercise a technological option that may shift the future PPC towards left (by depleting certain important resources, by causing irreversible damage to the waste receiving capacity of the environment), thus putting constraint on the choice of the future generation. Placed, in a different way, sustainability connotes the idea that such courses of action should be avoided that reduce the long run productive capabilities of the natural and environmental resource baser of the society (so that the PPC does not shift towards left).

Necessary and Sufficient Conditions for Sustainability

Sustainable development can be conceived as economic development that endures over the long run. Economic development, in traditional terms, can be, defined as GDP per capita or real consumption per capita. Alternatively, the traditional measures can be modified or extended to include some welfare indicators—health, education, quality of life, etc. The sustainable development (SD) is defined as at least non-declining consumption per capita or GDP per capita or any other welfare indicator.

What are the necessary and sufficient conditions for attaining SD? It is the capital stock that becomes important in this context. This is because capital provides the opportunity and capability to generate well being through the criterion of material goods and amenity and other services that give human life its meaning.

Now, capital (K) could be manmade (K_m) or given by nature (K_n). We can therefore write,

$$K=K_m+K_n$$

Statement 1 : The development satisfies the sustainability condition if it saves enough to offset the depreciation of its capital stock.

Let WSI be the sustainability index and S be the level of savings and Y be the level of output. Then statement 1 implies WSI>0 if $S>\delta K$ where δ is the rate of depreciation of capital stock.

Dividing through by Y (income), we get

$$WSI>0 \text{ if } S/Y > \delta K/Y = [\delta K_m/Y + \delta K_n/Y]$$

where δK_m is the depreciation of material capital and δK_n is the depreciation of natural capital.

Statement 2 : Manmade capital may have perfect substitution to natural capital.

If so, WSI>0 if $S/Y > \delta K/Y$ is the sufficient condition for sustainability. This is the essence of neoclassical growth economics and the condition is known as Very Weak Sustainability or Solow Sustainability conditions.

Statement 3 : There are some critical natural K_n^* (e.g., Keystone species and Keystone Processes) which are non-sustainable. Again, there is a lower bound stock limit (Z) to K_n to ensure ecosystem stability.

Sustainability is then defined as

$$\frac{S}{Y} - \frac{\delta K_m}{Y} - \frac{\delta K_n}{Y} = WSI > 0$$

With $K_n > 0$ and $\delta K_n^* > 0$

Since, growth is limited by these constraints and population goes on increasing, we need another condition for Sustainability, namely,

$\lambda > h$, where λ is the rate of technological progress and h is the growth of population.

This is known as modified Solow Sustainability condition or Weak Sustainability condition.

Statement 4 : (i) Some ecosystem functions are not amenable to meaningful monetary valuation. (ii) At least some of K_n is non-substitutable. (iii) Aggregate ecosystem structure (life-support capacity) does not come to economic calculation.

Hence, we shall preserve K_n along with K_n^*.

Therefore,

$$WSI > 0 \text{ if } \frac{S}{Y} - \frac{\delta K}{Y} > 0$$

and $\delta K_n \leq 0$, $\delta K_n^* \leq 0$,

This cannot be ensured by pure economic (market) consideration. This is ensured by moral (ethical) and cultural values. Introducing a concept of moral and cultural capital (K_c) the necessary condition for attaining sustainability is,

$\delta K_c \leq 0$

This is known as strong sustainability (ecological economics approach) and this is associated with the concept of 'decoupling' of the economy from the environment. (Waste material and waste energy are generated in a production process; waste energy cannot be recycled, but waste material can, up to a certain point. If waste material is recycled, the environmental impact of growth is reduced. Decoupling argues that waste per unit should be reduced at a faster rate than the economic activity grows.)

Statement 5 : Development is sustainable only if there is zero increase in the 'scale' of macro-economy. This is ensured if there is zero economic growth and zero population growth: The paradigm is known as steady state (Daly, 1991; 1992).

With all the condition under statement 4, we now add the condition that there is perfect complementarity between K_n and K_m. This is known as very strong sustainability or stationary state sustainability. [For a detailed discussion, see Victor, P. A. (1991): Indication of Sustainable Development: Some Lessons from Capital Theory; Ecological Economics, No. 6.]

Decoupling and Political Economy of Sustainability

Sustainability is fundamentally a matter of renewable resources. While non-resources are used they automatically become unavailable to future generation. The market economy has a mechanism by which the situation is taken care of. The market registers the gradual unavailability as the 'scarcity' and discourages the use of the concerned resource by placing it in an excess demand situation that increases its price. While this may increase the cost of production, there is the possibility that this would encourage the innovation of a substitute that depends more on renewable resources or decreases the use of the scarce resource in such a way that the cost is reduced.

The new technology reduces cost (enhances efficiency) by economising the use of scarce natural resource, utilising the (cheaper) renewable resources, economising the use of energy and recycling the waste and residual material (non-energy). Economic growth is decoupled from environmental impact if the waste per unit of economic activity is reduced at a faster rate than the economic activity grows.

As economies change, becoming less tied to natural resources and less polluting technologies are adopted the trade-off between environmental quality and marketed output would increase. There would thus increase the long run sustainability of the economy. This is described in Fig. 1.3.

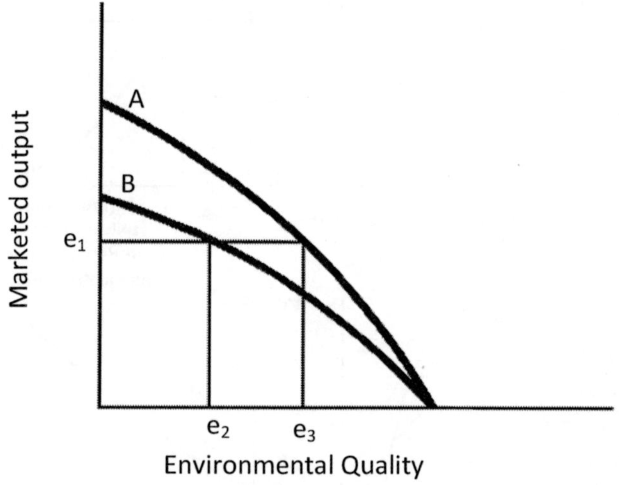

Fig. 1.3 : Production Possibility Curves under alternative situation.

Consider country A which is exposed to the same environmental endowment as B; but A has 'decoupled' growth from development in a better way so that B lies entirely within A. From Fig. 1.3 one understands that B can attain the same level of marketed output C_1 as A, with its own technology but only at a lower level of environmental quality (e_2). From the point of view of long run sustainability, A is better than B.

It is commonly believed that the developed economies posit a situation like A and B represents the status of the underdeveloped countries with respect to long run sustainability. However, the empirical evidences do not permit one to readily make such a conclusion.

Studies have been done to investigate the relationship between various environmental quality indices and the income levels attained in different countries. Several of the leading results are shown in Fig. 1.4. [Ref. Barry C. Field (1997), p. 408].

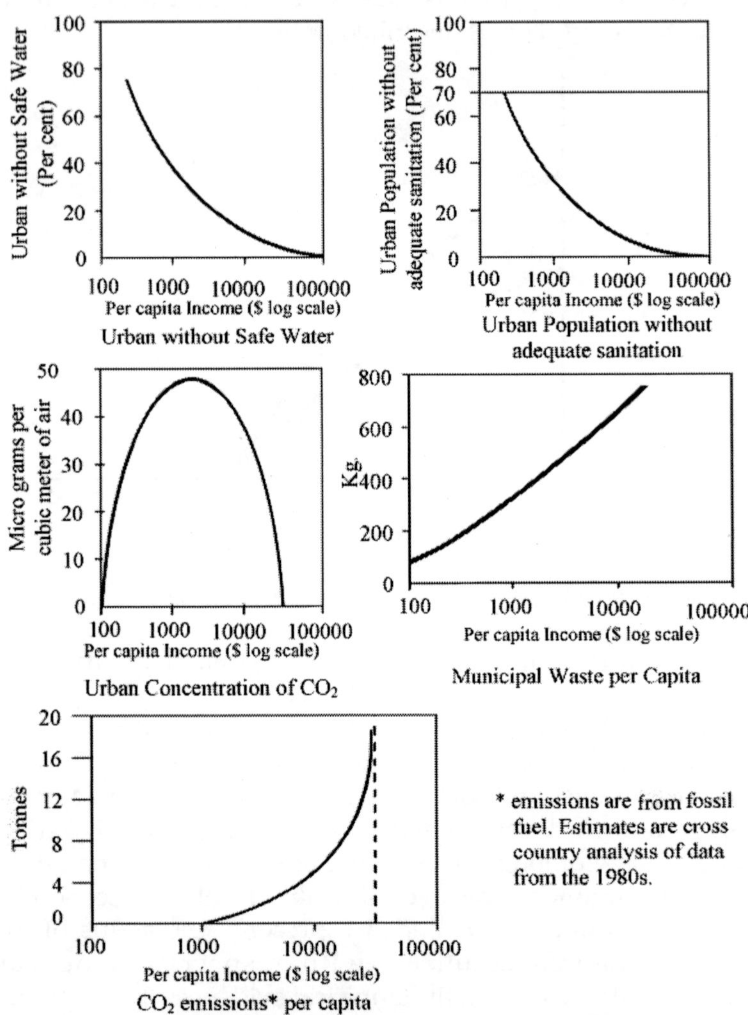

Fig. 1.4 : Environmental Quality Indices and Income Levels.

Studies thus point out that there does not exist a clear relation between the level of income and the indices of environmental characteristics. Some of them do perform better in the developed countries, some of them do not. This point is often missed while studying the political economy of ecological sustainability.

Optimal Growth and Sustainable Development

The optimal path of future consumption is traced out by setting up the problem in the following form,

$$\text{Maximise} \int_0^{\infty} U(C_t)e^{-\delta t}dt$$

where U is utility, C is real consumption per capita and d is the utility discount rate. Maximisation of this present value function can produce the paths of real consumption per capita over time, which may or may not be 'sustainable'.

Reducing the problem to,

$$\text{Max} \sum_{t=1}^{N} U(C_t)(1+s)^t$$

Subject to, $C_t + I_t = (1+r)\, I_{t-1}$

where U = Utility, C = Consumption per capita, I = Investment, r = rate of return on investment (marginal productivity of capital) and s = rate of discount.

Setting the Lagrange function,

$$L = \sum_{t=1}^{N} U\,(Cr)(1+s)'' + \lambda_t[(I+r)\, I_{t-i} - Ct - I_t]$$

$$= L\,(C_t, I_t)$$

The first order condition,

$$\frac{\partial L}{\partial C} = U'(C_t)(1+s)^{-t} - \lambda_t = 0 \qquad ...(1)$$

$$\frac{\partial L}{\partial C} = \lambda_t + I(1+r) - \lambda_t = 0 \qquad ...(2)$$

From (1),

$$U'(C_t) = \lambda_t(1+s)^t$$

Hence,

$$U'(C_0) = \lambda_0 \qquad ...(3)$$

From (2),

$$\lambda_{t+1}(1+r) = \lambda_t$$

Hence, $\lambda_0 = \lambda_1(1+r)$ or, $\lambda_1 = \dfrac{\lambda_0}{1+r}$

$$\lambda_1 = \lambda_2(1+r) \text{ or, } \lambda_1 = \frac{\lambda_1}{1+r} = \frac{\lambda_0}{(1+r)^2}$$

Similarly, $\lambda_1 = \text{or, } \lambda_0 = \lambda_t\,(1+r)^t$ ...(4)

Hence, $\dfrac{U'(C_t)}{U'(C_t)} = \dfrac{\lambda_1(1+s)^t}{\lambda_0} = \dfrac{\lambda_1(1+s)^t}{\lambda_1(1+s)^t} = \dfrac{\lambda_o}{(1+r)}$

Therefore, $U'(C_1) = U'(C_0)\,(1+s)^t\,(1+r)^{-t}$

Possibility 1: r = 0, s = 0.

$$U'(C_t) = U'(C_0)$$

Constant per capita consumption path (sustainable, because it gives non-declining per capita consumption).

Possibility 2: r = 0, s = 0.

$$U'(C_t) = U'(C_0)\,(1+s)^t$$

The optimal consumption path is not sustainable (consumption declines over time).

Possibility 3: r > 0, s = 0.

$$U'(C_t) = U'(C_0)\,(1+r)^{-t}$$

This is consistent with sustainable development (optimal consumption rises over time).

Possibility 4: r > 0, s > 0.

$$U'(C_t) = U'(C_0)\,(1+s)^t\,(1+r)^{-t}$$

(i) If r = s, Possibility 4 reduces to Possibility 1: Constant consumption.
(ii) If r > s, we get rising consumption over time.
(iii) If r < s, we get falling consumption over time.

We discuss this in Fig.1.5 with the assumption:

$$U'(C_t) = a\,C_t^b$$

The concept of sustainability also raise some fundamental question about how to assess the well being of the present and future generations. What should the present generation leave for its successors to maximize the chance that they would not be more worse-off than the present generation. The issue is more complicated because the future generation would not just inherit the pollution and depleted resource, but also would enjoy the fruits of the present generation's labour in the form of growth of education, skills, knowledge and physical capital.

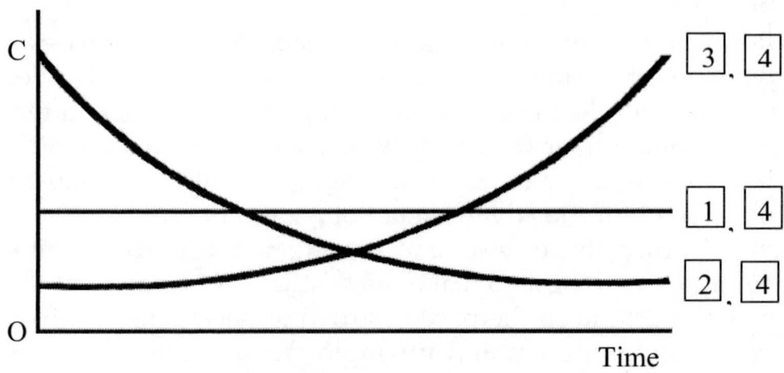

Fig. 1.5 : Consumption Path.

In resource study the factors of natural environment control the number, energy and abilities of inhabitants. Functional activities are guided by the efficiency and mental abilities of neighbours. In this sense some areas offer little resources for human use while others do in abundance. As for example, the tribal people of Khaira in South Munger, still lead a primitive life because the topography is rugged, the climate is unsuitable for raising crops and the forest environment creates favourable situation for hunting and forest gathering. In such an environment people spend much of their time and energy to satisfy a few primitive wants such as food, shelter, fuel, implements and personal adornment. In such an extreme environment inhabitants have little choice in the selection of functional activities. On the other hand, the people of Munger city have multitudinous occupations, good accessibility of roads, railways, waterways and airways. The industrial development, location of educational and social institutions, administrative establishment, medical facilities, etc. have given rise to dense population per hectare of land in comparison with Khaira, Sono and Chakai anchals of south Munger.

The strength of any nation, whether it be social, economic or political, depends mostly on the resources available at command and their proper utilisation. The evolution of civilisation centres command their spectacular growth in the field of social, economic, cultural and political advancements. History bears testimony to the fact that ancient civilisations flourished because they utilised available water resources; and wherever the growth of civilisations has been slow it has been in areas with less potentialities using the resources and these areas come under the category of underdeveloped countries. It is clear that the capacity of tapping the resources must be developed for the advancement of a country.

Today there is a race among the underdeveloped countries as to how to make the best use of resources at their command to keep the pace of growing necessities set by the growing population and unemployment. Unless these two major problems are solved properly, they will sap the vitality of the nation; and scientists and technologists of the world over must find out ways and means to combat the crucial problems being faced by the underdeveloped countries and must try to make them at par with the developed countries.

It is clear that a heavy responsibility lies on the scientists and technologists to uplift the underdeveloped countries by teaching them how to exploit the natural resources and make the best utilisation for the development of the country. The scientists by virtue of their various discoveries must direct the world to foster brotherhood and live in harmony.

The point is that we have not attached importance to the geography of resources in the university curricula. We do not realize that the planning of natural resources cannot be determined without a systematic introduction to the subject. We must know about our natural resources and thereby develop the art of their proper utilisation. This is necessary as the economic development of various nations has reached saturation point and some positive work in this field is most necessary for the survival of the human race. The purpose of this chapter is to shed light on some recent advancement in the concept of resources in introducing the whole book.

Meaning and Scope

According to Jones and Darkenwald, resource geography deals with "the productive occupations and attempts to explain why certain regions are outstanding in the production and exportation of various articles and why others are significant in the importation and utilisation of these things".

The demand for any item used by man is known as resources. In this sense anything in itself is not a resource but the need of such thing for a man makes them resources. In this way, an item is resource if in a certain time and space man has developed technological, cultural and physical abilities to use it. This shows the close relationship between the abilities of man and the resourcefulness of an item.

Generally hunting, fishing, forest gathering, industry, trade, minerals, water, land and soils are known as resources. But all these are the items which supply food, clothes, shelter, fuel, tools, materials of industries and luxuries as the physical base of natural environment, hence all these are known as resources. Besides these, other items which have an indirect influence in obtaining the resources are health, wants, knowledge, social adjustment, political stability, economic prosperity, national organisation and international cooperation, etc. are also known as resources. Commonly coal, iron, gold and silver are known as resources but the same serve the purpose of a man with the advancement of technological knowledge only, without which the resourcefulness of gold or silver cannot be estimated. The value attached to them in terms of capital, marketing, technological knowledge, sources of transport and favourable social, economic and political environment makes them resources.

To understand the resourcefulness of a thing, one has to understand the knowledge, habit and external and internal technological knowledge of persons including neutral stuff. There are two forces in physical environment : one is the natural element and other is the natural resistant force. The latter is generally harmful for the development of a nation. In this way, the progress of an economic landscape is determined by the above forces. Primitive people were surrounded by the neutral stuff of physical environment which has no direct influence on man. People in those days tried to fill their belly just like animals; and they had no courage to face the struggle actively. Such conditions are known as the passive adaptation of man which is just like the animal character of the present world.

Primarily the landscape was covered with forests but slowly with the occupation of land, exchange of little surplus with the neighbours and the growth of resources developed with the commercial activities of men. As for example, most of the early civilizations, viz., Egyptian and Mesopotamian in the old world and the Aztec, Melan and Incan in the new world have developed with the growth of subsistence agriculture. In early civilization the impetus for advancement came from European countries by occupying the professions of fishing, tending animals, tilling land, manufacturing, trading etc. With the decline of commercial powers of the Mediterranean, Phoenician, Cretan, Gressian, Carthaginian and Roman (from A.D. 1096 to 1270) revived the resource development in most of the American, African and Asian countries of the world.

The resources are dynamic in nature, as for example, many years ago coal was regarded as a blackstone. But nowadays, with the technical advancement of man, coal is regarded as an essential element of the present civilisations. Thus man has developed his technical abilities with the adaptation of nature in comparison with animals and the same is known as the "active adaptation of nature" for the growth of nations. In a natural environment according to the activities of men, the forest clearings give rise to permanent agriculture and in place of wild animals, domestic livestock are reared slowly. Later on fields and settlements developed.

According to Zimmerman and Michell, the knowledge of man is the greatest resource because the elements of physical environment have developed different cultural characteristics such as multifarious knowledge, love, social urbanisation and political organisation etc. which help in determining an item as resource. The limitation and barrier of human abilities cannot generate energy in itself and the same cannot restore resources. This is because the technical and cultural abilities of man are dynamic which have changed the use of resources in different manner.

According to Bowman, the geographical elements of environment are static but with the contact of man the characteristics of those elements have changed like cultural elements. In this way so far as the dynamic relationships of resource and cultural functions are concerned, there is no resource in itself but the same have been made by man as resource. Due to this fact the economic landscape to different nations are different in time and space which ultimately determines the potentialities of consumption and cultural advancement of the societies. Even in a country with the technical abilities of men and physical environment, the concept of resources is changing as for example the resources which are sufficient in peace time are not sufficient for a period of war.

The world is dynamic in its internal structure and so the age-old social, economic and political institutions must change with the advancement of age and time. The factors which one guided the development and utilisation of resources now find themselves incapable of ushering in an olden era, and they are now considered to be outdated. For instance, the yield which is produced with old techniques of production to cater the needs of the people now cannot keep pace with the time as there has been a tremendous growth in population and therefore, yields must be treble by using good manure and with the introduction of new crops along with modern scientific methods, otherwise we will hardly be able to feed the hungry mouths and to cater the needs of the people. A change in the old order or society is a prerequisite for a more efficient and higher utilisation of resources at a place which will provide proper solution to the fast multiplying problems.

The underdeveloped countries produce a large quantity of the world's raw materials yet they control an insignificant fraction of international trade in finished goods because their resources are being exploited by the market mechanism of the advanced countries. The underdeveloped countries have to face many knotty problems such as illiteracy, limited capital, underdeveloped industry and a primitive agriculture system. Even with limited resources at our disposal there has been a tremendous development after independence in India which managed to steer the ship out of the rough water. India has managed to go ahead with the development strategies with limited resources at its disposal. Poverty stands in the way but it does not hamper progress.

It is most striking to note that the developed countries overutilised the natural resources and this has had an adverse effect on the underdeveloped countries. On the other hand, the underutilisation of natural resources has kept the Indian economy in abject poverty. This unbalanced exploitation of natural resources may cause immense damage to the quality of human life and the life-supporting systems of ecology. In case the present trends of industrial and agricultural production, population growth and exploitation of non-renewable natural resources continue, this would cause immense damage to the earth's carrying capacity of human beings. The oil crisis also continues to be a sore problem in the economic affairs in most of the underdeveloped nations. The crisis of oil may diminish the rate of production of their natural resources. The value of oil production and prices has added new dimensions to resource development. The prices of raw materials in the international market have been artificially fixed at a lower level, whereas prices of finished goods have been inflated to suit the convenience of the developed countries. The prices in most cases do not give the true economic value of the products. After all, self-reliance is the goal of the underdeveloped countries and a rational policy of international trade can help to remove the obstacle in the way of development.

The responsibility for jointly safeguarding the natural resources would rest with the developing countries. This is the only way to eradicate poverty and its consequences.

India represents the paradox of poverty in the land of plenty. The reason is not far to seek. We have not been able to utilise our resources in an orderly manner. There are so many complex problems facing the Indian subcontinent like housing, medical facilities and per capita consumption of food, clothing, energy supply, minerals and other natural resources. The per capita income is very low.

We have not yet gone through the process of raising production in fields and factories to banish poverty from the land of plenty. We have not utilised modern techniques to raise production. We are still using the age old methods in spite of the fact that if we do not use the modern techniques we cannot feed everyone. The population problem is a baffling one. It has to be checked by using scientific methods. The illiterate masses must be taught that they have no moral right to increase the number of family members without finding resources for their proper maintenance. Illiteracy is the major cause which stands in the way of the progress of a country. It has to be fought at the war level and if we want to raise the status of the country, illiteracy has to be banished.

Lack of education, inadequate health care and the presence of irrational attitude towards the toiling masses and traditional social institutions are detrimental to the planned development. All these have threatened to thwart the forces of modernisation. Absence of knowledge about development, utilisation and conservation of our natural resources are the root cause of our inability to generate the powerful forces of modernisation. The central focus of this complex problem is that unless there is more for mass participations in the cause of advancement, there could be no prospect of rapid progress along a wide front.

Appraisal of Resources

The concept of resources includes physical, biotic, human and agricultural resources. This also includes mineral and man-made resources of settlements, industries, transportation lines and the development of scientific pursuits by acquiring technical abilities by people. Among all the resources, the human resource has great value as it formulates resources according to its choice. In any region, the vast reserves of coal and iron ore may have great economic value but the same is possible only with the help of technical abilities of man. Unless and until the inventions are made about any reserves by man, the same may not be regarded as resources because that is beyond the reach of man. Man as the creator and fashioner of resources occupies a dominant position in all economic and geographic studies. The relationship between man as the resource and environment can be divided into the following points:

(a) Man utilising the resources of nature as the creator of natural environment;
(b) Man utilising the resources of nature by virtue of his working in the natural environment; and
(c) Man utilising the resources of nature because human resources are the product of the physical environment.

Man is the dominant factor of resource geography as he produces, manufactures, distributes, consumes and also destroys resources. From pre-historic forest economy up to the present agricultural-cum-industrial economics, man occupies most of the habitable parts of the earth and uses its resources to satisfy his needs.

Thus the utilisation of resources depends on the favourable conditions of natural environment where each and every variable competes with the other in shaping and determining the cultural advancement and economic prosperity of the region. As for example, the production of wheat is not possible in highly wet areas. Similarly, the hot deserts are also not favourable for the cultivation of wheat in bulk. Only the reclamation of marshy land in wet areas and the provision of irrigation facilities in hot deserts can usher an era of favourable 'ecological balance' of environmental factors for the production of wheat in a particular region.

Resources and Culture

The culture of a man is the product of his physical environment and mental abilities and the increase of these two have given rise to the cultural advancement of a landscape. According to Karl Sauer, man's deformation of the pristine is assumed as essential. In a real sense cultural landscape is the result of dynamic and deformation of cultural abilities. In this way, culture in itself is the greatest influencing resource of a man. The table of other resources is also changing in time and space according to the cultural attainment which determines the intensity of the usefulness of an item as resource. In order to show the functional importance of this cultural hypothesis, Paul Seers has developed a formula of the physical resource of environment as land/population culture. According to this formula, the base of resource is divided by Population in a country which is the effect of a cultural landscape. As for example for primitive people coal, petroleum and other minerals are not regarded as resources but for a well-to-do person the same may be a resource.

Industrial Resources

Physical environment is a great source of human living which becomes the base for cultural advancement and the growth of industrial civilisations. On this basis three aspects have highly influenced resource development, such as: (1) technology, (2) social organisation, and (3) the knowledge of inhabitant. In any space and time, the level of industry fully depends on those natural resources which are utilised by the industry. Industries themselves are the product of the cultural landscape and the analysis of specific resources cannot be made keeping out of the cultural make-up of societies. The greater availability of resources, renewability and exploitability, etc. are the result of developed industrialisation. Underdeveloped, uncultured human society receives less resources from the physical environment; for example, petroleum was not known to primitive people as fuel oil but nowadays it is an essential commodity. In this way, the societies which have advanced in industrialisation are great source of the resource base (Figure 1.6).

In any region the question of the availability of resources depends on the technological advancement and the amount of resources available in that region. As for example, an area having hundred square kilometres may be an area of under-population if the fertility of soil and the production of crops are better in relation to the inhabiting population. Secondly, if the production of crops in that area is equal to the demand made by the people then it is said that the area is under optimum population. Thirdly, if an area has a higher number of people in relation to its productive capacity then it is said that the area is overpopulated. Fourthly, if the pace of population growth is rapidly increasing in comparison with the growth of production then it is said that the area is under population explosion.

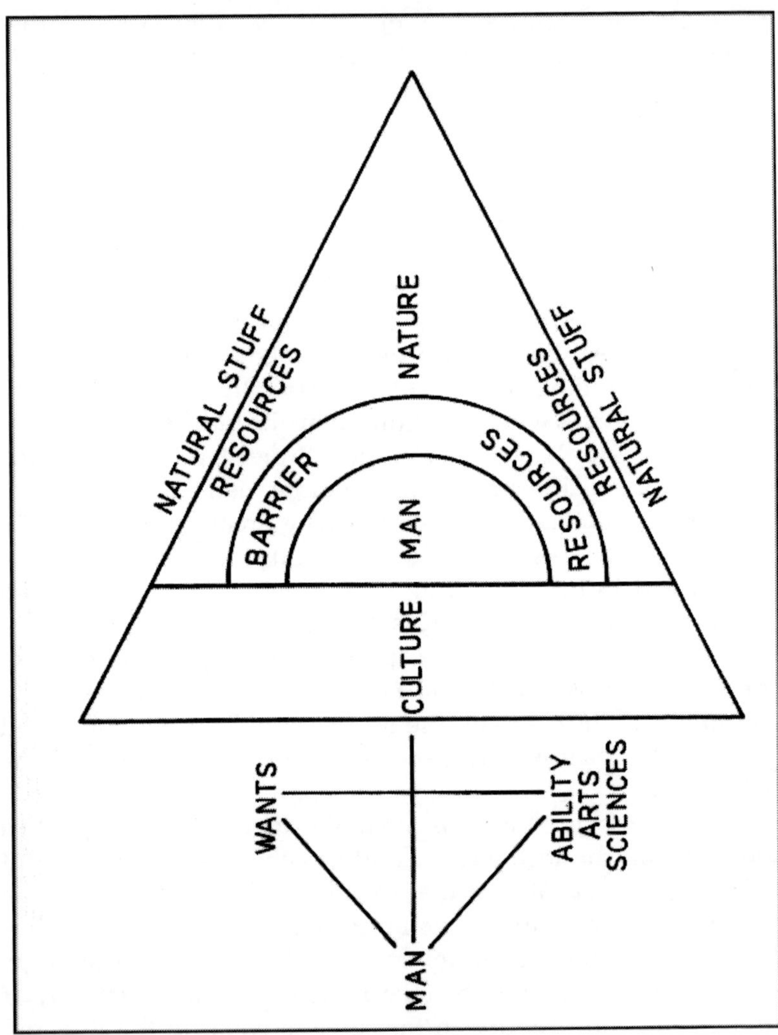

Fig. 1.6 : Mobilization of Resources.

Malthus has studied the relationship between population and resources at the end of nineteenth century and he used the terms "the means of subsistence" in lieu of resource. According to him, in any region population growth is always higher than resources. He suggested that the geometrical progression of population growth takes place in case "the means of subsistence" increase in arithmetical progression. In such a situation malnutrition will automatically bring famine conditions to the region.

In order to show the relationship between population and resources, E.A. Ackerman has classified the whole world into five groups:

(i) The United States Type : developed technology, a case of underpopulation.
(ii) The European Type : developed technology, a case of overpopulation.
(iii) The China or Egypt Type : less developed technology, a case of overpopulation.
(iv) The Brazil Type : less developed technology, a case of underpopulation.
(v) The Arctic Desert Type : less developed technology, a case of intensive overpopulation.

Resources are scarce in nature. Nowhere in this world, the desired level of resource development is possible unless a contentment prevails among the people for the fulfilment of their desires. Wants are unlimited and it is practically impossible to fulfil the desire of all persons at the same time. In areas where the growth of industry, agriculture, urban expansion and accumulation of capital, mineral wealth takes place, it is said that the growth of civilization has reached its peak, e.g., the areas of the Eastern Sea Board of North America. On the other hand, the forested rugged terrain, unfavourable climatic conditions of heavy rains and hot weather in the equatorial belt of the world bears some vulnerable diseases which makes the environment uninhabitable for human beings and as such this belt is the area of underdevelopment from the point of view of resource development.

With the advancement of scientific knowledge, the dimensions of resources are increasing, as now-a-days, the search for tidal power, solar energy, thermal power, atomic energy are fast changing the scene of resource development. Now the training in agricultural practices is given on the television. The use of the tractor, combine harvester, thresher, and the use of chemical fertilizers in the fields as well as the use of new dwarf varieties of seeds including the growth of agricultural resources beyond imagination. In some areas it is felt that the green revolution has taken place in the cultivation of wheat, rice and maize especially in northern India. Similarly, material culture has also changed because of the change of food habits. Clothing, advancement of technical abilities; and the old practices of cultivation through plough and oxen are not altered with the modern method of cultivation through scientific implements and the use of dwarf varieties of seeds, the use of synthetic fibres in place of traditional cotton cloth.

Classification of Resources

The resources can be classified into two groups—natural and human resources. (Fig. 1.7)

Natural Resources

This is the combination of elements available in nature. Keeping this in mind any element, creature, oxygen, nitrogen, rocks, minerals, landscape, soil, water, vegetation, livestock, etc. are regarded as resources due to their usefulness to man. The combination of a number of natural resources sometimes forms a composite resource as for example agricultural land, forest landscape, mining, etc. The agricultural landscape is the integrated form of land, soil, temperature, rainfall, etc. In this way, different elements and creatures of natural environment form resources both separately and in an integrated manner.

Physical environment is an integrated and composite unit of different materials and biotic resources which have formed different niches in space and time. Thus resources which have

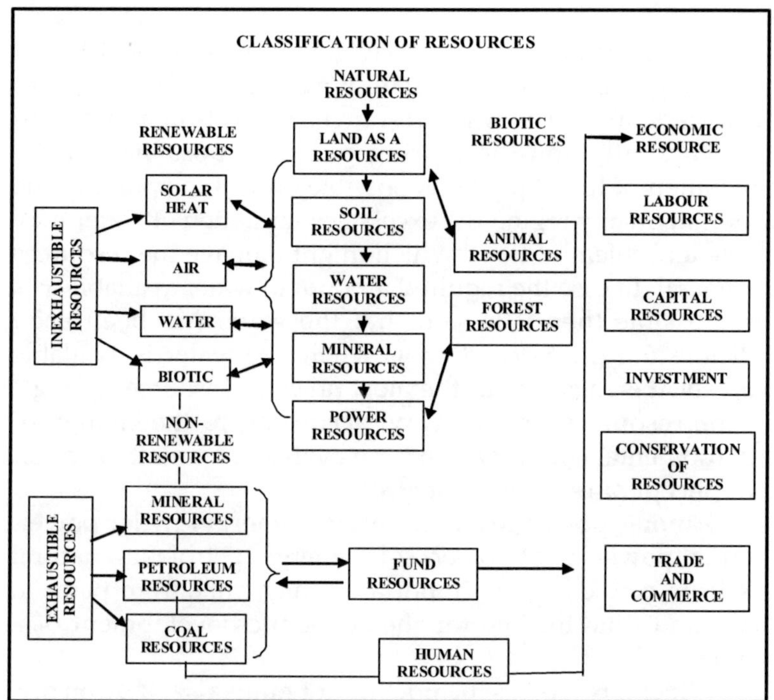

Fig. 1.7 : Schematic Representation of Classification of Resources.

formed different niches in space and time are dynamic in availability and use and their presence help directly in the activities of natural environment as a necessary part and parcel of geographical locations, size, accessibility, climate, landscape, features, soil, water, mineral resources, vegetation, livestock and birds including human resources.

The future demand of natural resources like flow and fund resources, water and mineral resources mostly depend upon the demand set by people on them for their judicious use. The flow and fund resources are always limited in quantity for a particular space and time. Thus the constant demand will pose a serious problem of population explosion on any kind of resources.

Water resource acts both as flow and fund resources. In case we allow the water to flow in channels, such water forms flow resources on which the navigation, rearing of fish and the sailing of boats for recreation purposes are possible up to the distant areas. The water resource, if it rests in a reservoir, in such a situation the rearing of fish is possible but the generation of hydel power is possible only at the point of the waterfall. In any region the excess accumulation of water may create a problem of flood which might damage the resources in general. In case the required amount of water availability is just possible then it is said that the water has been used judiciously. In hot desert areas where the water is available scarcely, it is regarded as the most precious resource amongst all the resources. In this way, according to the change of environmental conditions and the available amount of water, the concept of resource changes.

Among power resources atomic energy, hydel power, thermal power, tidal power, solar energy, petroleum, natural gas from cow dung are important which vary from place to place and time to time for the economic development of a region.

Mineral resources include host of minerals, viz., iron ore, copper, zinc, sulphur, limestone, gypsum, mica, etc. for the growth of material culture through the establishment of industries especially in areas where the power resources are available in bulk.

In this way, it can be said that the concept of resource is changing from place to place, time to time, man to man according to the task set by nature and the demand put by the society on a particular kind of resource.

Human Resources

Man as the creator, producer and consumer of natural resources, is a great resource in himself. Natural environment in itself is a dynamic element but without human use it has no importance. In this way, the activeness of man is necessary in order to procure benefit from the unseen resources of nature. The activeness of man is dependent on three things:

(a) Biological hereditary which includes race physiological behaviour, etc.

(b) Social, economic, cultural and political organisations which are acquired and adapted through organisational systems and traditional customs in different time and space as dynamic elements. In this sense living conditions, level of education, personal and family affairs, occupations and service position, social and economic status, citizenship and the consciousness of nationality, time, region, etc. are dynamic elements which affect population migration; and density, numbers, and behaviour in international relations which in turn, are important factors to affect the use of resources. This can be divided on the basis of several factors in which the following three are important:

  (i) The influence of Physical characteristics on natural resources;

  (ii) The influence of cultural factors on resources; and

  (iii) An integrated influence of natural and cultural elements.

The influence of physical characteristics of natural resources includes: (l) The total resource base and its influence; (2) The variability of the base of resources; (3) The characteristics and amount of each resource element; (4) The interrelationship amongst different resource elements; and (5) The locational attributes of resources.

In any space and time the base of human culture is the product of natural resources. The regions which have the greatest endowment of natural resources are the areas of highest technological advancement. Similarly, the availability of resources in greater amount will help in economic development. Recently, in the trade and industrial economic landscape, due to heavy influx of people on food items, raw materials, agricultural land, forest, minerals, water, houses, sources of transport, sources of recreation are developing, keeping in view the demand. In countries like U.S.A., U.S.S.R., China, India and Brazil, the bases of natural resources are vast, and in these countries the economic attainments are also substantial. During the Second World War when international trade was almost stopped, the wheels of the economy of every nation mainly depended on the natural resources base. But in peace time for the countries which have greater potentialities of favourable climatic conditions—soil, vegetation, livestock, birds and mineral wealth, the chances of prosperity are enough. In areas of single type of climate, the types of vegetations and crops grown are limited in number while in areas of temperate, sub-tropical and tropical climatic conditions, different types of crops and vegetables are available. Similarly, in a country where only coal is available, the chances of industrial development are less in comparison with the countries where different types of minerals, e.g., coal, petroleum, gas, iron and other minerals are available. With today's competitive economy, the demand for different resources is increasing according to the needs of the society (Figure 1.8). Due to high technological advancement, some items which are in demand these days, may not be so tomorrow or the demand may increase in place of the former.

Influence of Cultural Resources

The priority of resources is generally fixed by man. In any country, the total population, standard of living, wants of living, wants of society, fulfilment of demand of different resources, competitive demand, public opinion, political pressure, national and international political expediency, technological and scientific knowledge, availability of labour and capital resource ownership pattern, projects and legislation of the Government and the factors which are not equally distributed on the earth surface. Their amount, variety, frequency of occurrences are distributed in a explode; when discoveries so far out ran earlier estimates, it became a gluton of the market.

But neither technology nor extraction of earth's qualities is costless, many things are possible which are not currently economic. The United States possesses great deposits of oil, and Canada has great deposits of tar sands. From both it is wholly possible to extract enormous quantities of petroleum, but it has not proved economical to use either. Salt water can be made soft "to make the deserts bloom", but it is beyond our resources. Power from atomic sources is technically feasible, and seems to be on the verge of becoming economically bankrupt at least under some circumstances.

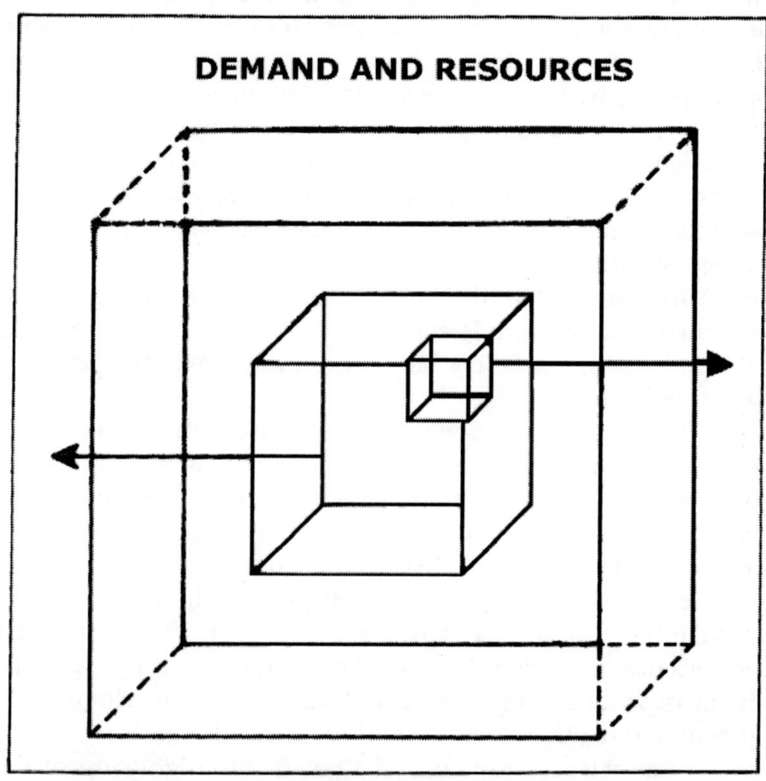

Fig. 1.8 : Phantom Resources Theory (After Zimmerman).

The goals that man seeks in resource use are equally important, and sometimes overlooked or inadequately considered. The attainment of the greatest possible income, measured in conventional dollar terms, may be sought for many purposes. This is an appropriate goal, and one for which the tools of economic analysis are well suited. But it is not the only goal, as most economists realize. Resources may be used for some ends which are not easily measured, or measurable at all, in monetary terms. One practical example of considerable quantitative importance today is the use of forests and other land areas for recreation. Others are the use of large quantities of minerals, labour and capital for defence purposes; and the shipment abroad of wheat and other food not needed at home to relieve distress and to assist the economic growth of the developing areas. It may be argued that, in each of these cases, the aim is to maximize something whether personal satisfaction not easily measurable in money terms, or national security, or something else. But, in some cases patterns of resource use seem scarcely to maximize anything, unless it be conformable with long established prejudices. Hence, the problem of goals became especially difficult when resource use in many countries and culture is considered. What may seem rational to Americans may be the height of rationality to the people of other culture.

These considerations about resources in general apply equally to soil. And, even in the United States, the objective of soil conservation programmes are not identical for everyone. Elaborate procedures have been developed for measuring the physical characteristics of soil—their depth, geologic history, size of constituent particles, slope, availability of plant nutrients, and many other features. Through practical experience and scientific research, a rich technology also has been developed for their use. Soils are used primarily by farmers and other private landowners, who seek maximum income as at least one goal; and who are thus highly responsive to economic considerations.

Each aspect of our definition of natural resources as applied to soil has undergone substantial change in the United States over the past three hundred years. Through erosion, different hydrologic relationships caused by forest clearings and other practices, irrigation, continued cropping and so on, the basic characteristics of the soil have been changed in many areas. Growth of large cities, development of transportation networks, creation for marketing institutions, and many other economic changes affected the economics of agriculture in still other ways. The goal of farming has changed to some degree, from provision of food and fibre for the farm family on a largely self-sufficient base to the production commodities for sale in the expectation, or hope, of a profit.

Resources are so enormously variable and diverse, in so many ways, that some grouping into broad classes seems essential. From a broad technological or management view point, resources can be grouped as shown in Table 1.1. Some resources are "flow resources" with a stream of uses possible without loss of the resource itself. These, in turn, are divisible into non-storable resources, of which sunlight is a good example, and storable ones, such as water from natural precipitation. In contrast are the "fund resources". All these were exhausted by use, but some can be renewed and some cannot. The "exhaustible" fund resources are typified by nature forests or soil fertility, either can be used and renewed, at least under many circumstances.

However, this is possible only within a range of reversibility. If all the trees of a species were cut, to use an extreme example, forests of the same species could not be re-established. Soil can be renewed when nutrients are used and even part of soil itself washed away or otherwise destroyed; but if the process goes too far it will prove impossible to reverse. Still other resources, the extraction of petroleum and mining of minerals for example, are "exhaustible and non-renewable", at least within the span of human planning.

Possibly we should add another class in exhaustible resources. For example, climatic factors, position of the earth's surface, basic geology, and perhaps other aspects or characteristics are so permanent and so difficult or impossible for man to change that they all are practically inexhaustible. The same might be said of the minerals dissolved in sea water. However, these can be considered, perhaps without too much violence, as flow resources of a non-storable kind.

Table 1.1 : General Characteristics of Natural Resources

Type of resource	*Relation of use in one period of time to use in following periods*	*Possibility of increasing supply in later periods by man's present activities*	*Management alternatives*	*Examples*
1. *Flow resources*				
a. Non-storable	None	None, or very use or non-use, Sunlight little		
b. Storable	Large, within limits of storage capacity	Considerable storage, non-storage, unstoring		Water stored in reservoirs or in soil
2. *Fund resource*				
a. Exhaustible but renewable	Large, within range of reversibility	Large, if necessary present steps taken	Maintain decrease or increase availability over time	Soil fertility tilling forest
b. Exhaustible and non-renewable	Complete and inverse; use now exactly reduces use later, and vice versa	Only by refraining from consumption now	Rationing over time, Substitute flow or renewable fund resources	Petroleum Peat and Muck

Source : John F. Timmons, *et al.*, Committee on Soil and Water Conservation of the Agricultural Board, Principles of Resource Conservation Policy, with some applications to Soil and Water Resources, Washington, (National Academy of Sciences—National Research Cuuncil, 1961), Publication 855, p. 8.

Wantrup uses a system of classification of natural resources (Table 1.2) which is similar to the foregoing in some respects, but differs significantly from it in other ways. While his major breakdown is closely related, his secondary breakdown for the stock resources or fund resources is on the basis of absence or presence of natural deterioration, and for the flow resources is inability or ability of human action to affect the resources. Still other classifications are possible; the problem is not simple, and different considerations may loom larger in the minds of some students than in those of others.

Table 1.2 : Wantrup's Classification of Natural Resources

1. Non-Renewable or Stock Resources
   - (i) Stuck not significantly affected by natural deterioration : metal ores in situ, Coal, Stones, Clays.
   - (ii) Stock significantly affected by natural deterioration : refined metal; subject to oxidations, oil and gas in case of seepage and blow-off, plant nutrients subject to leaching radioactive Substance in process of nuclear disintegration, surface water reservoirs subject to evaporation.
2. Renewable or Flow Resources
   - (i) Flow not significantly affected by human action: solar and other cosmic radiation, tides, winds.
   - (ii) Flow significantly affected by human action:
     - (a) Reversibility of a decrease in flow not characterized by a critical zone: precipitation, special locations that from the basis of site value, services from a species of durable producer of consumer goods.
     - (b) Reversibility of a decrease in flow characterized by a critical zone, animal and plant species, science resources, storage capacity of groundwater basins.

Source : S. V. Ciriacy-Wantrup, Resource Conservation Economic, and Policies (rev. ed.: Berkeley University of California Press, 1963), p, 42.

In any event, such classifications while logical and helpful in many ways, generally do not fit any particular beyond the reversible zone, sometimes within it. At the same time, these plant associations had the characteristics of flow resources, in the sense that important quantities of plant materials could be removed annually or at other intervals; and yet leave the basic resource fully capable of rejuvenation.

Principle of Resource Adequacy

In any country the efficiency of resource is measured by the number of population, the desired level of consumption and the total population, etc. Population and their living standard determine the total demand of resources. Whether the demand on any level can be fulfilled or cannot be determined by human knowledge, technology and the available physical resources. Man is not only the gatherer of resources, he also uses the scarce resources scientifically, increases their production and tries to lessen the pressure of human beings on a particular piece of land with a view to avoiding scarcity of physical resources. During the last two decades the growth of population and the economic development have taken place simultaneously but in order to raise the standard of people, family planning has been introduced in India. In this way, it can be said that the relationship between physical resources and population is a dynamic process.

The covariance analysis of these two variables has been presented by Ackerman in the form of an equation. Considering the demand and use of resources, the formula is as follows:

$$L = T +- S + R \pm S_{ev} - T_r$$

Where, L = Living standard of population:

T = Total base of resource characteristics and attributes of natural and physical resources (physical technology factor, administrative technology factor, resource stability factor);

S = Scale of economies;

R = Resource added in trade;

S_{ev} = Social, economic and political institutions and their characteristics; and

T_r = Techniques of resource use.

Conservation of Natural Resources

The conservation of resources has attracted the mind of scholars mainly after the Second World War due to population explosion, technological-industrial revolution and in order to fulfil the increased wants for a good living standard. During A.D. 1600 only a few minerals had been discovered such as clay, whetstone, millstone, iron ore, glass, tin, mica, salt, mercury, etc. But nowadays more than a hundred minerals have been discovered including petroleum, potash, nickel, aluminium, chromium, etc. In 1750, the total population of the world was only 66 crores but in 1975 this rose to about 400 crores. Hence, the resource requirements of these teeming millions also increases with the same rate.

A man can live in flexible manner in any environment which may be designated as an unspecialisation, but the evolutionary characteristics of human beings are of a different nature and the habitat and the scope of his activities are also not restricted. Man as a rational being, utilises the natural environment of the active ecological system. In such a manner that should be the dynamic elements of organic and inorganic world which sometimes used as resources, while at the other created barrier. In order to extract natural resources which have some economic value, a man determines the systems of organisation and space actively. The theory of the conservation of resources considers the base of physical balance of population and living standard of people which not only develops the present social, economic and political activities of man but also controls the future use of resources by less destruction, systematic utilisation and unexpectable exploitation.

The conservation of resources does not mean the use of resources either frugally or preservation. Conservation means the purposeful and planned use of resource and the protection of soil, forest and mineral resources from erosion and restoration. In this way, the conservation of resources always means the optimum use of resources which are intimately related with the population, living standard, technological advancement and cultural characteristics of the region which determines the productive capacity of resources also includes continuous, constant and sustainted use of resources. However, the rational use of resources as when and how to use, allocation, depletion, wastage, state of optimum use, planned integrated production come under the conservation of resources.

For planning and conservation of resources the consideration of the following aspects are necessary :

(i) The consideration of a region as a unit is necessary so that the knowledge of total resource base and their types, amount, characteristics and attributes of resources can be fully estimated for proper utilization.

(ii) In any space and time after considering the present resource need, the supply of the resources on lower prices should be immediately stopped in order to prevent destruction and misuse.

(iii) For the prosperity and production of resources, all possible methods of conservation must be adopted.

(iv) Those resources which are insufficient in quantity or exhaustible should be scientifically used.

(v) Such resources should be invented which are not exhaustible or which may take a long time to exhaust such as solar energy in place of coal and petroleum.

(vi) Resources complexes should be utilised in an integrated, balanced and in multiple ways.

(vii) Keeping in view the amount and availability of resources the study of a social, economic and political relations are highly needed.

(viii) By planning the national resources in a multitude of ways one should try to be self-reliant personally, regionally and socially.

(ix) In resource planning full cooperation between the government, public and private institutions is necessary.

(x) The planning and conservation of resources are necessary in order to make the general people good citizens.

Planning for Development

The objective of this study is to make plan for the better use of our resources. These could be rationality, planning for economic development, rise of productivity, rise of levels of living, social and economic equalisation, improved institutions and attitudes, national self-reliance, social discipline and participation of people in the task of nation building and so on, although it is difficult to fulfil the desired goals of development planning. Hence it is in this direction that we have to reorient our thinking and plans of action. And as we approach the desired goals, the changed social order will, in turn, accelerate the rate at which resources can be utilised or conserved.

The types of resources which need conservation depend on the availability of other resources, the final demand for them and the level of technology. For example, coal can he utilized for the production of power, fuel and has a potential for conversion into oil and other chemicals. Yet, to meet the demand for power, we have to use our coal reserves for quite sometime to come. But on account of its exhaustible nature, and its other uses, alternatives have to be found for the production of Power. These could be found through nuclear energy, or geothermal energy. And till a certain point of time, which we can call the threshold point, we must make the maximum use of our coal and after this point we must conserve our coal resource. The period between the present point of time and the threshold point is the time available to us for the development of alternative methods of power production which are not dependent on non-renewable natural resources. Other modes of conservation and efficient utilisation of resources could be through appropriate technological reforms, recycling of waste, matching of the objectives of organisations engaged in resource exploitation with the national objectives, reform of the social institutions and education of our people in new techniques of resource development and utilisation. In fact, survey, conservation and utilisation are three interlinked aspects of resources which cannot be considered in isolation without running the risk of drawing wrong conclusions. Geological and geophysical surveys enable us to identify the available resources, their concentration and their geographical distribution. This combined with the knowledge of distribution of demand for these resources across the country can help us in working out the economics of their extraction, establishing the need for conservation and developing the technology for their optimum utilization. It may be prudent to develop the technology of extraction which simultaneously tackles the problem of utilisation and conservation of resources. This requires an integrated approach to resource planning. For, otherwise there is a risk of unintelligent, overexploitation of resources as happened in the advanced countries, or under-utilisation of resources as in the developing countries. Economic and social advancement left to "natural forces", is highly unlikely to follow the desired direction. On the contrary, we may end up in a crisis situation as happened during the oil crisis from which it may be difficult and costly to retrieve lost ground and which certainly will impede our advancement. The author has chosen certain specific resources vital to our national development and would like to identify the issues stemming from the broad spectrum of the resources in order of their importance.

Sustainable Development of Resources and Environmental Issues

In recent years, economists have become increasingly aware of the important implications of environmental issues for the success of development efforts. We now understand that the interaction between poverty and environmental degradation can lead to a self-perpetuating process in which, as a result of ignorance or economic necessity, communities may inadvertently destroy or exhaust the resources on which they depend for survival. Rising pressures on environmental resources in developing countries can have severe consequences for self-sufficiency, income distribution, and future growth potential in the developing world.

Environmental degradation can also detract from the pace of economic development by imposing high costs on developing countries through health-related expenses and the reduced productivity of resources. The poorest 20 per cent of the world's population will experience the consequences of environmental ills most acutely. Severe environmental degradation, due to population pressures on marginal land, has led to falling farm productivity and per capita food production. Since the cultivation of marginal land is largely the domain of lower-income groups, the losses are suffered by those who can least afford them. Similarly, the inaccessibility of sanitation and clean water mainly affects the poor and is believed to be responsible for 80 per cent of diseases worldwide. Because the solutions to these and many other environmental problems involve enhancing the productivity of resources and improving living conditions among the poor, achieving environmentally sustainable growth is synonymous with our definition of economic development.

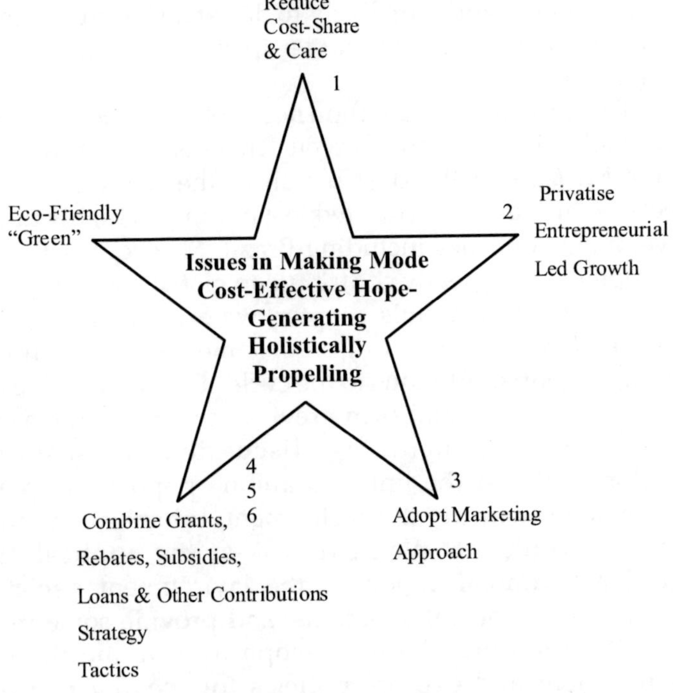

Fig. 1.9 : Holistic Model for Development.

Though there is considerable dispute concerning the environmental costs associated with various economic activities, consensus is growing among development economists that environmental considerations should form an integral part of policy initiatives. The exclusion of environmental costs from calculations of GNP is largely responsible for the historical absence of environmental considerations from development economics. Damage to soil, water supplies, and forests resulting from unsustainable methods of production can greatly reduce long-term national productivity but will have a positive impact on current GNP figures. It is thus very important that the long-term implications of environmental quality be considered in economic analysis. Rapid population growth and expanding economic activity in the developing world are likely to do extensive environmental damage unless steps are taken to mitigate their negative consequences.

The growing consumption needs of LDC populations may have global implications as well. There is increasing concern in the MDCs that the destruction of the world's remaining forests, which are concentrated in a number of highly indebted developing countries including Brazil, Mexico, Peru, and the Philippines, will greatly contribute to global warming and the greenhouse effect. In this chapter, we examine the economic cause and consequences of environmental devastation and explore potential solutions to the cycle of poverty and resource degradation. We begin, as in previous chapters, with a survey of basic issues, including discussions of sustainable development and the linkages among population, poverty, economic growth, rural development, urbanization, and the LDC environment. We next look at the applicability of traditional economic models of the environment, depict some typical environmental situations, and provide some relevant data. We then broaden our scope to examine the global environment and explore policies for seeking worldwide sustainable development. Pakistan's environment problems are this chapter's case study.

Environment and Development : The Basic Issues

Seven basic issues define the environment of development. Many grow out of the discussions in the preceding chapters. The seven issues are: (1) the concept of sustainable development, and linkages between the environment, (2) population and resources, (3) poverty, (4) economic growth, (5) rural development, (6) urbanization, and (7) the global economy. We briefly discuss each in turn.

Sustainable Development and Environmental Accounting

Environmentalists have used the term sustainability in an attempt to clarify the desired balance between economic growth on the one hand and environmental preservation on the other. Although there are many definitions, basically sustainability refers to "meeting the needs of the present generation without compromising need of future generations." For economists, a development path is sustainable "if and only if the stock of overall capital assets remains constant or rises over time." Implicit in these statements is the fact that future growth and overall quality of life are critically dependent on the quality of the environment. The natural resource base of a country and the quality of its air, water, and land represents a common heritage for all generations. To destroy that endowment indiscriminately in the pursuit of short-term economic goals penalizes both present and, especially, future generations. It is therefore important that development policy-makers incorporate some form of environmental accounting into their decisions. For example, the preservation or loss of valuable environmental resources should be factored into estimates of economic growth and human well-being. Alternatively, policy-makers may set a goal of no net loss of environmental assets. In other words, if an environmental resource is damaged or depleted in one area, a resource of equal or greater value should be regenerated elsewhere.

An example of environmental accounting is offered by David Pearce and Jeremy Warford. Overall capital assets are meant to include not only manufactured capital (machines, factories, roads) but also human capital (knowledge, experience, skills) and environmental capital (forests, soil quality, rangeland). By this definition, sustainable development requires that these overall capital assets not be decreasing and that the correct measure of sustainable national income or sustainable net national product (NNP) is the amount that can be consumed without diminishing the capital stock. Symbolically,

$$NNP = GNP - D_m - D_n$$

where NNP is sustainable national income, D_m is the depreciation of manufactured capital assets, and D_n is depreciation of environmental capital—the monetary value of environmental decay over the course of a year.

An even better measure, though more difficult to calculate with present data collection methods, would be

$$NNP = GNP - D_m - D_n - R - A$$

where D_m and D_n are as before, R is expenditure required to restore environmental capital (forests, fisheries, etc.), and A is expenditure to avert destruction of environmental capital (air pollution, water and soil quality, etc.).

In light of rising consumption levels worldwide combined with high rates of population growth, the realization of sustainable development will be a major challenge. We must ask ourselves, what are realistic expectations about sustainable standards of range. From present information concerning rapid destruction of many of the world resources, it is clear that meeting the needs of world population that is projected to grow by an additional 3 billion in the next 50 years will require radical and early changes in consumption and production patterns. We discuss these needed ranges later in the chapter.

Population, Resources and the Environment

Much of the concern over environmental issues stems from the perception that we may reach a limit to the number of people whose needs can be met by the earth's finite resources. This may or may not be true, given the potential for new technological discoveries, but it is clear that continuing on our present path or accelerating environmental degradation would severely comprise the ability of present and future generations to meet their needs. A showing of population growth rates would help ease the intensification of many environmental problems. However, the rate and timing of fertility declines, and thus the eventual size of world population, will largely depend on the commitment of governments to creating economic and institutional conditions that are conducive to limiting fertility.

Rapidly growing populations have led to land, water, and fuelwood shortages in rural areas and to urban health crises stemming from lack of sanitation and clean water. In many of the poorest regions of the globe, it is clear that increasing population density has contributed to severe and accelerating degradation of the very resources that these growing populations depend on for survival. To meet expanding LDC needs, environmental devastation must he halted and the productivity of existing resources stretched further so as to benefit more people. If increases in GNP and food production are slower than population growth, per capita levels of production and food self-sufficiency will fall. Ironically, the resulting persistence of poverty would be likely to perpetuate high fertility rates, that the poor are often dependent of large families for survival.

Poverty and the Environment

Too often, however, high fertility is blamed for problems that are attributable in poverty itself. For example, China's population density per acre of arable land is twice that of India, yet yields are also twice as high. Though it is clear that environmental destruction and high fertility go hand in hand, they are both direct out-growths of a third factor, absolute poverty. For environmental policies to succeed in developing countries, they must first address the issues of landlessness, poverty, and lack of access to institutional resources. Insecure and tenure rights, lack of credit and inputs, and absence of information often prevent the poor from making resource-augmenting investments that would help preserve the environmental assets from which they derive their livelihood.

Hence preventing environmental degradation is more often a matter of providing institutional support to the poor than fighting an inevitable process of decay. For this reason, many goals on the international environmental agenda are very much in harmony with the three objectives of development articulated.

Growth Versus the Environment

If, in fact, it is possible to reduce environmental destruction by increasing the incomes of the poor, is it then possible to achieve growth without further damage to the environment? Evidence indicates that the worst perpetrators of environmental destruction are the billion richest and billion poorest people on earth. It has even been suggested that the bottom billion are more destructive than all four billion people in between. It follows that increasing the economic status of the poorest group would provide an environmental windfall. However, as the income consumption levels of everyone else in the economy also rise, there is likely to be a net increase in environmental destruction. Meeting increasing consumption demand while keeping environmental degradation at a minimum will be no small task.

Rural Development and the Environment

To meet the expanded food needs of rapidly growing LDC populations, it is estimated that food production in developing countries will have to double by 2010. Because land in many areas of the developing world is being unsustainably over-exploited by existing populations, meeting these output targets will require radical changes in the distribution, use, and quantity of resources available to the agricultural sector. And because women are frequently the caretakers of rural resources such as forests and water supplies and provide much of the agricultural supply of labour, it is of primary importance that they be integrated into environmental programmes. In addition, poverty alleviation efforts must target women's economic status in particular to reduce their dependence on unsustainable methods of production.

The increased accessibility of agricultural inputs of small farmers and the introduction (or reintroduction) of sustainable methods of farming will help create attractive alternatives to current environmentally destructive patterns of resource use. Land-augmenting investments can greatly increase the yields from cultivated land and help ensure future food self-sufficiency.

Urban Development and the Environment

Demonstrated that rapid population increases accompanied by heavy rural-urban migration is leading to unprecedented rate of urban population growth, sometimes at twice the rate of national growth. Consequently, few governments are prepared to cope with the vastly increased strain on existing urban water supplies and sanitation facilities. The resulting environmental ills pose extreme health hazards for the growing numbers of people exposed to them. Such conditions threaten to precipitate the collapse of the existing urban infrastructure and create circumstances ripe for epidemics and national health crises. These conditions are exacerbated by the fact that under existing legislation, much and renders large portions of urban populations ineligible for government services.

Congestion, vehicular and industrial emissions, and poorly ventilated household stoves also inflate the tremendously high environmental costs of urban crowding. Lost productivity of ill or diseased workers; contamination of existing water sources, and destruction of infrastructure, in addition to increased fuel expenses incurred by people having to boil unsafe water, are just a few of the costs associated with poor urban conditions. Research reveals that the urban environment appears to worsen at a faster rate than urban population size increases of that the marginal environmental cost of additional residents arises over time.

The Global Environment

As total world population grows and incomes rise, net global environmental degradation is likely to worsen. Some trade-offs will be necessary to achieve sustainable world development using resources more efficiently, a number of environmental changes will actually provide economic savings, and others will be achieved at relatively minor expense. However, because man's essential changes will require substantial investments in pollution abatement technology and resource management, significant trade-offs between output and environmental improvements will occasionally become necessary. The poorer the country, the more difficult it will be to absorb these costs. Yet a number of issues including biodiversity, rain forest destruction, and population of growth, will focus international attention on some of the most economically strapped countries in the world. In the absence at substantial assistance to low-income countries, environmental efforts will necessarily have to be funded at the expense of other social programmes, such as education, health services, and employment schemes, that themselves have important implication for the preservation of the global environment.

Exactly what sacrifices need to be made and who should make them will continue to be matters of great controversy. Nowhere was this more evident than at the Second United Nations Conference on Environment and Development (UNCED)—the so-called Earth Summit—held in Rio de Janeiro in June 1992 and a follow-up conference in Kyoto, Japan, in 1997. Most cumulative environmental destruction to date has been caused by the developed world. However, with high fertility rates, rising average incomes, and increasing inequality in the developing world, this pattern is likely to reverse sometime in this century. It is thus unclear how the costs of global reform should be divided. Apportionment of responsibility for reducing environmental damage essentially hinges on the manner in which the question is framed. For example, if a limit is placed globally on levels of per capita pollution emissions, the approach would clearly favour lower-income countries that have much lower per capita consumption levels. Conversely, if international pressures try to limit the growth rate of per capita emissions or even to impose limits on the growth of national emissions, any movement in that direction would tend to freeze incomes in the developing world at a small fraction of those of their developed world counterparts.

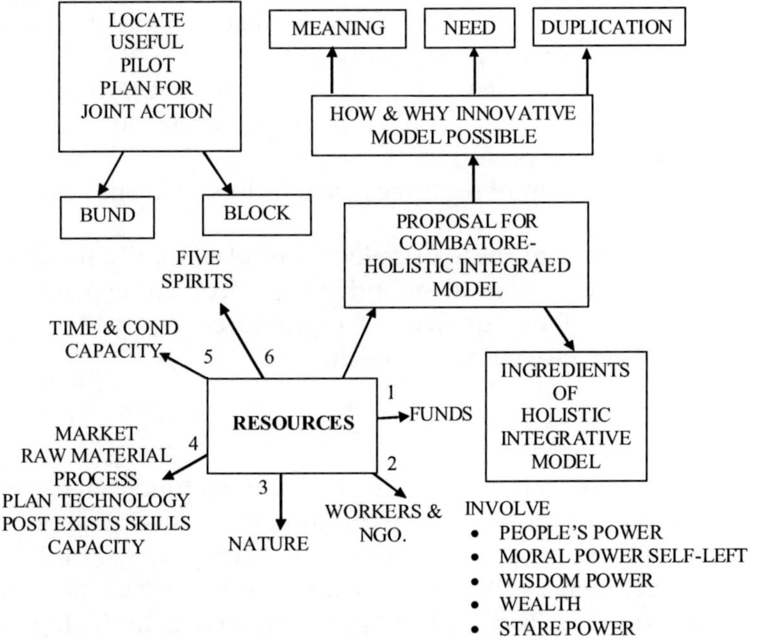

Fig. 1.10 : Analysis Series.

Statement of Problem

In this study problems to be tackled are as follows :

(i) How the location, growth and development or resources have taken place and what will be the theoretical background for their correct appraisal and planning?

(ii) How the geographical factors help in developing different types of resources?

(iii) What is the level of resource study in different parts of the globe with special reference to India?

(iv) What are the resources capable of development?

(v) What are the level of resource development in terms of water, forest, agriculture, industry, population and urban centres in the study area?

(vi) How the present standard of resource development be raised to provide decent living, social justice and economic preferment?

(vii) How different resources are interlinked in one system?

An answer to these questions would naturally result in identifying the problem areas and help in formulating planning scheme for balance growth of regional economy through integrated development approach.

Methodology

In the study of resource planning different methods have been employed after adopting from different sister disciplines such as economics, geography, statistics, demography, biology and other sciences. An interdisciplinary researcher need to be an expert historian, cartographer, artist, hydrologist, demographer, market researcher, sociologist, etc. Hence, his work is to assess the occurrences of different resources, level of their consumption, method of conservation with special reference to mapping, correlation and interpretation of comparative study, etc.

In the present study a systematic approach has been applied and a balanced development of resources and their planning in the district of Munger has been analysed on a firm statistical base. Relationships have also been established between different resources in the district while a large number of techniques have been applied in the preparation of maps and diagrams. The Choropleth and Isopleth techniques are most frequently used. Several maps have been prepared to illustrate the results of various statistical techniques applied to analyse and properly interpret the available data.

Sources of Data

The data has been collected from a number of sources and includes both published and unpublished maps, published literature, census data, unpublished data from official sources, field investigation and statistical analysis of relevant data. The following is an account of map utilised, data used and field work done :

(i) 1 inch = million scale map of Munger with anchal boundary made available by Mohammad Abbas, geographer and corrected by the author with the help of the map available from the Department of Census Operations, Government of India. This map has been used for the regional Study.

(ii) 1 inch = 1 mile topographical sheets of the Survey of India, Dehradun, was used for observing the drainage, types of settlement, village patterns in relation to different ecological conditions.

(iii) 1 inch = 16 miles cadastral map of the village, published by the Survey of India, Gulzarbagh, Patna, collected from the villagers in course of detailed field investigation for the socio-economic data, observing case study of resource development in different areas of the district. The fieldwork was completed during the period, November, 1979 to January, 1984.

(iv) The data available in published form include the Census figures for 1991, 2001 and earlier periods at district, anchal and village levels compiled by the Directorate of Census Operations, Government of India, Patna.

(v) Map of Dakra Nala, Surajgarha, Pump Canal Scheme and other irrigational projects were made available by the Department of Ganga Pump Canal Scheme, Munger.

(vi) A number of publications like Season and Crop Report, Bihar Through Figures, Bihar State Block Directory and Bihar Statistical Handbook, etc. published by the Department of Statistics and Evaluation, Government of Bihar, Patna have been consulted.

(vii) List of flood affected anchals and drought affected anchals have been collected from the Collcctorate Office, Munger.

(viii) During the year 1919 to 2001, the author visited the region several times in connection with the collection of relevant information, cross-checking and personal observation of significant elements. During the field-work a large volume of unpublished data and map were gathered from Government and private Institutions as well as from individuals.

(ix) While a large number of techniques have been applied in the preparation of maps, diagrams and models, the Choropleth and Isopleth techniques have been most frequently used. Several maps have been drawn to illustrate the results of various statistical techniques applied to analyse and properly interpret the available data.

(x) A large number of data have been made available in printed form and even greater amount has been collected in unpublished form from various sources, including Government Departments and fieldwork. An attempt has been made to analyse and interpret these datas scientifically and objectively by applying suitable statistical and cartographic techniques. File locational, spatial, hierarchial and functional studies of resources and their potentialities studied on the basis of theoretical models as this should be useful in regional planning and resource development.

Conclusion

Having gone through the important factors which affect the survey, conservation and utilisation of resources, it is suggested that the level of mass education and human health should be given due importance as without these no progress could be achieved. Proper attention to the development and right application of science and technology must be aimed to achieve this goal. State agencies, administrators and scientists should consider the problem of source conservation and planning, on a strong footing. The natural resources should be tapped and modern technology should be mobilized to achieve success in harnessing resources for balanced regional development.

REFERENCES

Ackerman, E. A. in K. N. Singh and J. Singh, "Bhoogol Ke Mul Tatwa", Tara Publications, Varanasi, 1973.

Ahmad, K. S., "Environment and the Distribution of Population in India", Indian Geographical Journal, 1941.

Chatterjee, S. P., "Minerals and Mineral Products", Calcutta Geographical Review, 1942-43.

Ciriacy-Wantrup, S. V., Resource Conservation and Policies (rec. ed, Berkeley : University of California Press), 1963, p. 42.

Dayal, P. "Aspects of Industrial Location in India," Transactions of Indian Council of Geographers, Vol. I, Dec. 1964.

Dayal, P. "The Energy Resources of India", Geology and Geography Section, 64th Session of Indian Science Congress, 1977.

Darling, F. F., "A Wider Environment of Ecology and Conservation", Deadalus, 96 (4), 1967, p. 1003.

Freeman. T.W. "Geography and Regional Planning", National Geographer, Vol. V (Planning Number), 1962.

Husley-Tanslcy Report, "Conservation of Nature in England and Wales", Her Majesty's Stationery Office, 1947.

Jain, J.K., "Water Resources of the Arid Region of Rajasthan", Indian Journal of Geography, 1967.

Khasnabis, R., "On the Economic Problem of Sustainable Development".

Krishnamurthy, N., "Papanasam Hydro-electric Project", Journal of the Madras Geographical Association, 1940.

Mandal, R. B., "Nature of Population Geography", Dimensions in Geography, (ed. V.N.P. Singh and R.B. Mandal), Associated Book Agency, Patna, 1975.

Malthus, T.R., A Summary, View of the principle of population in Encyclopedia Britannica Supplement, 132-3, reprinted G.S. Drmko *et al.*, "Population Geography : A Reader", New York, 1970, McGraw-Hill, pp. 44-70.

Raja, M., "Is India Over Populated", Geographer, 1950.

Stamp, L. Dudley, "Applied Geography", Op. Cit.

Stamp, L. D., "Man and the Land", 1955.

Zimmerman, E. W., "World Resources and Industries", Harper and Row, New York, 1961, p. 402.

2

Review of Literature, Approaches, Principle and Hypothesis Testing

Overview

The resource potentialities have been studied and recorded by the specialists of various disciplines such as economists, geologists, geographers, botanists and zoologists. In this field efforts of individuals, institutions and agencies have also been made extensive studies of the country's natural wealth and collected enormous information on the quantity and quality of different categories of resource potential.

In discussing the history of resource use Zobler has divided it into four stages: initial scarcity, abundance, changing abundance and lastly again that of scarcity. The stage of initial scarcity exists for want of the knowledge of resources and technology, the ultimate stage of scarcity is reached when all the resources are near exhaustion. These stages described by Zobler may be considered as ideal and may not be attained in all countries in the same form. Duncan is of the opinion that in India there was a long period of abundance when some vital resource fronts were opened and utilised in a better way.

The subject of agricultural resource has drawn attention of economists and geographers from early decades of the present century. Renne has rightly said, that among, all enterprises 'agriculture is strictly the most fundamental enterprise'. According to J.L. Buck "land utilisation is the satisfaction, which the farm population derives from the type of agriculture developed, the provision for future production and contribution to national needs. Fox has drawn attention to different meaning of land use, as a formal concept and land utilization, as a functional concept. In 1930, Stamp began to organize the Land Utilization Survey of Britain, in which he devised a method of land classification and field investigation. The classification set out by Fagg and Hutchings was better suited to the study of small areas. Similarly, an alternative approach employed by W.D. Jones and V.C. Finch, Jones and P.W. Bryan involved the assignment of four or six digits in the form of a functional notation to separate categories of land use. A further development of land use studies has been in the direction of quantitative description, especially area measurement. Much of the early works were undertaken in the United States by J.M. Trefethen. J.P. Latham, B.J.L. Berry, Berry and A.M. Baker and P. Haggett and Board have compared different sampling schemes for the collection and analysis of land use data. V.R. Singh suggests the adoption of qualitative and quantitative land use studies with special reference to India. J.T. Uyanga explicated that the small farmer limited perception for lack of funds in farming is a dominant constraint to his freedom of choice. R.B. Mandal suggested the method of land capability classification and is of the opinion that land capability should be tested in relation to net sown area and the land capability index required for certain land use. A. Kumar and G.P. Sharma in their recent study presented the temporal dimensions of land use changes from 1951-52 to 1972-73, R.C.P. Singh emphasized the use of linear programming, land reforms and conservation of land for the maximization of agricultural production.

The first systematic research conducted on population geography of India has been the doctoral dissertation of G.S. Gosal under the able guidance of G.T. Trewartha. Chatterjee described in detail the distribution of rural and urban populations in seven macro regions of India. Ahmad had earlier studied the influence of physical and cultural environments on the distribution of population in India as a whole. There are several published articles concerning the distribution and density of population on State, geographic regions, and district level study. Among these mention may be made of Kuriyan's study on Kerala, Verma's study on Punjab, Prasad's on Bihar, Ananta Padmanabhan's on Madras, Chatterjee's on West Bengal and Krishnan's on Orissa. Niliya Nanda published a study on 'Distribution and Spatial Arrangement of Rural Population in East Rajasthan'. Some research publication on Population Growth for areas small than a State are also available. In this category, Chatterjee and Ganguli's study of Naida and Tirunelveli districts, Ramachandra's study of Malnad and Krishnan's study of the border districts of Amritsar and Gurdaspur are noteworthy. R.B. Mandal emphasized that population studies are collections of systematic knowledge of the massing of people on the earth surface and the investigator's duty is to set the things in a right direction. According to V.N.P. Sinha towns act as centres of population concentration which attract rural population immensely and accelerate the pace of rural-urban migration in Ghana.

A systematic regional study of forest wealth was probably first attempted by Hart (1932), who laid stress on the utility and importance of the forest resource to man in context of the Indian economy with a detailed analysis of the forests of the Madurai district, as a case study.

The vegetation of the upper Doab of Uttar Pradesh has been analysed by Mukherjee who has pointed out the geographical distribution of the sal forests, scattered woodlands and scrub vegetation.

Bose contributed a paper on plant geography of Damodar Valley and recorded altitudinal variation of sal in the Upper Damodar Valley from 300 metres in the valley bottom to the plateau at an altitude of 600 metres. Shafi stressed the role of forests in the national economy and focused attention on the need for proper planning based on problems of forests utilization. Pal gave a description of the coniferous forests of the Himalayas, the sal forests of the foothills and the scrub forests of Bundelkhand. Karnik contributed two papers on regional studies of the forest wealth of the Tapti Valley and Khandesh in the Deccan Plateau. Sinha made a systematic regional analysis of the types of forests in Orissa. The work of

National Atlas Organisation may be mentioned here for its forests and land use map in which the distributional pattern of the types of vegetation of India has been clearly shown. Work on Assam's forests has been done by Bor, Stracey and Das, Jacob, Purkayastha, Das, Rowntree, Saikai, Rajkhowa and Rao.

Industrial locations are the milestone of progress and prosperity of the nation. They are like yardsticks to measure the economic growth and cultural advancement. Economists and geographers are generally interested in the spatial variations of industrial locations with respect to the availability of raw materials, labour supply, besides transportation, communication, capital, site, situation and climatic conditions in a particular region. Thomas, in his presidential address to the Calcutta Geographical Society, strongly pleaded for a careful planning of industries in India. Ganguli presented his survey of the Bengal-Bihar industrial region where a marked concentration of mining and metallurgical industries was a significant aspect of India's industrial growth. Karan identified the patterns of the industrial landscape of Jamshedpur. Krishnan taking the example of iron and steel industry emphasized the role of geographical factors such as proximity to raw materials and sources of power in the location of mineral based industries. Hameed focused attention on the cause of Telangana's industrial backwardness and suggested measures for removing the existing regional imbalances in economic development of Andhra Pradesh. Gananathan and Bhanumati presented their survey of the manufacturing industries in and around Poona. Dayal analysed the role of geographical factors in the location of iron and steel and cement industries of India. Durrani examined the locational factors in his study of the industrial development of Rajasthan. Sengupta presented her valuable study of the industrial growth in Hooghly region of West Bengal. Bhardwaj reviewed the industrial growth of the Punjab since 1947. R.B. Mandal discussed about the meaning and scope, concepts, principles, approaches, methods of study and hypothesis testing with special reference to industrial studies. Ishwari Prasad classified forces behind the location of industries. C.R. Pathak emphasized that the growth of urban centres acts as catalyst for industrial development in India. G.N. Singh dealt with the problems of industrial growth, trends in favour of dispersal of industries to underdeveloped areas over the development of small scale and ancillary industries, and the suggestions about planning with special reference to Chotanagpur plateau.

Geographical studies of the transport system of any country have been mainly attempted on a regional basis. W. Linden has remarked, "if the objective of the state is to achieve the optimum economy by regional planning one of the most effective ways to achieve the objective is by controlling transportation. According to Belosouv, "no regional economic location scheme can be considered complete or can be properly evaluated if it does not provide transport links and does not envisage a transport network necessary to achieve the planned development."

Sourairanjan's study of the Buckingham canal represents the first systematic attempt to understand the navigational problem of an inland waterways system in India. Sinha's study of the transport problems of Orissa indicated the inadequate development of road transport in that State. Mukherjee emphasized the need of a new rail link between Calcutta and Bishnupur in view of the inadequate transport requirements of Allahabad, situated on the confluence of the Ganga and the Yamuna and attracting a large inflow or traffic. Patterns of railway traffic flow in Madhya Pradesh were identified by Lakshmi who correlated the State's railway system with its economy and worked out the density of passenger and freight traffic flows. Mayer indicated the role of personal transportation in planned development of a region. Morlol's studies on transport centre influence on poverty and transportation linkage. Richards worked out the technological change in plantation transport system. Shanmuganandan outlined the conceptual development in transportation geography.

India possesses abundant wealth of marine and inland fishes but the geographic study of this important resource is rather inadequate. Sundar Raj presented a description of the pearl fisheries of South India. Banerjee, in his paper on fisheries of Bengal, discussed different sources of the supply of fish in Bengal. Mukherjee, in his study of the river fisheries of Bengal, has given a detailed account of the distribution of fisheries by emphasising the conditions of fishermen and the need of appropriate measures for the conservation and preservation of this resource. Hora contributed valuable papers on the fisheries of India.

About a dozen specialized studies of the distribution and occurrences of various minerals in the country have been made by the Indian Geographers. Chatterjee published a series of papers which contain a detailed account of the geographical distribution of all the important ferrous and non-ferrous mineral deposits of the country. Basu made a study of Darjeeling coal and emphasized its utilization. Negi described the mineral resources of Kumaon and Garhwal region of Uttar Pradesh and pointed out that the main problem regarding full utilization is the inaccessibility of deposits. Pasupatinathan attempted a brief survey of the distribution of mineral resources of India. Mishra described the role of mineral resources in economic development.

Water resource has been a subject of special interest and attention to the Indian geographers who have also studied the problems of flood and river projects. A good account of water resource is contained in Majumdar's book, 'Rivers of the Bengal Delta'. Banerjee published an outline of the devastations caused by floods in the Damodar Basin and discussed briefly the ways of controlling the floods. Mukherjee attempted a survey of the electric installations in pre-independent Bengal. Chaudhuri has attempted a survey of the power zones of West Bengal with special reference to the sources and power requirements of the State. Jain described the surface and underground water resources of Rajasthan. Sinha examined the installed capacity of electric power in Bihar and the total demand for electricity at the end of Fourth Plan Period.

Urban geography is the study of urban places which evolve, grow and exist as services centres, largely for the surrounding area. The earlier geographic studies on the growth of urban places were mainly confined to South India and were published in the journal of the Madras Geographical Association. Mention may also be made of the works of Subrahmanyan, Srinivasachari on the growth of Madras city, Singh worked on Jaunpur, while Mishra and Tiwari on Mathura, Singh on Meerut, Gupta's study about the ancient temple town of Tribeni Bansberia. Thacker analysed that India need not imitate blindly the more advanced countries but evolve its own patterns of industrial and urban development which will suit out social conditions and economic resources. Mathur and Sundram studied about the growth foci in regional strategy through spatial analysis. Mallick worked on the development of small towns for balanced urbanization and economic growth.

With increase in surplus labour in the agricultural sector or farm sector it is necessary in the interest of economic development to find alternative employment for some of the labour now engaged in peasant agriculture. Redundant labour with no marginal productivity is thought to exist as disguised unemployment in the agricultural sector. If demand increases in the industrial sector, the transfer of this labour into industry can hold down, maintain profits, stimulate industrial investment, the marginal productivity of labour becomes much higher in the non-agriculture sector. As a large proportion of population is engaged in the low productivity sector, the level of per capita income of the people is also low. This eventually result in a lower expenditure pattern on industrial products compared to the expenditure incurred on food and other essentials, thus limiting the market for the industrial products and the opportunity of profitable employment of labour, capital and enterprise in the production on non-farm product, there is thus little scope for the growth of the secondary and tertiary sector, not only there is a heavy pressure of population on land but this pressure constantly increases at a high rate with man-hour applied in agriculture may yield zero marginal output.

Models of growth particularly those of Norks, Lewis, Ranis and Fai analyse the process of economic growth of underdeveloped economy in terms of movement of labour from primary sector to the secondary and tertiary sector on the assumption that the marginal productivity of the last man-power on most of the cultivable holding is zero and hence this marginal labour can be transferred to the industrial sector without retarding agricultural production. On the other hand, transfer of surplus labour to the industry of capitalist sector can produce capital goods and add something to the total production.

Nurkse believed that the degree of disguised unemployment varied in the range of 25° to 40° in various overpopulated countries. In his model we find the existence of surplus manpower in the agricultural sector, which according to him, may be effectively used for the construction of capital projects like rail, road, irrigation etc. Employment generated through these projects will raise both output and capital and wages per worker, technique remaining the same. This contention resembles the Lewisian theory of capital-widening process of growth. Lewis' two sector model of growth analyses the process of capital accumulation, output and employment extension by adopting the labour intensive technique of production in a labour surplus economy. The Lewis-Nurkse model of growth favours the adoption of the labour intensive technique of production in a surplus labour subsistence economy, because the supply of labour being abundant is supposed to be perfectly elastic at the subsistence wage-rate. The adoption of the labour intensive technique of production in the industrial sector will employ surplus labour of agriculture. But after the disappearance of the surplus labour from the subsistence sector of the economy, the capital intensive technique of production should be put into operation. Lewis has prepared industrial to agricultural expansion because of his contention that private owned industry has the character of saving and the flow of output should be distributed in favour of the saving class.

In Rains and Fei model the rate of growth of industrial employment and output depend mainly on the strength of two forces—capital accumulation and technological change. The greater the rate of capital accumulation, the greater will be the rate of expansion of industrial employment and output. Moreover in an economy with substantial surplus labour, innovation can also make a significant contribution, especially if they are sufficiently strong and labour absorbing, thus making possible maximum use of the retatively abundant factor. As this process goes on the underdeveloped economy will sooner or later exhaust its surplus labour and will be able to reach a significant turning point in the development process. A quick rate of industrial investment will lead to a rapid increase of the marginal physical productivity of labour which in turn will further stimulate a rapid rate of labour absorption.

In this way we find that every country wants to increase the relative size of the industrial sector. Some may wish to employ the entire population increase outside of agriculture. Indeed where rural overcrowding is extreme it may be necessary to aim at absolute reduction in agricultural employment. In that case still more rapid growth of the industrial sector is required. It is quite natural, therefore , that any effective development plan must take into account a long run target for increase in the relative proportion of labour force in industry. However, there is a limit to the extent of labour absorption in industries, especially in underdeveloped countries.

Khusro has empirically observed in the context of the Indian economy that the non-agricultural sector also suffers from the problem of overpopulation and unemployment. Therefore, it should not be wise to transfer the surplus labour from the agricultural sector. He firmly believes that the need for increasing the supply of food and the need for employment can be simultaneously met by employing the agricultural population within the agricultural sector, on capital construction. The surplus labour can also be employed within the agricultural sector by expanding auxiliary industries like dairy farm, animal husbandry including poultry farm. However, the process of industrialization must be stimulated in order to raise output, capital and employment. Therefore, his strategy of surplus labour transfer is logically based on the theory of balanced growth.

Again, if we assume that the income elasticity of demand for food is less than unity, the transfer of surplus labour from agriculture will give rise to marketed surplus of food to some extent. However, this amount is likely to be less than the demand for food by the transferred workers in the non-agricultural sector. The reasons for this are obvious. When the surplus labour is transferred from farms, those who are left on the farms may consume more than before so that the surplus of food will be less than the previous consumption of the departed labourers. On the other hand, those who are transferred to the non-agricultural sector will get a higher income and therefore may tend to consume more than before. Besides, there will be some cost in providing civic amentities and social overheads for the transferred workers in the non-agricultural sector. This has to be financed by the agricultural surplus. There will also be some cost involved in transferring food which Nurkse has mentioned and some cost on transfer of men, which he did not mention. All these will eat into the surplus.

Apart from this, there are certain practical difficulties in the way of mobilising the concealed surplus of food from the agricultural sector to meet the food requirement of transferred labourers in the industrial sector. As Tripathy points out, "the mobilization of surplus manpowar is not practical through fiscal manipulation. It can be achieved only through organisational and structural change."

The problem of mobilising surplus manpowar would require disentangling of the passive working labour force from farms so that it could be released for utilisation in the industrial sector. The difficulty in mobilising such labour lies in the fact that in the agricultural sector both the passive and effective labour force exist side by side. It is, therefore, quite likely that in the absence of any re-organisation of the rural economy the release of the marginal workers would lead to a decline in the effective labour force causing a reduction in agricultural production.

Moreover, the remaining on farm would be expected to improve the standard of their performance after the transfer of the surplus labour force. As a matter of fact "There is a considerable scope for improvement in the performance of labour in the agricultural sector without any additional input of capital, what would be required is a complete reorientation in the norms of economic behaviour and profit."

However, the two sector growth model given above does not cover a large class of two sector-growth models. Since the total production of the economy is divided into production of consumption goods and investment goods, the planning authority has already decided the allocation of total demand between consumption and investment in determining the general rate of development. This requirement on the demand side is not necessarily matched by the supply side. Since the growth of an economy depends on realised investments, it is highly desirable to pay closer attention to the supply of investment and consumption goods. Thus the natural and simplest division of the economy into two sectors is that between the consumption and investment goods sector.

Mahalanobis and Feldman have also explained the capital goods sector and consumer goods sector in their model. They have not taken into account the agricultural sector separately which Lewis, Nurkse and Ranis and Fei models have explained. In Mahalanobis and Feldman model, total investment is being divided into two portions, one being used to increase the production of basic capital or investment goods, that is K-sector and the other to increase the production of consumer goods that is C-sector.

So that $K + C = 1$

They lay great emphasis on large investment in the capital goods sector in order to maximize the long term growth of the economy as a whole. They strongly advocate the development of capital or investment goods industries so as to maximise the long-term growth of consumer goods industries, and place reliance upon cottage and small industries for supplying consumer goods. In the first phase of industrialisation, employment will be generated in the consumer goods industries. The progress of heavy industries will accelerate the pace of economic growth.

Given priority to the development of the K-sector, the domestic supply of investment goods would become more and more important. Machinery and equipment for the development of consumer goods sector will then be available within the country. In the early phase of development, however, the country has to depend undoubtedly on imports of capital goods which will gradually decline and ultimately there will be no import of capital goods. This of course does not mean that the country will not purchase capital goods from other countries. She could also manufacture and export capital goods as well.

Mahalanobis' and Feldman's stress on the heavy capital goods industries is very sound for an underdeveloped country like India. But investments on large scale capital goods and heavy machine building industries require strong and sound economic position of the country concerned. It must have satisfactory foreign trade position with no balance of payment difficulties, high level of food production etc. In the initial stage consumer goods sector should be adopted, because it is less capital intensive and easier to start off with for a poor country. Consumer goods sector will also solve the unemployment problem by adopting labour intensive techniques.

Malenbaum is not in favour of capital goods industries in the backward areas, because in his opinion, if industrial development continues along the lines of large scale industries, there will be need for huge amount of foreign loans which would bring about an increase in the burden of interest. But industrial development is a process and is directed to the establishment and expansion of modern manufacturing, mainly through large scale industries. Emphasis on light manufacturing, re-organisation of handicraft and cottage industries is only due to the shortage of capital in the less developed countries.

Hoffman analyses that in the early stages industrial development, consumer goods industries are of major importance, their net output being on an average five times as large as that of capital goods industries. In the third stage, the net output of the two groups are approximately equal. In the fourth stage, the consumer goods industries are left far behind by the rapidly growing capital goods industries.

Chenery has classified industries into three groups. He has given a separate category to intermediate goods, because, according to him, their growth nature is somewhat different from the other groups. He is of the view that machinery and equipment sector is most sensitive to resource endowment, a large portion of it can be supplied more economically by import when a country has comparative advantage in the product of the primary sector.

The pattern of industrialization is affected by changes in technology and consumer's habit, capital requirements and so on. In the early phase of development, specialised machinery is rarely used. The small size of market and the frequent changes in the design of products and the production process does not justify heavy investment in capital goods industries.

For the development of the economy, however, capital goods sector should be given greater emphasis, because investment devoted to the capital goods will have a growth inducing influence which investment in the consumer goods sector will lack. It will contribute to the cumulative expansion of the investment potential of the economy. Such industries are of fundamental importance because they have a notable effect on the character and pattern of industrial development of an underdeveloped country. The expansion of the capital goods sector in an underdeveloped economy is the key to the process of capital formation and economic expansion. As the marginal propensity to consume in an underdeveloped county like India is much higher than the propensity to save, the only alternative is to raise the marginal rate of saving by expending the capital goods sector. In other words, the expansion of this sector determines the marginal propensity of the economy to save and invest.

But the growth of a particular industry will not in itself stimulate economic development. It is only a balanced development of various industries which will stimulate development through the process of mutual support, by creating demand for each other's goods. Where a major target is set up for a particular industry, the planning authority should see to it that the various common duties and raw materials which are essential for its production are available. To illustrate, if the planning Authority sets steel target production at 6 million tonnes, it should plan at the same time the availability of adequate quantity of various raw material such as iron ore, coal etc. necessary for its production. The finished steel product shall form an input for several other industries. Therefore, prudent planning also required that the fulfilment of target in capital goods industries should be related to its utilisation in consumer goods industries. All this is necessary because we know that economic activity is closely interdependent.

In analysing the sectoral growth models one objective is thus obvious, and that is industrialising the underdeveloped county. A question may here arise as to why industrialisation is essential for economic development of a less developed country. Economists like Colin-Clark have pointed out that economic growth is positively correlated with the number of working population engaged in the secondary and tertiary sector and low per capita income is associated with high proportion of population engaged in the agricultural sector. Economic growth therefore entails the movement of resources from a low productivity agricultural sector to a high productivity sector that is industry.

Industrial sector plays an important role in economic development because it makes all the idle and dormant productive resources active. In the broader sense it may mean the use of power, tools, large capital investment, sophisticated technology and organisation, all implying an extended division of labour and exchange of goods in money economy. Decreasing returns is completely ruled out in industries due to improvements in techniques and by the frequent introduction of new invention and improved machinery and equipment.

Industrialisation has become one of the greatest world crusade of our time. Late Pt. Jawahar Lal Nehru said, "Real progress must ultimately depend on industrialization." That is why most of the underdeveloped countries are moving towards industrial development. Industrialisation is technically so superior that it tends to raise materially the level of wages, reduces the hours of work and raises the living standard. While giving agriculture due importance, it is realised that agriculture and industry are closely linked in the process of economic growth, but for achieving rapid economic growth, manufacturing industries have to play a pivotal role. The rapid development of organised industries not only stimulates the growth of small and village industries, but also helps in mechanising agriculture, accelerating its growth. Industrialisation, thus, is an aid to agriculture. The pace of agriculture production is likely to increase with industrial development. Demand for foodstuffs and raw materials is likely to be increase during the process of industrialization. In this way we find that the industrial sector has to play a more significant role than agricultural sector in accelerating the tempo of economic growth.

The development of industries, however, is not possible without the help of necessary infrastructures. This comprises of basic services without which primary and secondary productive activities are not feasible. Economic infrastructure includes investments in transport, communication, power, irrigation, etc. Expansion of social overhead capital involves expenditure on education, health, housing and various other welfare schemes. The availability of adequate overhead facilities brings about external economics to other industries, lowers their capital co-efficient and by improving the efficiency of general investment make possible a more rate of economic growth. It serves as a prerequisite to growth.

Industrialisation needs a well disciplined system of transport and communication. The potentialities of infrastructural growth are enormous but they need to be properly utilised and exploited. The role of transport and communication in economic development is comprehensive. It ushers in an era of industrial awakening in the county and has subsequently accelerated the pace of national advance. It enlarge opportunities for social, and potential changes and builds up the economic fabric of the country. It is considered as the vanguard of industrial growth.

Mandal discussed about the meaning and scope, concepts, principles, approaches, methods of study and hypothesis testing with special reference to industrial study. Prasad classified forces behind the location of industries. Singh, dealt with the problems of industrial growth, trends in favour of dispersal of industries to underdeveloped areas over the development of small scale and ancillary industries, and the suggestion about planning with special reference to Chotanagpur plateau.

Power is an important over-head for agricultural and industrial development particularly for achieving a particular tempo of industrial development. This makes necessary the spread of power in order to promote village and small scale industries and increase agricultural as well as industrial production. The increase in the total installed capacity of electricity in the country has been a key factor to its industrial advancement in the country.

People with good health stand better to support an industrial crusade. Improvement in medical facilities and spread of education accelerates the pace of economic development. Education becomes one of the means of vertical social mobility in a technical world. The sweep of education throughout the world has enhanced the need for industrialisation. Therefore, the service sector plays a dominant role in modernising the industrial and agricultural sector.

Lewis' concept of balanced growth shows that income elasticity of demand for industrial precuts is greater than that for agriculture and income elasticity of demand for service is greater than that for agriculture and industry. However, there is one important question involved and that is whether infrastructure investment should precede a growth of industries or a growth of industries should come after industries have grown. This question has not been settled at the theoretical level so far, but many developing countries have experienced bottlenecks in the development of industries without the appropriate development of infrastructure. In fact, development of infrastructure provides the base for industrial construction.

Nevertheless, the agricultural sector cannot be neglected. If agricultural sector is neglected and agricultural output is not raised substantially, agricultural prices will rise. One can hardly cite the instance of a country where the absence of any one of the sectors has been a cause of successful development of the country. Both the sectors need simultaneous encouragement for a well-planned policy of economic development. Only the degree of sectoral adjustments could vary to achieve the highest level of growth.

The sector in the model taken together must cover the whole economy. For example if a model consists of two sectors, a mere subdivision of the economy into agriculture and manufacturing industry is not complete. Such economic activities as trade, transport, education, health power etc. must be included in the model grouping those activities all together as non-agricultural.

Since the developing countries wish to accelerate the rate of development by an intensive utilisation of sectoral resources, the question of the strategy for economic development acquires importance in the planning of development policies. The planning Authority follow the strategy of balanced and unbalanced growth to solve the resources allocation problem among the different sectors of the economy.

Both the balanced and the unbalanced growth are based on the assumption of complementarity between different branches of economic activities. If economic activities in different branches were independent of each other, than both the balanced and the unbalanced growth doctrines would collapse. If the development of one sector does not affect the development of other sectors, the doctrine of balanced growth will become meaningless. On the other hand, if the unmatched growth of one sector does not set any force in action to lead to the expansion of other sectors, then there will be no particular advantage from unbalanced growth. However, there is one basic difference and that is of degrees.

Though balanced and unbalanced growths are based on the same assumption, unbalanced growth theory concentrates on development of a few industries. The strategy of unbalanced growth seems to have been derived from the Feldman model of economic growth which has constituted generally to the kingpin of Soviet planning model and which has emphasised the importance of the expansion of the investment goods sector in accelerating the growth of an economy. This inevitably involves a process of disproportionate development implying a proportionately much higher growth rate for the capital goods sector. This is found to create disequilibrium and great stresses and strain in the growing economy. Prof. Hirschman emphasise the concentration on investments at growing points which generate the highest backward and forward linkage effects, since resources in underdeveloped countries are limited, they cannot be spread over a large number of sectors because in this way they will be practically wasted.

The theory of integrated area development has very recently entered in the field of economics in recent years; a number of economists have evinced keen interest in integrated Area Development Concept. A massive programme of agricultural and rural development is a sensible course of resource development in country. Bihar government has worked out on integrated rural development programme in Bihar (1989-90). Virendra Agrawal suggests the need of evaluating policies and programmes for the rural poor. During the 55th Central Advisory Council Meeting on the 22nd July 1986, the Government of Bihar has organized a symposium on integrated Resource Development of the Bihar state. An integratel resource survey of development planning for Bihar has also been taken in hand by the Government with the help of Geological Survey of India, Calcutta also.

The most outstanding feature brought out by the examination of the work done in the field of resources are informative in character and contain useful data on the amount of resource potential and their regional distribution. Plan oriented studies of resources suited to the economic needs the country focused attention on the objective analysis of the whole process of resources creation and utilization which are strikingly absent. A sound feature that emerges from the report is absence of any systematic regional coverage. A third feature is the relatively inadequate emphasis on problems of conservation. It is, therefore, obvious that there is still a lot to be done to promote the study of resource of the country.

Approaches

In any scientific enquiry of resource appraisal that is customary to have certain approach of study which deals with the sample, regional, general including past, present and future conditions of resources. In the Munger division, the availability of various types of resources, their variable demands set by the people and differential levels of their development have given rise to varying approaches of resource analysis which are as follows:

Retrogressive Approach

This is the fundamental approach of studying resources from inception in a particular geographical environment up to the present distribution and future conditions. Generally, in retrogressive approach past conditions through historical method has been analysed which throws sufficient light on the present and future conditions too. It is true in this respect that past is key to the present. Any resource has its history of development along with the present pattern of distribution and future conditions without which the advancement of resources cannot be measured either quantitatively or qualitatively.

Retrospective Approach

In retrospective approach generally present conditions are studied in detail with due emphasis on the past and also reported on the 'Urbanisation' published in the Draft Five Year Plan.

The theory of integrated area development has very recently entered in the field of economics and geography. In recent years, a number of economists and geographers have evinced keen interest in Integrated Area Development Concept.

A massive programme of agricultural and rural development is a sensible Course of resource development in a country. Bihar Government has worked out an integrated rural development programme in Bihar (1979-80). Virendra Agrawal suggests the need of evaluating policies and programmes for the rural poor. During the 45th Central Advisory Council meeting on the 2nd July, 1976, the Government of Bihar has organized a symposium on Integrated Development of the Munger division. An integrated resource survey and development planning for Munger has also been taken in hand by the government with the help of Geological Survey of India, Calcutta.

The most outstanding feature brought out by the examination of the work done in the field of resources are informative in character and contain useful data on the amount of resource potential and their regional distribution. Plan oriented studies of resources suited to the economic needs of the country focussed attention on the objective analysis of the whole process of resource creation and utilization which are strikingly absent. A second feature that emerges from the report is the absence of any systematic regional coverage. A Third feature is the relatively inadequate farming operations lies there. In this case, the farmers desire to minimize the expenses. In the farming system, it may be wiser to use biotic resource in between three fields in order to minimize the cost.

Prospective Approach

In prospective approach of resource appraisal generally the potentiality and the projection of future conditions of resources are stressed, keeping in view the past and present trends of the growth of resources. In view of the land development it is essential to analyse the future conditions of resources. Almost all the disciplines as economics, geography, sociology, etc. consider resources not only as a general trends of growth but also future conditions due to the growing population and the growth of biotic, human and other resources are of varied nature and hence the study about resource through eco-system has been considered to be the best one.

Sample Approach

In sample approach the investigator tries to choose individual conditions of resource development on smaller levels of territorial unit after selecting only a few from a large observation through purposive, systematic, nested or clustered sampling basis or as it suits the investigator. The study emphasizes on problems of conservation. It is, therefore, obvious that there is still a lot to be done to promote the study of resource geography of the country.

Regional Approach

This approach includes a large economic or physical regions of the study area through which the generally analysed in a nutshell to improve the conditions of the resource appraisal in a larger region.

Principle Approach

In principle approach the generalised conditions of resource such as Von Thumen's theory of resource development and land development around a settlement and the physical and economical optimum and limits of McCarty and Lindberg can be taken into account. Thus any sort of generalisations of any resource in any region comes under the principle approach.

Genetic Approach

In genetic approach the origin and evolution of resources in different historic periods have been analysed. The investigation about the original conditions of resources is necessary in view of their future development. It is impossible to plan for the future without knowing the past conditions. In this way, the genetic approach has been considered to be the best way of resource appraisal for the past, future trends. In resource appraisal the analysis of the present situation in depth automatically gives great impetus towards past and future conditions. Resource development have their own history of growth and development through different phases of technological advancements and production characteristics of the society and the door of progress towards past, present and future conditions of resources.

Predictive Approach

In predictive approach mainly the estimation about the future conditions has been made after considering synoptic situations of the past. Predictive approach is generally applied in almost all scientific disciplines because there is no resource to escape out without predicting the future conditions of resources in any planned economy. Although any prediction or projection is fraught in itself but still there is some hidden merit of which the planning of the State and the Central Government and the former depend.

Prescriptive Approach

In prescriptive approach just like a doctor's prescription about the disease, the investigator tries to prescribe certain modifications in the existing rules of resource development for the good production of resources and developing the resources on desired lines for planned development. It is customary to suggest something in any geographic and economic studies but prescriptive approach of study will solve certain aims of this project in view of planned development of resources in Munger division.

Eco-system Approach

In eco-system approach the ecological balance of different factors like population, land, water, climate, vegetation, drainage lines, relief feature, etc. control the resource development in such a way that each and every type of resource could not be grown with equal production level. In different regions due to the varying eco-system conditions the general conditions of resources in a village is known as sample approach, because in the Munger division a large number of villages are distributed and hence the sample approach has been considered to be the quickest, cheapest, simplest and the best method of research analysis in any part of the globe or any branch of scientific investigation.

Systems Analysis Approach

With the consideration of the process of feed-back mechanism and circular motion of inherent benefit, keeps the resources in such a circular fashion among varying interest of individual, society, fields, technological conditions that each and every resource have to specialize certain specific geographical regions.

Quantitative Approach

Through quantitative approach the estimation and calculation of statistical inference, models, formulae, cartographic analysis of maps, diagrams, charts, mathematical and geometrical solutions of resource development are generally emphasised.

Principles of Resource Planning

Economists and other social scientists have studied resource planning at different levels. Resources are of four types :

- (i) Natural resources like crops, vegetation;
- (ii) Manmade resources, like transport, and communication, industries, housing, establishment and health centres;
- (iii) Biotic resources like wild lives, birds and catching of fish; and
- (iv) Human resources like population distribution, concentration, trends of growth, age-sex structure.

In the beginning man knew nothing about other resources except the importance of human resources. But with the increase of knowledge and dimensions of technical knowledge, the scope of resource study have enormously increased. Hence, economists have been concerned with the linkage of different resource use and their stages of development in order to meet the changing needs and character of resources in different parts of the globe.

The principles of resource study may be summarized as follows :

(i) Principles of comparative advantage,
(ii) Law of diminishing return,
(iii) Principles of economic rent,
(iv) The maximization of production,
(v) Minimization of effort and input,
(vi) Optimization of crops protective space,
(vii) Optimization of the quality of produce in relation to environmental factors and demand.

The principle of comparative advantage is useful for locational study of resources. This may be stated under these two assumptions :

(a) that only natural resources have been used by man whose economic gains are higher, and
(b) that the enterprise of natural resources and human resources are determined in terms of comparative advantage of network.

This principle is mainly useful in comparing the potentiality of resource development for differential enterprises, so that the highest income may be incurred from a particular region. In this sense sometimes the agriculture produce may be more paying than industrial produce goods. This is determined through the difference in demand and cost of agricultural and industrial produce of goods. In areas of land cultivation the rate of production show that the areas of highest hectares may not be the areas of highest yield.

Another principle is the 'law of diminishing return.' The decreasing returns for each additional unit of input is called the law of diminishing return. At this point farmers usually decide that it is not worthwhile to continue increasing input on additional hectares of land in the margin of production.

The theory of economic rent (Ricardo, 1817) is one of them. It arises due to the difference in land, costs of production and transport, storage and improvement in other facilities. Economic rent never affects price but prices affect it. The idea of "economic rent" is useful in understanding the operation of production factors. It is a useful technique to estimate marginal productivity by means of a production function designed to examine productivity in terms of varying inputs of labour and capital. The Cobb-Douglas production function, originally designed to examine quanta of industrial production in terms of carrying inputs of labour and capital, is generally used in the form $P=L^aA^b$, where P is the production, L is the input of labour, A is the area of land to be used, a is the marginal productivity of labour and of land in b proportion to total production.

Agricultural resource is so designed that the maximization of production is possible. Due to the nutritive quality of staple food for men and economic importance, the maximization of agricultural product is necessary. This may be possible by increasing the cultivable land, farm input and technological advancement. The successful operation of this principle is almost a prime necessity in view of providing adequate food to the rapidly increasing human number.

In planned development of resources, the principle of minimization of input is highly needed in changing social make-up of the society for better living conditions of toiling masses and guideline for the regional development and planning.

The principle of optimization of crops protective space is a function of economic value of produce, delicacy of grown methods of cultivation and for time taken in harvesting different levels of protection are necessary for different crops.

The last principle of resources planning is the optimization of the quality of produce in relation to environmental factors and demand. The quality of agricultural produce can be modified with the use of better seeds, fertilizers and irrigation, even if the environmental conditions are not so much favourable and the demand is also lesser. In economic geography the total demand is of less importance in relation to the international market of produce in the present-day world.

Hypothesis Formulation

Before dealing with the hypothesis formulation, it is necessary to know about the general framework for carrying out any research project in the field of resource study. In view of its presentation, some of the steps are as follows :

1. Selecting the problem of research;
2. Stating the problem;
3. Thinking through the purposes;
4. Formulating the hypothesis;
5. Determining the evidence needed;
6. Assembling the evidence;
7. Processing and analysis of evidence; and
8. Presenting the results.

All these are the different steps of research and in this sense one should not go ahead on second step, without clearly understanding the first. Thus hypothesis testing is a series of actions in which the previous step is an undertaking. But a research activity might not complete all the steps, such as it might stop with the formulations of hypothesis only.

Hypothesis shows the relationship between two or more variables. Thus a statistical hypothesis is a formidable answer to a question about relationships. Generally, research hypothesis has been tested against null hypothesis. The hypothesis $X=X_1$ is tested against the alternative $X=X_1$. Where X and X_1 are the means of two obverse variables, whereas = is a sign of not equal to. Hypothesis originate from the observations of events. They provide a bridge between the theory and observed real world situation. This is because researchers generally ask questions about the relationship between variables in light of the existing theory. After the formulation of hypothesis it is advanced for testing and drawing inference.

The testing of hypothesis may proceed by drawing deductive or inductive logic of inferences. In deductive research all events have been explained by citing conclusions which are based upon theoretical principles. In inductive research relationship between two variables is illustrated on the basis of data gathered through experience. This is because we have few principles and laws in economics and therefore, an economical research is mostly inductive. Both theoretically and by common knowledge it could be suggested that the population of the Munger division should have close relationship with the area under different crops between 1999-2000. The problem can be restated by hypothesis that increase in population (dependent variable) in 1999-2000 is dependent on the increase in area under different crops (independent variable). This simple relationship is illustrated in a scatter diagram (Figure 2.1).

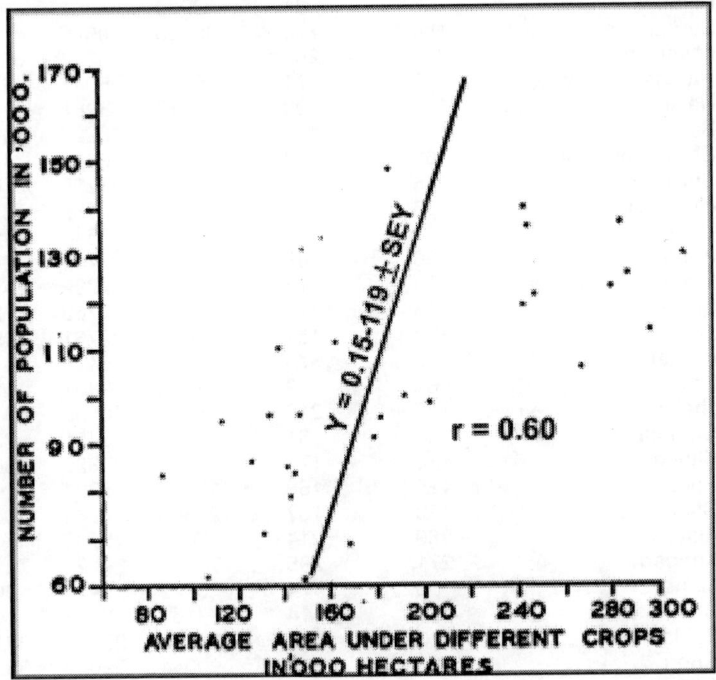

Fig. 2.1 : Munger Division : Relationship between Population and Average Area under Different Crops.

Table 2.1 : Munger Division : Relationship between Population and Average Area under Different Crops (2001)

Name of anchals	*Average area under different crops in '000 hectares(X)*	*Population in '000 (Y)*	x^2	y^2	*xy*
Munger	91	298	8281	88804	27118
Bariarpur	96	92	9216	8464	8832
Jamalpur	51	182	2601	32761	9282
Dharhara	107	104	11449	10816	11128
Kharagpur	236	108	55696	11664	25488
Asarganj	39	60	1521	3600	2340
Tarapur	50	85	2500	7225	4250
Tetiha Bambor	91	58	8281	3136	5278
Sangrampur	48	79	2304	6241	3792
Barahiya	147	123	21609	15129	10081
Pipariya	63	31	2268	961	1953
Surajgarha	184	225	33856	50625	41400
Lakhisarai	238	261	56644	68844	62118
Ramgarh Chowck	75	71	5625	5041	5325
Halsi	114	99	16416	8281	11286
Barbigha	132	115	17424	13225	15188
Sheikhopur Sarai	68	59	4624	3481	4012
Sheikhpura	188	165	35344	27225	31028
Ghat Kusumbha	96	37	9216	1369	3552
Chewara	78	62	6084	3844	4836
Ariari	96	88	9216	7744	8448
Islamnagar Aliganj	139	113	19321	17689	15707
Sikandra	153	113	23409	12769	17289
Jamui	138	181	19044	32761	24978
Barhat	60	74	3600	5476	4440
Lakshmipur	67	99	4489	9801	6633
Jhajha	137	216	18769	46656	29592
Gidhaur	37	61	1369	3721	2257
Khaira	130	177	16900	31329	23010
Sono	125	168	15625	28224	21000
Chakai	163	187	26569	34969	30481
Alauli	260	214	67600	45796	55640
Khagaria	271	295	73441	87025	79945
Mansi	55	74	3025	5476	4070
Chautham	131	114	17161	12996	14934
Beldaur	133	174	17689	30276	23142
Gogri	138	243	19044	59049	33534
Parbatta	224	192	50176	36864	43008
Total	4649	5097	717406	879357	726395

Source : Compiled by Author.

With dependent variables on vertical axis and independent on the horizontal, to know the distribution and magnitude of one variable with the other comparisons may be made so that their correspondence can be estimated.

In order to test the hypothesis an example has been given in Table 2.1.

X = 120, Y = 134

To get the correlation values the formula for Pearsonian correlation coefficient is as follows :

$$r = \frac{\Sigma xy - \frac{(\Sigma x)(\Sigma z)}{N}}{\sqrt{(\Sigma x)^2 - \frac{(\Sigma x)(\Sigma y)}{N}(\Sigma y)^2 - \frac{(\Sigma y)^2}{N}}}$$

Where r = Pearsonian correlation coefficient
Σx = Sum of all values of independent variables.
Σx^2 = Sum of the value of independent variables.
Σy = Sum of all values of dependent variables.
Σy^2 = Sum of square of all dependent variables.
Σxy = Sum of products of all x and y values.
N = Number of units.

With the help of Table 2.1 the product moment correlation coefficient calculation is as follows :

$$\mathrm{r} = \frac{\Sigma \mathrm{xy} - \frac{(\Sigma \mathrm{x})(\Sigma \mathrm{y})}{\mathrm{N}}}{\sqrt{(\Sigma \mathrm{x})^2 - \frac{(\Sigma \mathrm{x})(\Sigma \mathrm{y})}{\mathrm{N}}(\Sigma \mathrm{y})^2 - \frac{(\Sigma \mathrm{y})^2}{\mathrm{N}}}}$$

$$= \frac{-}{\sqrt{\qquad \frac{}{\qquad}}\sqrt{\qquad \frac{}{\qquad}}}$$

$$= \frac{102817}{\sqrt{(195688)}\sqrt{(148637)}} = \frac{102817}{170549} = 0.60$$

r = 0.60

We get the value of product moment correlation coefficient as *r*=0.60. This shows that the relationship is very close in between area under different crops and population. The result has been tested by students. 't' test by the formula:

$$t = \sqrt{\frac{r^2(n-2)}{1-r}}$$

$$= \sqrt{\frac{(0.36)(38-2)}{1-0.6}}$$

$$= \sqrt{\frac{(0.36)(36)}{0.4}}$$

$$= \sqrt{\frac{12.96}{0.4}}$$

t = 32.4

HO : According to Null Hypothesis *(HO)* : There is no correlation between area of land under different crops and population in the Munger division in 1995.

Hi : According to alternative hypothesis (*Hi*) : There is correlation between area of land under different crops and population in the Munger division in 1995.

Table-Value of 't' at 24 degree of freedom *(df)* for 95% level of significance is 1.71 whereas the calculated value of to = 32.4. Therefore, HO (Null hypothesis) is rejected in favour of *Hi* (research hypothesis).

The regression of y upon X has been calculated by the method of least squares according to the formulae :

Y= a + bx

Where Y is the dependent variable and x is the independent variable. The value of band a are given by the following formulae :

$$b = \frac{N.\Sigma XY - (\Sigma X)(\Sigma Y)}{N.\Sigma X^2 - (\Sigma X)^2}$$

$$\text{and } a = \frac{Y - b.x}{N}$$

Where X and Y parameters are the same as mentioned earlier.

In the present case, the value of y = a+b has been estimated as :

Y=0.15X – 119

The accuracy of this estimate of y can be tested by calculating the standard error of estimate. It is a measure of the error to be expected in estimating the value of X by means of the computed value of y.

The formula is as follows :

$$\text{Standard Error (Sey)} = \frac{1-r^2}{\sqrt{N}} = \frac{1-0.6}{\sqrt{38}} = 0.07$$

The standard error of estimate gives the probability of error of two-thirds of all cases on either side of the regression line. Thus the final equation takes the form :

Y = a+bx +Sey

Y = 0.15x – 119 ± 0.07

Need of Hypothesis Testing in Research

To improve the methods of research, economics hypothesis may be stated and tested. In economics the techniques of mapping, photo interpretation, remote sensing and fieldwork enriched the analysis of spatial distributions of crops. All these enable the investigator to observe and record accurately so that hypothesis can be tested against real situations available in the Munger division.

Recently several new techniques have been adopted for hypothesis testing in economics. In 1960s regression techniques were quite common but today factor analysis is very popular among economists. Recently, multidimensional scanning has also been introduced to investigate into the problems related to the choice between alternative locations. These techniques have been borrowed from social sciences to economics to explore the relationship implicit in spatial organisation.

The need of hypothesis testing is to improve the theoretical constructs of the economic discipline. Thus in order to know the exact relationship between theory and occurrences, hypothesis must be developed by those who have sufficient knowledge of the past generalizations published in previous literature must be tested under different cultural circumstances.

Lastly, it can be said that in hypothesis formulation answerable questions should have been examined with the help of field observation, mapping and review of literature. All these help in the refinement of propositions and maximization of research efforts. Hypothesis formulation also represents the initial step in a well established approach of research.

REFERENCES

Ananta Padmanabhan, N., 'Density of Rural Population and Terrain Types in Madras State', *Bombay Geographical Magazine,* 1957, p. 5.

Banerjee, A.K., 'Fisheries of Bengal', *Calcutta Geographical Review,* 1942.

Barker, R. 1997. *And the Waters Turned to Blood.* New York: Simon and Schuster.

Basu, B., 'Darjeeling Coals : Utilization Possibilities', *Calcutta Geographical Review,* 1943.

Belosouv, I.I., 'Transportation and the Formation of Economic Regions', *Soviet Geography,* Review and Translation, Nov., 1964, p. 19.

Berry, B.J.L., 'Sampling, Coding and Storing Flood Plain Data', *United States Department of Agriculture Hand Book,* No. 237, Washington, DC., 1962.

Bor, N.L., 'The Relief Vegetation of the Shillong Plateau, Assam', *Indian Forester,* 1938, 64.

Bryan, P.W., 'A Technique for Recording Land Utilization in City and Rural Surveys' *Geography,* Vol. 16, 1930, pp. 211-15.

Buck, J.L., 'Land Utilization in China', London, 1937.

Chatterjee, S.P., 'Physical Features and Population Distribution in West Bengal', *Calcutta Geographical Review,* 1961, p. 23.

Chatterjee, S.P., "Minerals and Mineral Products", Calcutta Geographical Review, 1942-43.

Chatterjee, S.P., 'Regional Pattern of Density and Distribution of Population in India', *Geographical Review of India,* 1962, p. 24.

Chatterjee, S.P. and A.T. Ganguli, "Population Distribution in Two Districts of India", Calcutta Geographical Review, 1943, p. 5.

Chaudhari, M.R., "Power Resources of West Bengal", Geographical Review of India, 1965.

Das, H.P., Agar perfume Industry in Assam, Gauhati University (Science) Journal, 1959.

Dayal, P. "Aspects of Industrial Location in India", Transactions of Indian Council of Geographers, Vol. I, Dec. 1964.

Duncan, C., Resource Utilization and Conservation Concept Economic Geography. Vol. XXXVIII, 1962, p. 39, 119.

Durrani, P.K., "Industrial Development and Locational Factors in Rajasthan", Annals of Social Science, 1965, Vol.1, p.1.

Fagg, C.C. and G.E., Hutchings, "An Introduction to Regional Surveying", Cambridge, 1930.

Fai, G.K. and Rains, G., "Innovation Capital Accumulation and Economic Review", vol. LIII, p. 248, 1963.

Fieldman, G.A., "A Soviet Model of Growth", in Eassys in Economic Growth, Oxford Univ. Press, p. 223, 1979.

Fox, J.W., "Land Use Survey : General Principles and a Newzealand Example", Auckland, 1956.

Ganguli, B.N., "Iron and Steel Industry in Bengal Bihar Industrial Belt, Calcutta Geographical Review, 1949, Vol. II, pp. 3-4.

Gupta, K.L., "Tribeni-Bansberia, Geographical Review of India", 1960, p. 22.

Haffman, W.G., "The Growth of Operational Industrial Economics", Manchester Univ. Press, 1988.

Hagget, P. and C. Board, Rotational and Parallel Traverses in the Rapid Integrating of Geographic Areas, Annals of the Association of American Geographers, 54, 1964, pp. 406-410.

Hameed, S.A., "Industrial Pattern of Telangana : A Geographical Analysis", Indian Geographical Journal, 1962, Vol. 37, p. 2.

Hart, W.E., "Forest of Madurai District", Journal of Madras Geographical Association, 1932.

Hora, S.L., "Presidential Address", Zoological Society of India, 1951.

Jacob, M.C., 'Regeneration of the Bonsum Amare Forests of Assam', Indian Forester, 1952, 71.

Jain, J.K., "Water Resources of the Arid Region of Rajasthan", Indian Journal of Geography, 1967.

Jones, W.D. and V.C Finch, 'Detailed Field Mapping in the Study of the Economic Geographers of an Agricultural Area', Annals of the Association of American Geographers, Vol. 15, 1926, pp. 148-57.

Karan, P.P., 'Patna and Jamshedpur,' Geographical Review of India, 1952, 14(2).

Karnik, C.R., "Flora of Tapti Valley", Bombay Geographical Magazine, 1956.

Khusro, "Economic Development with no Population Transfer", Asia Publishing House, Bombay.

Krishan, Gopal, 'Regionalism in Growth of Population in Punjab's Border Districts of Amritsar and Gurdaspur', 1951-61,' Geographical Review of India, 1968, 30, pp.1-17.

Krishna, M.S., 'Geographical Control of Mineral Industries', Bulletin of the National Geographical Society of India, 1952, No.16.

Kumar, A., and G.P. Sharma ,'Changing Pattern of Land Use Categories, Land Use Efficiency in Haspura Anchal', in Dimensions in Geography, (eds.) V. N. P. Sinha and R.B. Mandal, Associated Book Agency, Patna, 1979, p. 217.

Kuriyan, G., 'Population and its Distribution in Kerala', Journal of Madras Geographical Association, 1938, p.13.

Lakshmi, B., "Pattern of rail traffic flow in Madhya Pradesh", the National Geographical Journal of India, 1967, vol.13, p. 2.

Latham, J.P., "Methodology for an Instrumental Geographic Analysis", Annals of the Association of American Geographers, 1963, pp.194-211.

Lewis, W.A., "Development with Unlimited Supplies of Labour", The Manchester School, 1958, p. 132.

Majumdar, S.C., "Rivers of Bengal Delta", 1943.

Mahalanobis, P.C., "The Approach of Operational Research to Planning in India", Asia Pub. House, Bombay, 1963.

Mallick, U.C., "Development of Small Towns for Balanced Urbanisation and Economic Growth", Place of Small Towns in India, 1979, p. 55.

Mandal, R. B., 'Nature of Population Geography', Dimensions in Geography, (Eds. Sinha and Mandal), Associated Book Agency, Patna, 1979, p. 6.

Mayer and Kohen, Readings in Urban Geography, Central Book Depot, Allahabad, 1969.

Mishra, S.D., A Note on the Socio-Historical Geography of Mathura, The National Geographical Journal of India, 1958.

Morlok, E.K., 'The Effects of Reduced Fares for the Elderly on Transit System Routes', The Transportation Centre North Western University, 1971.

Mukherjee, B.N., A Rail-Road development Between Howrah and Bishnupur, Observer, 1958, p. 4.

National Atlas of India, Forest and Land Use Map, 1957, p.11.

Negi, B.S., The Mineral Resources of Kumaon and Garhwal, Geographical Review of India, 1960. Part I., The Forest Wealth of Uttar Pradesh, The Geographer, Aligarh, 1950, 3(1).

Nehru, J.N., "Speeches", New Delhi, March 1963.

Nurkse, R., "Problem of Capital Formation in UDC", New York, 1953.

Pasupatinathan, K., 'Minerals of India' Geographical Teacher, pp. 1965-66.

Pathak, C.R., Spatial Variation in Urban and Industrial Growth in India, Dimension in Geograpy, (Eds.) V.N.P. Sinha and R.B. Mandal 1979, pp. 433-450.

Prasad, I., Population Distribution and Dynamics of Supply in Bihar, Geographical Outlook, 1956, p. 2.

Prasad, Ishwari, "Location of Industries Dimensions in Geography (Eds.) V.N.P. Sinha and R.B. Mandal 1979, pp. 407-431.

Purkayastha C.C., Working Plan for Shillong Pine Forest, Shillong, 1948.

Raja, M., Is India Over Populated: Geographer, 1950.

Rajkhowa, S., Forest Types of Assam, Indian Forester, 1961.

Ramchandran, R., 'Population Trend in the Malnad', Deccan Geographer, 1965, 3.

Rao, A.S., The Vegetation of Khasi and Jaintia Hills : Proceedings of the Pre-Congress Symposium of Meghalaya and Eastern Himalayas, International Geographical Union, Department of Geography, Gauhati University, 1968.

Renne, R.R., Land Economics, Harper and Brothers, New York, 1947, p. 26.

Richard, M., Technological Change in Plantation Transport System : A Caribbean Example, 1930-70, The Journal of Tropical Geography, Vol. 41, December, 1975, pp. 1-8.

Saika, M.L., Grasslands of Assam, Symposium on Vegetation Types of India, 1955.

Sengupta, P., Some Aspects of Industrial Growth of Hooghly Region, The National Geographical Journal of India, 1958, pp. 79-139.

Singh, G.N., Problems and Trends of Industrial Growth in Bihar, Dimension in Geography, (Eds.) V.N.P. Sinha and R.B. Mandal, 1979, pp. 471-478.

Singh, R.C.P., Agricultural Land Use Planning in India, Dimensions in Geography, (Eds.) V.N.P. Sinha and R.B. Mandal, Associated Book Agency, Patna, 1979, p. 317.

Singh, V.R., The Research Methodology in Rural Land Use Analysis in India: A Systematic Approach, in Dimension in Geography (Eds.) V.N.P. Sinha and R.B. Mandal, Associated Book Agency, Patna, 1979, p. 216.

Sinha, B.N., 'Forest Economy of Orissa', Indian Geographer, 1958.

Sinha, B.N., Transport and Communication Problems of Orissa, The National Geographical Journal of India, 1957, Vol. 3, p. 2.

Sinha, R.P., Power Development in the State of Bihar, Abstracts of the International Geographical Union Congress, 1968.

Sinha, V.N.P., Changes in the Geographic Patterns of Population in Ghana, Dimensions in Geography, (Eds.) V.N.P. Sinha and R.B. Mandal, 1979, pp. 118-139.

Srinivasachari, C.S., Growth of Madras City, Journal of Madras Geographical Association, 1934, p.14.

Stamp, L. Dudley, Land Use and Resources : Applied Geography, Penguin Books, 1960.

Stracey, P.D. and H.P. Das, 'Assam's Economy and Forest', Shillong, 1949.

Thacker, M.S., "India's Urban Problem", University of Mysore, 1965, pp. 1-49.

Thomas, L. 1974. The Lives of a Cell: Notes of a Biology Watcher. New York, Viking, p. 153. UNDP (United Nations Development Programme). 1999. Human Development Report 1999. New York, Oxford University Press. [Essential reading and reference.]

Tiwari, A.K., Mathura: A Study of Site and Evolution, The National Geographical Journal of India, 1963, Vol. 9, pp. 41-50.

Trefethan, J.T., A Method for Geographical Surveying, American Journal of Science, 32, 1936, pp. 454-64.

Uyanga, J.T., Measuring the Perception of Farming problems by Small Scale Farmers in a Nigerian Locality, in Dimensions in Geography, (Eds.) V.N.P. Sinha and R.B. Mandal, Associated Book Agency, Patna, 1979, 216.

Verma, S.D., 'Density and Pattern of Population in Punjab', the National Geographical Journal of India, 1956, p. 2.

Zobler, L., "An Economic Historical View on Natural Resource Use and Conservation", Economic Geography, Vol. XXXV1II, 1962, p. 191.

3

Profile of the Region

Resource appraisal of an area cannot be elaborated without knowing the geographical and economic background. Keeping this in mind resources are of various types including land, water, air, and other. But the ecological condition of their origin and growth varies considerably in different parts of the Munger division. Such conditions are the product of the geological, physiographic, climatological, floral and agricultural-cum-socio-economic conditions of the region under study. These conditions vary considerably from area to area because the southern portion is of rugged topography having forest infested hills and valleys, but the northern portion is almost fluvial alluvium deposits which is deposited in pleistocene period due to the filling of the fordeep which formed after the rise of Himalaya in Eocene, Miocene and Pliocene periods.

Location

The Munger division is located in between 24°22' and 25°49' north latitudes and between 85°36' and 86°51' east longitude covering an area of 7884.76 square kilometres. The nearest neighbour in the north are the district of Saharsa, Darbhanga and Samastipur, in the east lie the districts of Banka, Bhagalpur and Madhepura, in the west lie the districts of Begusarai, Patna, Nalanda and Nawada, and in the south the districts of Giridih and Deoghar (Figure 3.1). The river Ganga flows through the division from west to east and divides it into two unequal parts. The northern smaller part is a flat, alluvial plain traversed by the Burhi Gandak river, which flows through it from the north-west to south-east. The southern portion is also to a great extent alluvial but the general level is higher, and the surface, more undulating. A large area particularly at Jamui and Kharagpur District is composed of hills and valleys covered with forest tracts and bushes.

Climatologically, the Munger division lies in temperate regions where the effect of north-east and south-west monsoon rains are highly felt. But economically it lies in the transitional belt of a granary land in the north up to the forested and mineralised belt of Chotanagpur scrap belt in the South.

The division is subdivided into five districts namely, Munger, Jamui, Lakhisarai, Sheikhpura and Khagaria having thirty-eight anchals (Fig. 3.1 and Table 3.1).

Geology

Geologically, the Munger division lies in the southern fordeep of Tethys sea which filled up by the Himalayan or Indo-Brahm river during Pleistocene period. This lies in a saucer-shaped basin filled up by the south flowing rivers originating from the Chotanagpur plateau. The master stream of river-Ganga flows from the region having meandering courses engraved in the soft alluvial soil with its inception in Tertiary Era. But the southern part of the Munger division is composed of hard Gondwana and Dharwar systems of rocks newly covered by the alluvium in which the deposition of mica, bauxite, and other deposits are mainly found. The location of Pirpahari near Munger Jamalpur ridge, Kharagpur and Sheikhpura hills are the fine examples of lier found in the district.

Important rocks found in Munger are quartzite, slate, phyllite, lime-stone, peridotite, biotite, pyroxine, mica schects in the lower portion of the basal rock. But in the north Khadar of Silica deposit are found due to annual flooding in the river which are the rule. Near Chotanagpur scrap belt mainly in areas of Jamul, Chakai, Jhajha and Lakshmipur the coarse alluvium soil mixed with coarse sand are the rule which is known as bhangar soil or older alluvium.

Table 3.1 : Some Relevant Information of Munger Division (1961-2001)

Name of Districts and Anchals	Area in sq. km.	Total population					Density of population					No. of villages 2001	No. of towns 2001
		1961	1971	1981	1991	2001	1961	1971	1981	1991	2001		
1	2	3	4	5	6	7	8	9	10	11	12	13	14
Munger	242.00	148,693	173,931	215,727	254,105	297,741	614	719	891	1050	1230	137	Munger
Bariarpur	160.52					92,406					576		
Jamalpur	85.45	125,646	146,636	179,074	210,931	181,751	574	670	817	963	1133	126	Jamalpur
Dharhara	282.85	53,307	62,836	72,132	84,964	104,037	188	229	256	301	368	186	Bariarpur
Kharagpur	306.79	100,203	199,116	160,241	188,748	108,409	265	315	423	499	353	146	Kharagpur
Asarganj	59.48					59,507					1000		Asarganj
Tarapur	71.04	67,779	83,811	104,804	123,449	84,694	519	642	801	802	1192	149	
Tetiha Bambor	101.38					57,734					569		
Sangrampur	85.94	60,205	71,060	68,123	80,242	78,518	420	496	476	560	914	119	
Barahiya	241.06	88,090	96,547	92,689	169,178	122,741	309	369	354	647	509	48	Barahiya
Pipariya	58.39					31,114					533		
Surajgarha	389.66	108,701	136,757	166,162	199,129	224,587	260	327	397	476	576	196	
Lakhisarai	283.30	112,244	140,492	188,990	228,829	261,871	324	406	546	661	924	153	Lakhisarai
Ramgarh Chowck	104.27					71,112					682		
Halsi	151.06	64,908	84,247	102,659	142,299	90,800	236	360	373	452	602	101	
Barbigha	94.36	85,521	100,766	121,066	145,085	115,460	572	674	809	970	1224	98	Barbigha
Sheikhopur Sarai	55.12					58,833					1067		
Sheikhpura	185.18	89,356	106,983	127,420	152,700	165,018	355	425	505	606	891	112	Sheikhpura

Name of Districts and Anchals	Area in sq. km.	Total population					Density of population					No. of villages 2001	No. of towns 2001
		1961	1971	1981	1991	2001	1961	1971	1981	1991	2001		
1	2	3	4	5	6	7	8	9	10	11	12	13	14
Ghat	92.59					36,504					394		
Kusumbha													
Chewara	115.06					61,563					535		
Ariari	146.65	57,442	68,362	80,425	96,381	88,124	281	334	393	471	601	88	
Islamnagar	172.89					113,250					655		
Aliganj													
Sikandra	184.01	99,218	126,830	154,319	188,15	112,517	279	357	435	530	666	135	
Jamui	173.91	67,604	85,368	103,425	126,075	181,227	389	391	594	725	1042	81	Jamui
Barhat	232.16					74,385					320		
Lakshmipur	251.77	91,733	111,676	138,135	168,386	98,865	168	205	346	371	393	138	
Jhajha	427.39	90,165	110,846	131,492	160,289	215,551	211	259	314	375	504	197	Jhajha
Gidhaur	71.11					60,670					853		
Khaira	418.71	78,295	96,225	117,269	142,951	177,008	183	255	274	334	423	102	
Sono	392.27	67,269	86,103	100.072	122,012	168,289	171	220	255	311	429	250	
Chakai	774.04	78,501	95,386	113,924	138,873	187,034	101	123	147	179	242	600	
Alauli	274.47	95,240	114,272	128.810	165,444	213,845	347	416	469	603	779	45	
Khagaria	262.36	115,472	137,534	174,962	224,721	295,480	440	524	667	857	1126	49	Khagaria
Mansi	70.01					74,299					1061		
Chautham	163.97	74,885	91,063	109,174	140,223	113,761	320	389	466	599	694	41	
Beldaur	223.56	67,107	78,296	92,124	118,324	174,456	300	350	412	529	660	29	
Gogri	250.31	85,854	98,755	146,849	188,613	243,303	420	483	718	922	972	39	Gogri
Parbatta	241.04	102,146	121,282	121,739	1156,361	192,212	356	423	424	545	797	85	

Source : District Census Hand Book, Munger, 1961, 1971, 1981, 1991 and 2001.

Fig. 3.1 : Munger Division : Location and Administrative Units.

In Fig. 3.2 the distribution of minerals in the Munger division is shown (Table 3.2).

Table 3.2 : Production of Minerals in Munger Division (1996-2000)

Name and production of the minerals	*Year*			
	2002-03	*2003-04*	*2004-05*	*2005-2006*
Ordinary Stone (in cubic feet)	13,079,500	18,045,500	20,000,00	32,154,538
Sand (in cubic feet)	2,215,300	2,610,000	2,800,000	1,462,453
Slate (in sq. feet)	4,500	2,000	2,000	1,000
Brick (in number)	22,300,000	24,000,000	28,000,000	217,961,608
Silica (in million tons)	42,000	50,000	35,000	23,409
Mica (in kilogram)	2,278,900	228,400	346,000	4,123,500
Soap Stone (in million tons)	390,406	312,221	4,984,547	510,206

Source : Mining Officer, District Mining Office, Munger, 2007.

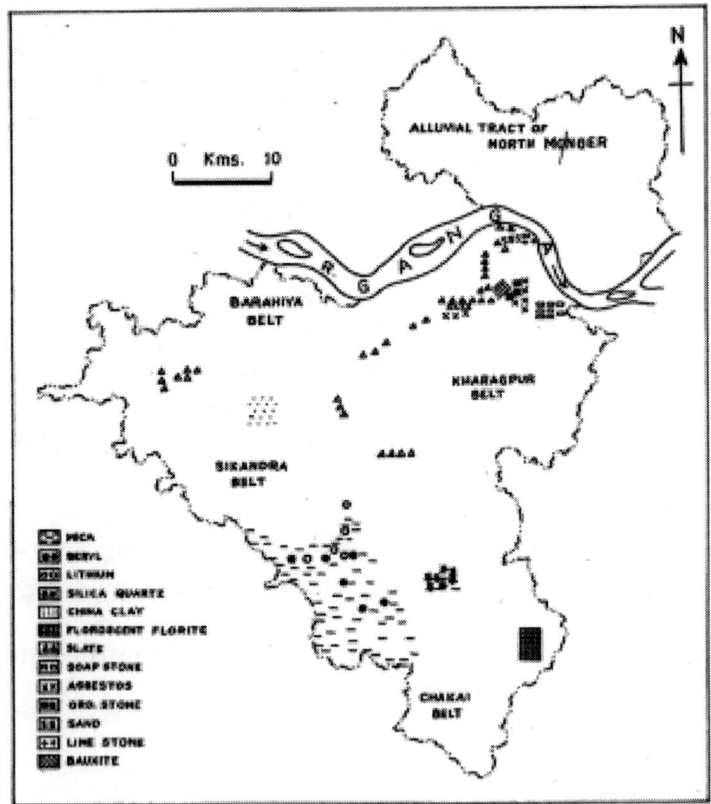

Fig. 3.2 : Munger Division : Occurrences of Minerals.

Bauxite: Bauxite and laterite occur on the top of a few laterite capped hills, e.g., Khapra, Maruk, Maira and Thaidi of which the Maruk hill is most important. Both pisolitica and massive bauxite ore generally occur through excavation at 350 to 400 metres depth. Geological Survey of India carried out preliminary investigation in these four hills, and inferred reserve of 1.5 million tons of medium to high grade bauxite has been estimated there.

Beryl: Occurrences of beryl in association with mica pegmatites are widely known in the Munger division. Some of the important localities being Bijaia, Lewa, Asarhua, Chatkarm.

Building Materials: The most important building material is granite. Fresh as well as jointed varieties suitable for making slabs are found. Besides, there are occurrences of travertine, Kankar, and chalk-concretions which are suitable for lime on small scale.

Clay/Fire/Clay: Extensive occurrences of clay occur at the top of Khapra, Maruk and Mtairu hills. Deposits of white clay occur in the west of Pirpahar hill, 4.8 kilometres east of Munger. Some of these may be suitable for ceramic, textile or refractory industries.

Felspar: Good quality of Felspar mainly microcline occur in mica pegmatite at a number of localities of which Borgi and Marwa are important.

Kyanite: Pockety segregations of bladed Kyanite have been noted in association with quartz vein. These blades measure as much as 20 cms. × 3 cms. in size.

Mica: The Bihar mica belt which extends through Munger is an important source of Mica. Important deposits are located at Behra, Harni, Mamva, Bongi and Donta. Rocks of good quality ruby mica ranging in size from 2 cms. to as much as 30 cms. across are found. Besides the occurrence of lepidotite at Bijaia is also worth mentioning.

Mineral water: The hot springs of Bhimbundh and adjoining areas are important for their medical value. Their potentiality as a source of geothermal energy is under study.

Quartzite: Extensive reserves of quartzite suitable for use as furnace lining and for the manufacture of Silica bricks occur in the Kharagpur hills about 11 kilometres from Jamalpur. Important occurrences are at Barhat, Bhimbudha and Kemal.

Other mineral occurrences in the division include asbestos at Goria Khoghat and at the Pirpahar hill, Graphite at Baghmari, Corcendum in the Jamui hills, Bermiculite and lead ore in the neighbourhood of Chakai. Besides there are minor occurrences of ilemenite uranium ore, tin ore and sillimenite.

Physiographically, the Munger division may be divided into five parts after considering the contour heights from mean sea level. Topographic conditions, rock structures and patterns of drainage lines are shown in Fig. 3.3.

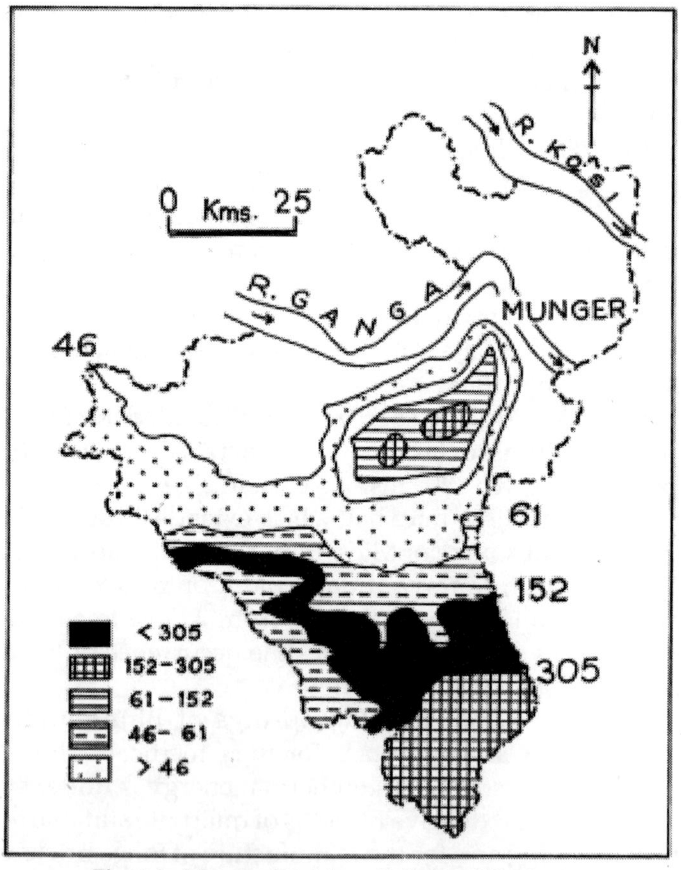

Fig. 3.3 : Munger Division : Relief and Drainage.

In Munger division 305 metres of contour line covers part at the top of Kharagpur hill. In Sikandra, Sono, Khaira and Jhajha anchals the area is 305 to 152 metres high and the similar position is found below the top of Kharagpur hill. The area above 152 to 61 metres from mean sea level is found north of Sikandra and Jhajha, and the fringe area of Kharagpur hills. The area above 61-41 metres high from mean sea level is found north of Sheikhpura, Sikandra and Jhajha area and around Kharagpur, Jamalpur and southern part of Munger anchals. Beyond this almost all areas of Barahiya Tal, Ganga riparian tract and the Gandak-Kosi basin are below 46 metres high from mean sea level.

The northern part is a saucer shaped region of North Bihar where almost all major rivers concentrated to meet with the Ganga. For this fact, the location of riverine flood plains, chaurs, marshy-land and diara are the rule.

South Munger comes under the stable shield of granitic and metamorphic rocks of Archaean age, the oldest system recognised by geologists, rises gradually towards southern boundary. These shields are ridges of hard rocks of granite-family, but the general character of the country is that of an old peneplain dissected by steep gradients of rivers which cut their valleys downwards, offers evidence of fairly recent uplift and this also has been responsible for younger stage relief features. Further in the South, there are several types of schists and gneisses, as for example mica schists, phyllite and slate. Mica schists composed chiefly of muscovite, hiotite, quartzite and felspar with subordinate magnetite and chlorite. Slate and phyllite are best developed in Kharagpur hills. The soil cap of schists and slate are fairly thick. The gneiss and schists are sometimes capped by laterite which is residual product of weathering of these rocks. A large development of laterite is seen in Kharagpur hills, where it has resulted from the alteration of ferruginous slate and phyllite and ferruginous quartzite. Figure 3.4 shows the river geomorphology in the environs of Munger especially its northern part where the mighty river Ganga flows from west to east in a semi-circular fashion projecting its head towards the north. The middle part of diara is called Taufir where swales are divided into numerous branches with intervening sand bodies. The unequal distribution of sand bodies is seen in this diara tract where the sign of arrow shows the direction of the movement of flood water and flow of sand along with running water especially during floods. A ridge is found just north of Munger near Kastharni ghat. In north-western part (Fig. 3.3) Burhi Gandak river makes a confluence with the Ganga and similarly the Kosi river makes a conflucnce in north-eastern part with the Ganga.

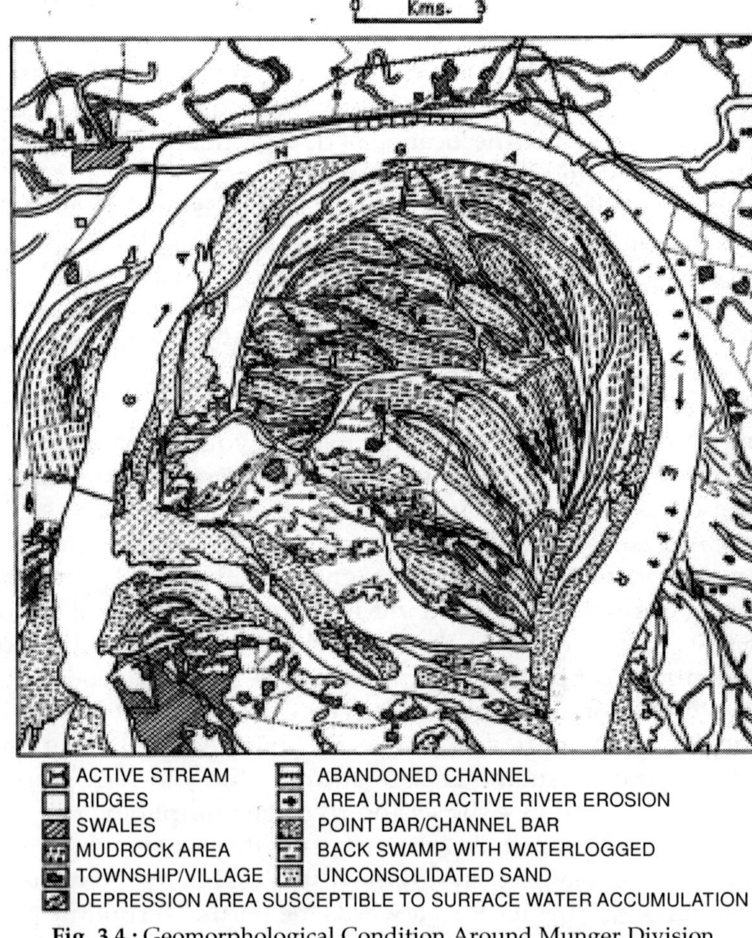

Fig. 3.4 : Geomorphological Condition Around Munger Division.

North of the Ganga quite a large number of abandoned channels are found. In this way, it is quite clear that the entire area is divided into a small piece of land with a variegated picture of landforms especially due to the erosion and deposition of sand by the flowing water of the Ganga in flood. The main feature of this area is active strong ridges of hard stones, swales in the sandy diaras, hardy granite rock near Kastharni in Munger and abandoned channel north of the town of Munger. Areas under active erosion are shown with the sign of arrow. The dotted area is a sand bar, broken strip is waterlogged swampy ground, which is found in the diara areas.

There are several peaks rising to the height of 450 metres and the highest point is 542 metres above MSL. In south low ranges and isolated peaks are the rule. These seem to be the extension of outlying parts of the Vindhyan system of rocks. The description of hills are as follows:

***Kharagpur Hills*:** It is an important and extensive hill range in South Munger which covers 676 square kilometres of area. This hill forms, a distinct watershed, the country to west being drained by the river Kiul and that to the east by the Man and other streams. This forms an irregular triangular block extending from Jamalpur to Jamui railway station. This area consists of a number of steep ridges rising from the low ground on all sides, with scarp face of massive quartzite in some places. In between these rocky and irregular hills, some valleys are fairly extensive with patches of cultivation around. In valleys of hills several hot springs are found as for example Bhimbundh, Sitakund, Rishikund, etc. These hot springs are probably due to the deep rooted thermodynamic action and the existence of active volcano in this region during cretaceous period for which the Chinese traveller Hieun Tsiang had recorded in the Seventh Century A.D. "that by the side of the Capital and bordering on the Ganges river is the I-han-mountain, from which are belched forth masses of smoke and vapour which obscure the light of the Sun and Moon". Bhim bundh is associated with the Pandavas while Sitakund in the memory of Sita. Rameshwar hot spring is also a sacred bathing spring. Near south-western fringe of these hills of Shringirikhi hot spring is found which is a sacred place associated with Shringirishi. There is a beautiful five mouthed fountain and crystal clear bathing kund also. There are several peaks rising to a height of about 450 metres and the highest point is Maruk (488 metres of about sea level). It is a table topped hill covering with deep layers of laterite. Among other peaks the height of Maira is 458 metres, Manikasthan 567 metres and Abhainath 482 metres.

Gidheshwar Hills: South-west of Kharagpur hill lies the Gidheshwar hills. It is a continuation of hills of Nawadah district and covers an area of 210 square kilometres forming a compact cluster between Khaira and the western boundary of Munger. In the east there is a cliff overlooking Khaira and the Kiul river, and this area continues in the south into the rocky valley of the Kiul river. On the south in the village of Sakdari there is a hot spring "Panchbhur" which is surrounded by precipitous walls of rock. The "Panchbhur" is river originated from the five splendid springs and hence the name of the streams. The water of this spring is acidic, probably a continuation of the base of the sulphuric hot springs of Kharagpur hill. Long ridge of the mountain is known as hill range. The highest peak is Ekgora having a height of 544 metres.

***Satpahari Hills*:** In south-western Munger there is a high hill named as Satpahari beyond which the Kiul river breaks through the range by a narrow gorge. The hill is 542 metres above mean sea level and 360 metres above the country at its base. There are also hills in the level alluvial plains near Sheikhpura which abruptly rise in the south but in the north crags are almost precipitous. There are small, isolated stony hills of fair size on both sides of Lakhisarai-Jamui road south of Titachat in the plain.

Climate

Climate of this area can be described as transitional in between parching heat of the west and moist of the east. The hot weather starts in the beginning of March and lasts till the middle of

June. At this time stray cases of 'Loo' are seen in summer and dew fall in winter. In the evening the temperature falls down and the night is not so oppressive. The monotony is broken down, there is strong gale and storm, which is known as Kal Baisakhi. The mercury drops down considerably only to rise again after a few days.

The monsoon usually breaks in the third week of June and lingers till September and sometimes till October. Though the temperature falls down, the heat is oppressive in the day, while it is stuffy and sultry during the night. The average rainfall being 19.36 centimeters, 33.78 centimeters, 29.93 centimeters and 24.30 centimeters in June, July, August and September respectively.

The cold season starts from November and linger, till February. In November morning and evening are very cold, in December and January the temperature goes down and the night is very cold. But in February the night is still cold but the day becomes progressively warmer until the hot weather sets in March.

Table 3.3 : Mean Monthly Temperature and Humidity of the Munger division (2008)

Name of months	*Mean daily maximum temperature in °C*	*Mean daily minimum temperature in °C*	*Mean relative humidity per cent at 8 hrs.*	*Average monthly rainfall in cm.*	*Mean No. of rainy days*
1. January	23.22	11.16	24.44	2.35	0.5
2. February	28.83	13.33	13.88	0.46	1.4
3. March	31.72	17.77	12.77	0.97	1.1
4. April	35.88	22.00	15.00	3.43	1.3
5, May	37.11	24.41	20.50	10.81	3.3
6. June	34.43	25.72	27.22	23.40	10.9
7. July	32.61	26.16	30.55	26.20	14.6
8. August	31.83	26.55	31.11	22.15	14.6
9. September	31.88	25.50	28.88	13.92	10.5
10. October	31.11	22.61	24.44	0.20	2.3
11. November	27.83	16.55	22.77	0.10	0.6
12. December	23.55	11.41	23.33	-	0.2

Source: 'Working Plan'—District Forest Office, Munger, 2009.

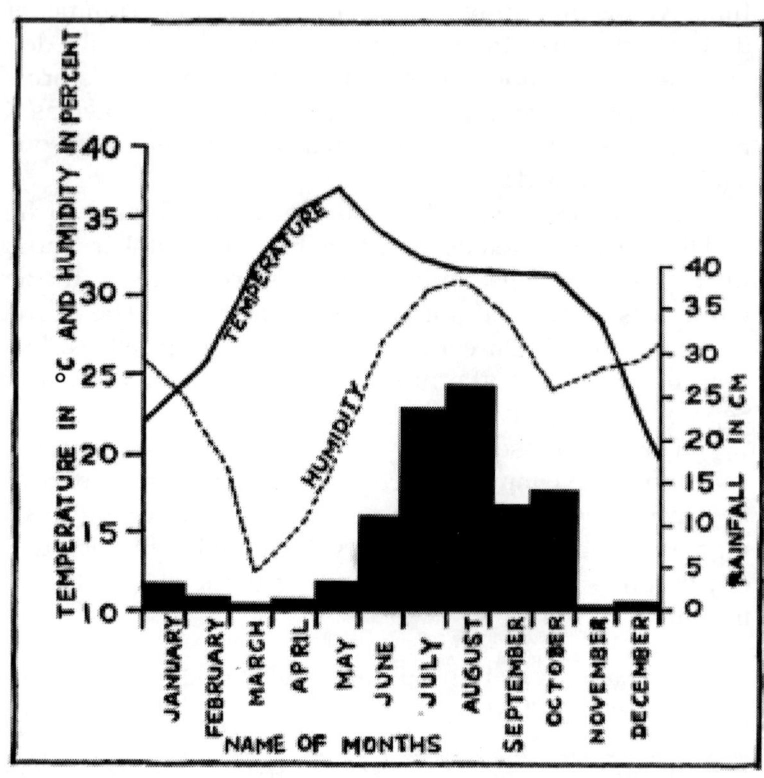

Fig. 3.5 : Munger Division : Temperature, Humidity and Precipitation Graph.

Figure 3.5 and Table 3.3 show the distribution of temperature, humidity and precipitation graph of Munger. The average annual temperature is 25°C in the Munger Division. The highest temperature (45°C) is recorded in the month of May every year. The lowest winter temperature was recorded 11°C in 1980. In day time generally the temperature is 5°C high in comparison with the night. The variation of winter and summer temperature is around 30°C. According to the range of temperature the summer months are warmer and the winter months are colder. In summer months of May and June the local wind 'loo' is quite prevalent in Munger.

In Munger division the relative humidity is high but the absolute humidity is low. The relative humidity is about 90 per cent in July, August and September but it is lowest especially in the month of March. In rainy season the dampness is quite felt and this leads to waterlogging in low lying areas which is hazardous for health. This is because the mosquito breeding on such stagnant water quite often give rise to Malaria and Kalazar epidemics. From March to July due to occasional rain, the humidity increases slowly and slowly in April, May, June and July, but just after rainy season at the time of retreating monsoon to October and November, due to excessive damp ground the temperature decreases.

The distribution of rainfall is quite uneven from January to December in the Munger division. About 80 per cent rain falls from June to October and the highest in the month of August, 262 mm. (Table 3.4). In November and December, the rainfall is just 54 mm and 6 mm respectively in the month of January with the onset of westerlies about 112 millimetres rain falls but it decreases towards March. Beginning from March up to July it increases towards the highest amount of 303 millimetres in Munger recording station.

The monthly distribution of rainfall in the Munger division is quite uneven (Fig. 3.6). As for example, in the month of January Munger block receives 11.8 millimetres rainfall whereas it is 21 millimetres in Kharagpur and 23 millimetres in Sono. These areas are forest covered mountainous parts of the district. Alauli anchal of North Munger receives just 5 millimetres rainfall but 13 millimetres in Tarapur and Sangrampur anchals during the same period. The highest amount of rainfall 334 millimetres receives Ariari anchal and the lowest amount of 98 millimetres in Surajgarha anchal. In the month of August the highest rainfall is 331 millimetres in Lakshmipur and Jhajha anchals.

Whereas the lowest amount of 94 millimetres falls in Barbigha anchal of South Munger. In total, the highest amount of 1,293 millimetres rain falls in Jhajha whereas the lowest 644 millimetres in Barbigha anchal of South Munger. In other anchals more or less the same condition prevails.

Table 3.4 : Blockwise and Monthly average rainfall in Munger division (1961-2001)

(in mm, 2001)

Name of Districts and Anchals	*Jan*	*Feb*	*Mar*	*Apr*	*May*	*June*	*July*	*Aug*	*Sep*	*Oct*	*Nov*	*Dec*	*Annual rainfall*
Monghyr													
Munger	11	9	9	3	3	157	323	180	220	93	5	2	1,048
Bariarpur	10	10	9	4	3	157	323	176	220	92	5	2	1,045
Jamalpur	10	10	9	3	31	158	315	170	225	90	5	2	1,028
Dharhara	11	3	1	5	29	149	307	170	213	92	4	2	1,000
Kharagpur	21	12	5	38	47	127	227	235	253	93	3	2	1,063
Asarganj													
Tarapur	13	13	20	4	32	146	201	257	182	65	6	24	963
Tetiha Bambor	13	13	24	39	35	154	200	258	182	61	2	20	1003
Sangrampur	13	13	28	40	34	150	200	256	180	60	2	20	997
Lakhisarai													
Barahiya	-	-	-	-	-	-	-	-	-	-	-	-	1,000
Pipariya	-	-	-	-	-	-	99	171	229	98	-	-	601
Surajgarha	-	-	-	-	-	-	97	170	233	93	-	-	595
Lakhisarai	-	-	-	-	15	76	291	208	138	39	-	-	767
Ramgarh Chowck	10	5	9	5	31	157	257	298	180	39	10	5	1,008
Halsi	9	5	8	5	33	155	259	294	183	43	10	5	1,007
Sheikhpura													
Barbigha	-	-	-	-	-	55	232	94	168	95	-	-	644
Sheikhopur Sarai	-	-	-	-	-	56	235	94	170	95	-	-	649
Sheikhpura	18	8	12	6	30	149	289	319	260	55	8	-	1,144
Ghat Kusumbha	17	8	12	6	30	147	286	320	261	55	8	-	1,141

Chewara	-	-	-	-	22	96	335	206	189	48	5	-	902
Ariari	-	-	-	-	23	94	334	205	186	46	5	-	894
Jamui													
Imamnagar Aliganj	18	9	11	17	22	104	256	260	188	32	8	2	927
Sikandra	20	8	10	15	20	100	246	261	198	30	8	1	917
Jamui	10	11	21	19	33	170	271	255	177	40	8	3	1,017
Barhat	18	7	15	12	25	109	241	265	191	35	6	1	925
Lakshmipur	20	8	10	15	20	100	246	261	198	30	8	1	917
Jhajha	19	12	21	13	51	166	323	331	275	65	14	3	1,293
Gidhaur	21	11	15	15	32	104	301	298	206	41	11	2	1057
Khaira	-	-	-	-	-	-	-	-	-	-	-	-	1,000
Sono	23	5	17	14	26	150	-	-	218	49	15	-	1,020
Chakai	10	5	16	27	44	139	-	-	140	67	-	-	-
Khagaria													
Alauli	5	1	4	23	54	104	207	220	179	68	8	1	954
Khagaria	10	5	16	27	44	139	-	-	14	67	7	4	943
Mansi	-	-	-	-	-	-	-	-	-	-	-	-	1,000
Chautham	-	-	-	-	-	-	-	-	-	-	-	-	1,000
Beldaur	-	-	-	-	-	-	-	-	-	-	-	-	1,000
Gogri	8	7	16	22	43	186	237	233	212	67	9	22	1,040
Parbatta	7	9	26	4	37	158	243	266	207	53	40	04	1050

Source: Statistical Office, District Statistical Office, Munger, 2001.

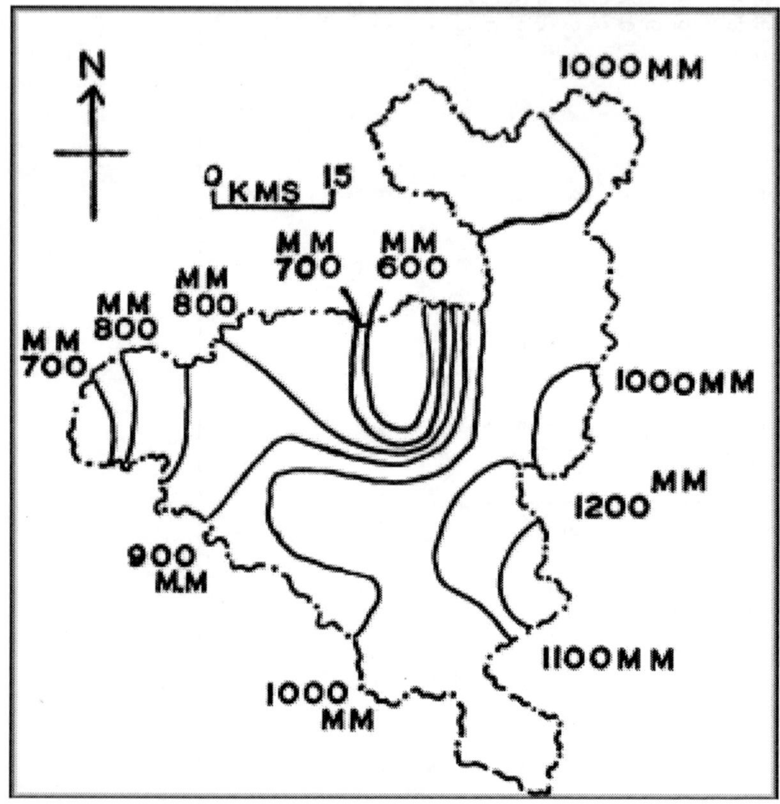

Fig. 3.6 : Munger Division : Distribution of Rainfall.

Drainage

The Munger division is rich in natural drainage lines as the master-stream of Ganga runs from Barauni in the west up to Gogri in the east which covers a distance of 42 kilometres. Beginning from the west up to 43 kilometres near Surajgarha this river flows towards north-east with a sharp bend towards north near Munger. The river Burhi-Gandak flows south-east after originating from the Tarai area of Champaran district and enter into the Munger Division west of Khagaria and ultimately meet with the Ganga west of Khagaria town. It is a navigable river with a tributary of Balan which flows in the region north of Khagaria. The river Bagmati also traverse the area north of

Khagaria and ultimately join with the Tiljuga river near Chautham. Kamla river enters near Khagaria and joined by the Bagmati in the lower riches. It is known as Ghugri river which flows towards south to meet with the river Ganga. In the southern part of the district Kiul, Barnar, Allai, Anjan and Ajai rivers are flowing from Chotanagpur plateau and ultimately meet with the Ganga.

Thus in South Munger the principal rivers are Kiul and the tributary Man. Besides this there are also a number of smaller streams which come down from hills during rains, but subside as rapidly as they rise. For the greater part of the year there are sandy water course with little or no current and exhausted by the demand for irrigation before they reach the Ganga. Figure 3.7 shows the relationship between forest and drainage lines towards east near about the area of Jamui-Bhagalpur road. This map shows the intensity of drainage lines and the distribution of dense forest with special reference to 1″= One mile Toposheet Number 72 K. This area is located in between 25° N to 26° 3′ N and 26° 22′ E to 86° 30′ E longitude. The important rural settlements are Khirbohina, Basmara, Nimorn, Jogia, Buhduta, Kohbara, Karmania, Dhamankara, Chaukia, Gaight, Banpukharia, Kendwatari, Baghra, Gabbi, Gujra, Sarma and Tetaria. Thus it is an area of high drainage density from where minor rivulet radiate from south to north in accordance with the nature of slope height and shape of the ground.

Soil and Natural Vegetation

Soil is the greatest resource of the region on which the production of crops, site for home and the construction of roads, railways and institutions are possible. Soil is the life-line of the agricultural system of a region. North Munger is covered by alluvial soil while the north-eastern part is covered by the alkaline soil due to influence of the Kosi river. In Khagaria district, the new alluvium or khadar soil is an important asset to agriculture while in the south older alluvium or bhangar is found with coarse-texture especially in Jhajha,

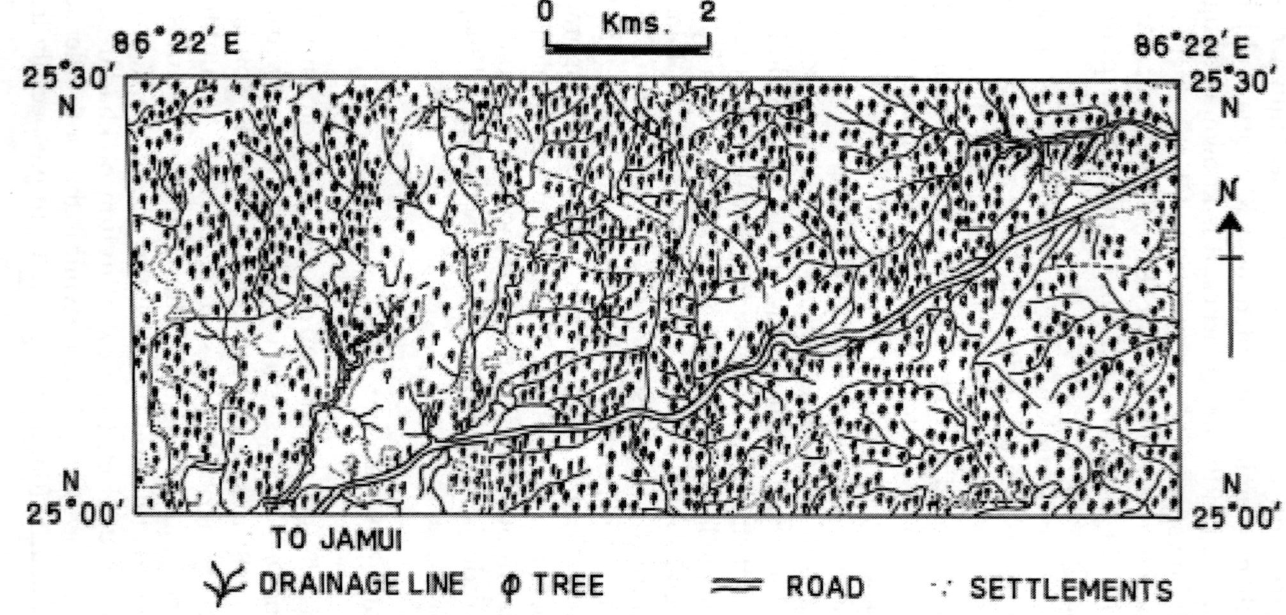

Fig. 3.7 : Munger Division : Drainage Pattern in Relation to Forest Resource.

Sono, Chakai, Jamui, Khaira and Lakshmipur anchals mixed with forest soil. P.H. Value ranges from 7-11 in the soil of Khagaria district, 7-5 in Southern plains of Munger, but in the forested terrain of south it ranges from 3-5 only (Figure 3.8).

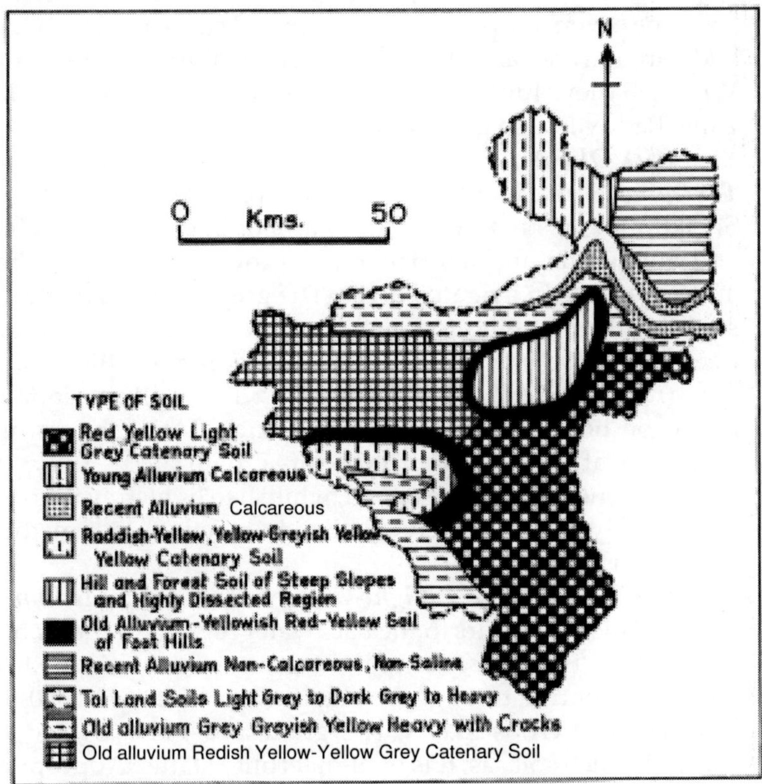

Fig. 3.8 : Munger Division: Soil Association.

In the Munger division newer and older alluvium including the coarse texture sandy soil is found in the plain land whereas forest soil is prevalent with proficiency of laterite in Kharagpur, Jamui and Gidhaur hill areas. The following soil groups are found in the Munger division :

(i) ***Young Alluvium, Calcareous:*** This association is found in Alauli, Khagaria and some portions of Chautham anchals located north of the river Burhi-Gandak. The major crops are maize, wheat, chillies and tobacco. The soil is white to olivegrey in colour with dark grey soil in chaurs or tals, showing slightly to moderately alkaline reaction (pH 7.4-8.6) and are moderate to high in fertility status.

(ii) Recent Alluvium, Non-Calcareous, Non-Saline: This soil zone covers the part of Chautham and entire Beldaur, Gogri and Parbatta anchals. The soil is low to medium fertility. Maize, Wheat, Barley, Jute and Khesari are important crops of this zone. Paddy is also grown in this area.

(iii) Old Alluvium, Grey to Greyish Yellow, Heavy-Texrtured Soil Crackings: This soil zone exists in Barbigha, Sheikhpura anchals, Ariari and parts of Lakhisarai, Barahiya and Surajgarha anchals. The crops grown are paddy, wheat and sugarcane in irrigated areas. The grey coloured neutral to slightly alkaline soil is medium heavy textured with wide and deep polygonal cracks. They possess medium fertility status.

(iv) Recent Alluvium, Calcareous: This soil association is found on both banks of the river Ganga. Maize, sugarcane, arhar, wheat and other Rabi crops are grown successfully. The soils are generally light textured, whitish to light grey in colour and slightly to moderately alkaline in reaction with medium to high fertility status.

(v) Tal Lands Soil, Light Grey to Dark Grey Medium to Heavy: This soil occurs in the backwater regions of the Ganges throughout its course. It is light grey to heavy in texture, showing neutral to slightly alkaline reaction (pH. 7.0-8.0) Bumper Rabi crops like wheat, gram, peas, khesari, etc. are grown in such soils as, it is of high fertility status. Appropriate drainage measures will render the soil suitable for Kharif cultivation.

(vi) Reddish Yellow-Yellow Greyish and Yellow Catenary Soils: This soil association is found in a very small part of Khaira and Sono anchals. The soil is slightly alkaline. Major crops are kodo, til, kulthi and paddy.

(vii) Red Yellow and Light Grey Catenary Soil: This association is located in Chakai, Jhajha, Sono, Khaira, Jamui, Lakshmipur, Tarapur and Sangrampur anchals. Major crops of the area are kodo, til, kulthi, gond and paddy.

(viii) Hill and Forest Soils of Steep Slopes and Highly Dissected Region: This association occurs in the Kharagpur, Dharhara, and Jamalpur where the plateau region begins. The soil is very shallow, gravelly and stony and deserves to be put under forest to check further deterioration due to erosion.

(ix) Yellow, Red-Yellow and Black Catenary Soils of Rajmahal Trap: Soils under this association occur in some parts of Chakai and Khaira anchals especially in areas of metamorphic rocks. It is medium to heavy ill texture, neutral to alkaline in taste and have good fertility status. Paddy, gram, til, kodo, kelai etc. are the chief crops grown in this region.

(x) Old Alluvium, Yellowish Red-Yellow Soils of Foot-hills: This soil association is found in a thin strip of the Khargpur hills where maize, arhar, kulthi, paddy and gram constitute the major crops. The soil is generally deep, light textured, acidic, and poor in fertility.

Only natural environment of the study area have been discussed above because these are the governing factors of resources. Without considering the physical attributes and their political units it may not be possible for an investigator to take a certain level of data for any territorial unit over a physical landscape. In this sense the locational dimensions, geological conditions, soil and vegetation have been discussed in different subheads, so that the impact of these basic infrastructures may open up a new line of thinking for resource appraisal in the Munger division.

REFERENCES

Ahmad, Enayat, Bihar : A Physical, Economic and Regional Geography, 1960, p. 27.

Coupland, H., Final Report on the Survey and Settlement Operations of Munger (North), 1920, p. 16.

Dayal, P., "The Bihar Plain—A Regional Study", Transactions of the Council of Indian Geographers, Vol. V, 1563, p. 2.

Guha, S.K. and P. Nag, 'Studies on the Hot Springs of Munger District', Bihar, 1971.

Jacob, V.C., 'A Note on the Hot Springs of Bhimbandh Area, District Munger', Geological Survey of India, 1962, p. 2.

Kayastha, S.L., "Conservation of Natural Resources in Himalaya : A Vital Need"; Seminar on Himalayas, New Delhi, December, 1965.

Prasad, S.D., District Census Hand Book, Munger, Vol. IV, Part-viii, 1961, p.

Prasad, S.D., 'Census of India, 1961, Vol. IV, Part-I-A-(i)', 1967, pp. 5-8.

Roy Choudhary, M.K., 'Geology and Mineral Resources of the States of India'; Part V, Bihar, 1973, p. 1.

4

Human Resources and Sustainable Development in the New Millennium

The active and potential workers are known as human resources. In this sense it is a fact that workers of the city of Munger are more valuable due to their superiority in skill in comparison with workers of the rural areas who are called unskilled labour for a job in the urban institutions on high wages. On the other hand, blindmen, old persons and other sorts of non-workers have less potentialities to form resource because they are liabilities on a worker or national economy.

The science which deals with human resources is known as "demography" or population studies. This term was first used by Guillard in 1855. Demography has been derived from two Greek words, Demons + Graphien, where "Demons" means people and "Graphien" means to describe. Population studies deals with the analytical interpretation of population dynamics and composition which covers a wider area.

Population is the greatest resource among all the resources on the earth. This is because man is the creator, surveyor and destroyer of resources, who occupies a dominant position both in economic and geographical studies. This relationship between man and environment can be divided into three different groups :

(a) Man utilizing the resources of nature as the creator of natural environment;
(b) Man utilizing the resources of nature by virtue of his working in natural environment; and

(c) Man utilizing the resources of nature because human environment being the product of physical environment. In this way, the number of persons at a place is the focal point of resource investigation. Without man universe is resourceless because only those elements are considered as resources which are needed and utilized by man.

The concepts of population study are basically related to the following points :

(a) The use of natural resources, i.e. environment, soil, space, time, energy, goods, techniques, money, information, and other sources for the optimum population distribution over earth's surface.
(b) The spatial distribution, density, change and growth of population depend upon the economic conditions and availability of food in any region.
(c) The demographic attributes as the movement of people, fertility, mortality, age-sex structure and literacy play an important role in economic growth of inhabitants themselves.
(d) People on particular place and time provide an opportunity to make the best use of available resources for the production of food and labour in other jobs.
(e) Population planning for regional economic development and welfare of the people.

The human characteristics fall into three groups :

(a) absolute numbers.
(b) 1. Physical characteristics : age, sex, race, morbidity, intelligence;
 2. Social characteristics : marital status, family household, residence, literacy, education, language, religion, nationality and ethnic group;
 3. Economic characteristics : industry, occupation, income.

(c) Population dynamics : fertility, mortality, migration and change.

The purpose of this study is to highlight population as a resource in the Munger division in close connection with the demographic aspects.

The Sustainable Development Paradigm

The concept of "sustainable development" transcends the classical development paradigm, and consists of the following two components :

(1) Sustainable Human Development, and
(2) Environmental Sustainability

1. Sustainable Human Development (SHD)

SHD represents an evolution of the classical concept of development : its emphasis has moved from the material well-being of states to the well-being of individual human beings.

While the classical approach was based on three factors of production, namely land, capital and labour (human beings), the new paradigm of SHD places people at the center, as the principal actor and the ultimate goal of development.

By enhancing human capabilities to expand choices and opportunities for men, women and children, SHD creates an environment in which human security is guaranteed and individual human beings can develop their full potential and lead a life of dignity and freedom.

A. Human Development

The concept of "human development" was developed by UNDP in 1990, and is the process of enlarging people's choices. The Human Development Indicator (HDI) is an indicator of the degree of human development enjoyed in respective countries. It comprises of the following components :

(1) Longevity (life expectancy at birth)
(2) Knowledge (adult literacy rate, gross enrolment ratio)
(3) Decent standard of living (GDP per capita measured in PPP$).

B. Human Poverty

Poverty has many dimensions and has been defined in various ways. The following are five basic definitions of poverty :

(i) **Income poverty** = Lack of minimally adequate income or expenditures. Around 1.2 billion people live on less than US$ 1 a day (1993 PPP US$), and 2.8 billion on less than $ 2 a day.
(ii) **Extreme poverty** = Inability to satisfy even minimum food needs.
(iii) **Overall poverty** = Inability to satisfy essential non-food as well as food needs.
(iv) **Relative poverty** = Poverty defined by standards that can change across countries or over time. An example is a poverty line set at one-half of mean per capita income implying that the line can rise along with income.
(v) **Absolute poverty** = Poverty defined by a fixed standard. An example is the international one-dollar-a-day poverty line which is designed to compare the extent of poverty across different countries.

The Human Poverty Index (HPI) was introduced in the Human Development Report, 1997 in order to measure the extent of human poverty in a community. It measures the extent of deprivations in developing countries in the following three dimensions :

(1) A long and healthy life—vulnerability to death at a relatively early age, as measured by the probability at birth of not surviving to age 40.

(2) Knowledge—exclusion from the world of reading and communications, as measured by the adult illiteracy rate.

(3) A decent standard of living—lack of access to overall economic provisioning, as measured by the percentage of the population not using improved water sources and the percentage of children under five who are underweight.

C. SHD and Human Security

The concept of "human security" emphasizes the security of individual human beings, in contrast to the traditional concept of security focused on the security of States. In the post-Cold War world, the reduction of the risk for a nuclear conflict has been followed by the dramatic increase of the threats to human security in the form of gross human rights violations, intra-states conflicts, poverty, environmental degradation, illicit drugs, transnational organized crimes, infectious diseases, the outflow of refugees and antipersonnel landmines. The lack of human security hinders the achievement of SHD and, vice versa, when there is no SHD, the security of human beings is endangered. Conflict prevention and peace-building are among important means in ensuring human security.

D. SHD and Human Rights

Human rights and SHD share a common purpose to secure the well-being, freedom and dignity of all people everywhere. Human rights and SHD are interdependent and mutually reinforcing. SHD is essential for realizing human rights and human rights are essential for the achievement of full SHD.

By expanding choices and opportunities for men, women and children, SHD address their economic, social, cultural political and civil rights.

SHD requires 7 freedoms :

(1) Freedom from discrimination (by gender, race, ethnicity, national origin or religion).
(2) Freedom from want (to enjoy a decent standard of living and well-being).
(3) Freedom to develop and realize one's human potential.
(4) Freedom from fear (of threats to personal security, torture, arbitrary arrest and other violent acts).
(5) Freedom from injustice and violations of the rule of law.
(6) Freedom of thought, opinion and to participate in decision-making and form associations.
(7) Freedom for decent work without exploitation.

E. Gender Empowerment

In order to secure SHD, measures must be taken to promote the political, economic and social empowerment of women and gender equality and to ensure their participation in all phases of development, from planning to implementation. UNIFEM has played an important role in this regard.

The Gender Empowerment Measure (GEM) examines whether women and men are able to actively participate in economic and political life and take part in decision-making. It is a composite index using variables constructed to measure the relative empowerment of women and men in political and economic spheres of activity. The final GEM value is given by three indices measuring economic participation and decision-making, political participation and decision-making, and power over economic resources.

F. SHD and Governance

A crucial aspect of SHD is governance (accountability, transparency, corruption, election, participation, democracy, free media, access to information, human rights, rule of law).

G. Impact of New Technologies on SHD

With regard to the impact of new technologies on sustainable human development, the following issues need to be considered :

(i) Participation
(ii) Knowledge
(iii) New Medicines
(iv) New Crop Varieties
(v) New Employment and Export Opportunities

2. Environmental Sustainability

(i) Sustainability of ecosystems

The sustainability of the Earth ecosystem is threatened by human conduct. The Earth's existence itself will be jeopardized by the continued exploitation of natural resources and manufacturing and industrial activities, already affecting negatively the delicate balance that existed in the ecosystem.

(ii) Biodiversity

Biodiversity signifies the existence of a variety of species at different levels of the eco-system. It also means the existence of diversity within individual species that need to exist in balance. Sustainability of their existence depends on preservation of the delicate balance among them that are linked to each other.

(iii) Climate change

The slow but steady changes taking place in the environmental conditions as a result of gradual climate change can have an impact on the living conditions of many parts of the world, as lakes began to dry up and forests have started disappearing. The most recent negotiations in the COP 7 may signify the beginning of concerted actions that the international community is taking to stablilize carbon dioxide and other greenhouse gas emission in the atmosphere at a level that would prevent dangerous interference with the climate system.

(iv) Environmental Ethics

It is incumbent on us human beings to consider how we can conduct our human affairs in order to realize sustainable human development, while, at the same time, preserving the integrity of the Earth, and to recognize the interdependent relationship that exists between the Earth and the human beings. Towards this end, it is imperative to commit ourselves to fundamental environmental ethics that should govern the conduct of human affairs.

International principles constituting environmental ethics are set forth in four key documents : The 1972 Stockholm Declaration on the Human Environment; the 1982 World Charter for Nature; the 1992 Rio Declaration on Environment and Development; and the 1998 Earth Charter.

Low Cost Practical Consultancy and Sustainable Human Development

The concept of consultancy has been unknown to rural development programmes. Traditionally consultancy has been a high cost business affordable for only organised sector with high investment. But now time has come when rural development programme do have benefit of services of experts. Retired people, freedom fighters, volunteers from other organisations, experimenter have role to play in this process. What is required is acceptance of the fact that everybody has important role to play and working in isolation, individualistic, nuisance mindsets have accelerated the process of collapse. It is here not only man as a part of the Society but also society itself should have system of low cost practical consultancy. Due to problem of CTB, law and order problem was rampant, stock exchange failure became common. It is essential that new investment potential with stability and security are identified in small towns and rural area. The field of consultancy is bound to open up. How do we go about it? This question which this project tries to address itself. This is golden time to implement various reports of international and national bodies. Lessons of various efforts need to be suitably adjusted with the local conditions. This project by adopting evolutionary, humanistic approach to the whole developmental process did try to come out with implementable plan of action. Needless to say there cannot be soft options.

Regional Conditions and Patterns of Growth

The Munger division has a population of more than 51.44 lakhs (2001). It has been distributed over 7920 square kilometres and for the fact the density of population comes to 650 persons per square kilometres. The total population in 1951 was only 20.55 lakhs while it was 22.68 lakhs in 1961 and 27.45 lakhs in 1971, 33.14 lakhs in 1981 and 40.41 lakhs in 1991. This shows that the growth of population is not so alarming but it is moderate in nature. More than 14 per cent people live in urban areas and the rest in rural areas.

Before 1889, the census was conducted by collectors only, which was simply an estimate of population. But after 2001 census records of population were available in a published form. Since then the population growth seems to be steady but in 1921 due to various diseases, famine conditions and influezna the growth of population in that census counted very low in almost all parts of India for which the Munger division was not an exception.

During 1921-51, the growth of population was slow but after 1951 census it was slightly high due to the availability of medical facility, control over various diseases, good agricultural harvest with the use of chemical fertilizers, improvement of seeds and the provision of irrigation have immensely helped in the growth of population, as all these have immensely changed the environmental conditions of the region.

The population of the district has almost doubled itself over a period of hundred years since 1872. The bulk of increase has, however, taken place after 1921. In fact, the increase in 1921-31 was itself more than the net addition to population during the preceding five decades. The highest growth was 21.40% in 1961-71 decade including the population of present Begusarai district in Munger. The growth of population between 1971-81 was 20.75% and between 1981-91 was 21.93% while the growth of population between 1991-2001 was 28.14%. The rate of population growth was 2.66% in 1951-61 in Barahiya anchal. The highest growth rate was 38.30% in Jamui anchal during the same period. The other anchals of high population growth were Beldaur and Chautham (27.54%), Khagaria (25.01%), Jhajha (23.36%) and Gogri (22.33%). The anchals of very low growth of population were Kharagpur (12.13%), Parbatta (10.56%), Khaira (11.37%), and Chakai (14.25%). All these give an idea that areas of high population growth were having urban location of greater achievement in agricultural production.

The decadal change of population has been calculated on the basis of formula :

$$\text{Decadal Growth Rate} = t\sqrt{\frac{Pi}{PO}} - 1 \times 100$$

Where t is the number of years interval, Pi is the last census population and PO is the previous census population.

Table 4.1 : Growth of Population in the Munger Division (1872-2001)

Year	Population	Absolute variation	Percentage variation
1872	1,812,986	—	—
1881	1,969,950	156,964	8.66
1891	2,036,021	66,071	3.35
1901	2,070,191	34,170	1.68
1911	2,135,000	64,809	3.13
1921	2,029,965	105,035	4.92
1931	2,287,154	257,189	12.67
1941	2,568,544	277,390	12.13
1951	2,849,127	284,583	11.10
1961	3,387,082	537,955	18.85
1971	2,745,180 + 1,147,429	505,527	21.40
1981	3,314,806	569,626	20.75
1991	4,041,866	727,060	21.99
2001	5,144,665	1,129,799	34.71

Source: Census of India, 2001 and Census Handbook, Munger.

During 1961-71, the highest population growth were in Halsi anchal (29.96%). The lowest growth was observed in Gogri anchal (15.24%). The other anchals of medium population growth were Sono (27.90%), Jhajha (22.93%), Tarapur (23.65%), Surajgarha (25.82%), and Lakhisarai (25.16%). In 1961-71, the highest growth in rural areas were mostly due to non-adoption of family planning measures by the rural folk, improvement in medical facilities and the development of agriculture.

Density of Population

The average population density is 696 persons per square kilometres in the Munger division. The highest population density is 1230 for Munger anchal and the lowest density is 242 persons in Chakai anchal of South Munger.

In between these two extremes, the density of population has been divided into four categories according to the median and quartile values. Where Q1 is 346, M is 435 and Q3 comes to 594 persons per sqaure kilometres.

The zone of highest population density (over 1000) comprises the Gangetic riparian tracts of Munger, Jamalpur, Asarganj, Tarapur, Barbigha, Shekhopur Sarai, Jamui, Khagaria and Mansi anchals. This high density is the result of easy accessibility and early establishment of human habitation in the region of fertile soil with 3 to 4 seasonal crops in a year. In some of the anchals there is a high population density which is due to the home of the sedentary agricultural population where the saturation point seems to have been reached long ago. This high density region is interspersed with the area of medium density where the wet and lowlying paddy fields are separated by sandy lands and swampy chaurs. Here high density co-exists with great agricultural insecurity on account of the production of single rice crop annually and the periodic flood.

The zone of medium population density has 601 to 972 persons per square kilometre. This is available in Sangrampur, Lakhisarai, Ramgarh Chowk, Halsi, Sheikhpura, Ariari, Islamnagar Aliganj, Sikandra, Gidhaur, Alauli, Chautham,

Beldaur, Gogri and Parbatta anchals in the Munger division. Rice cultivation is prevalent in this region which naturally has a high potentiality to support dense population.

The third zone of semi-sparse population density is located in Bariarpur, Tetiah Bambor, Barahiya, Pipariya, Surajgaraha, Chewara and Jhajha anchals in the Munger division. Here due to heavy rain, forest covered rugged terrain, malarious climate, marshy land and frequent change in course of the Kosi, the semi-sparse population density is found.

Areas of sparse population density comprises Dharhara, Kharagpur, Ghat Kusumbha, Chewara, Barhat, Lakshmipur, Khaira, Sono and Chakai anchals of south Munger. The average population density is 359 persons per square kilometre. In this zone the vast expanse of Kharagpur and Sheikhpura hills having a cover of dense forests restrict high population density.

Share of Male and Female Population

The percentage of male and female population is almost equal throughout the Munger division (2001). The highest percentage of male population is 54, found in anchals of Munger, Bariarpur, Asarganj, Tetiah Bambor and Parbatta. The lowest value of male population is 52 per cent found in anchals of Surajgaraha, Lakhisarai, Ramgarh Chowk, Halsi, Barbigha, Shekhopur Sarai, Sheikhpura, Ghat Kusumbha, Chewara, Ariari, Islamnagar Aliganj, Sikandra, Lakshmipur, Jhajha, Khaira, Sono, Chakai and Beldaur anchals.

So far as the female population is concerned the highest value of 48% is available in anchals of Surajgaraha, Lakhisarai, Ramgarh Chowk, Halsi, Barbigha, Shekhopur Sarai, Sheikhpura, Ghat Kusumbha, Chewara, Ariari, Islamnagar Aliganj, Sikandra, Lakshmipur, Jhajha, Khaira, Sono, Chakai and Beldaur and the lowest value of 46% is found in anchals of Munger, Bariarpur, Asarganj, Tetiah Bambor and Parbatta. The causes of such a distribution of male and female populations are industrial development in the environs of Munger and Jamalpur and the cultivation of maize in North Bihar. All anchals having rice cultivation have a higher percentage of female population (Table 4.2).

Table 4.2 : Density and Male-Female Ratio of Population in the Munger Division (2001)

Name of Anchal	*Per km²*	*% of Male*	*% of Female*	*Sex ratio*
Munger	1230	54	46	859
Bariarpur	576	54	46	861
Jamalpur	1133	53	47	874
Dharhara	368	53	47	881
Kharagpur	303	53	47	877
Asarganj	1000	54	46	885
Tarapur	1192	53	47	894
Tetiha Bambor	569	54	46	863
Sangrampur	914	53	47	883
Barahiya	509	53	47	899
Pipariya	533	53	47	894
Surajgarha	576	52	48	937
Lakhisarai	924	52	48	908
Ramgarh Chowk	682	52	48	941
Halsi	602	52	48	941
Barbigha	1224	52	48	918
Shekhopur Sarai	1067	52	48	934
Sheikhpura	891	52	48	907
Ghat Kusumbha	394	52	48	934
Chewara	535	52	48	906
Ariari	601	52	48	930
Islamnagar Aliganj	655	52	48	926
Sikandra	666	52	48	930
Jamui	1042	53	48	900
Barhat	320	53	48	888
Lakshmipur	393	52	48	920
Jhajha	504	52	48	910
Gidhaur	853	53	48	900
Khaira	423	52	48	935
Sono	429	52	48	933
Chakai	242	52	48	925
Alauli	779	53	47	893
Khagaria	1126	53	47	879
Mansi	1061	53	47	868
Chautham	694	53	47	894
Beldaur	660	52	48	918
Gogri	972	54	46	866
Parbatta	797	53	47	885
District Average	696	53	47	903

Source : Census of India, 2001.

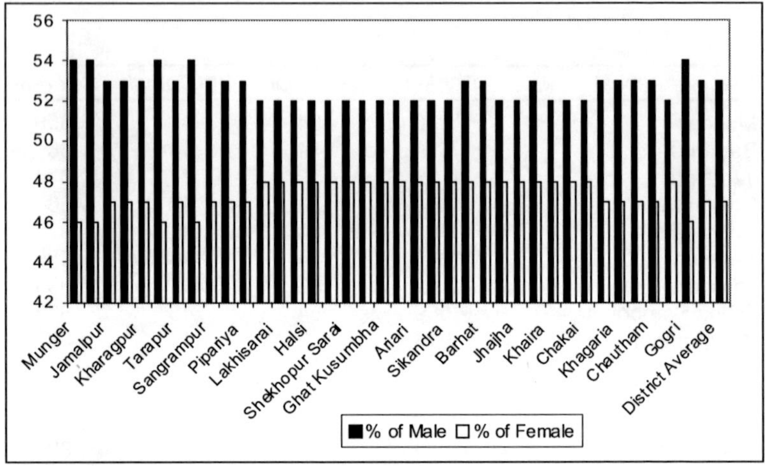

Fig 4.1 : Munger division : Ratio of Male-Female (2001).

The ratio of male and female population is unbalanced in the Munger division. On an average there is 903 female for 1000 male in the division. The anchals of Ramgarh Chowk, Halsi, Shekhopur Sarai, Ghat Kusumbha and Khaira have more female in comparison with their male counterpart, whereas in other anchals there are lesser number of female in comparison with male (Table 4.1), whereas the lowest ratio of female population is 859 in Munger and the highest ratio of female population is 941 in Ramgarh Chowk and Halsi anchal (2001).

Rural Urban Population

In the Munger division 29% population was urban, whereas 71% was rural (2001). So far as anchal level pattern is concerned, the highest 63% was urban in Munger anchal corresponding to only 37% as rural. Similarly, in Jamalpur anchal 53% was urban and 47% rural. In almost all other anchals rural population dominated over the urban. This shows that even in 2001 census, the Munger division was least urbanized (Table 4.3).

Table 4.3 : Percentage of Rural and Urban Population (2001)

Name of the Anchal	*% of Urban population*	*% of Rural population*
Munger	63	37
Jamalpur	53	47
Bariarpur	46	54
Kharagpur	15	85
Lakhisarai	30	70
Sheikhpura	26	74
Barbigha	33	67
Barahiya	32	68
Jamui	37	63
Jhajha	16	84
Khagaria	15	85
Gogri	13	87
Asarganj	10	90
District average	29	71

Source : District Census Handbook, Munger, 2001.

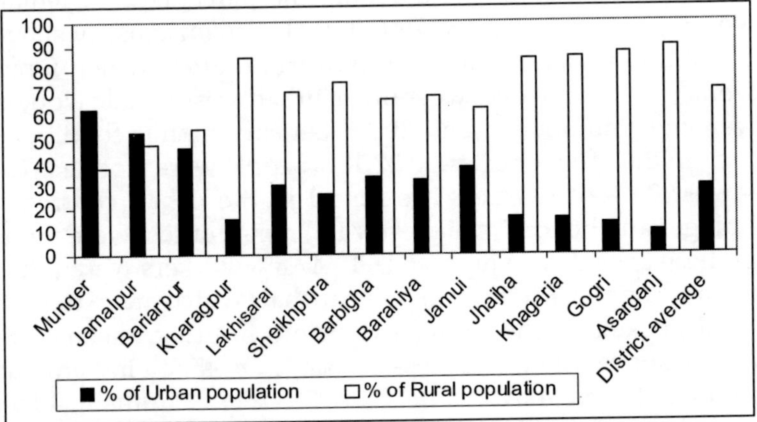

Fig 4.2 : Munger division : Percentage of Rural and Urban Population.

Per Capita Net Sown Area

It is distressing to note that the district of Munger has the least acreage of per capita NSA. Here 0.17 hectare is the highest value which has been calculated in Beldaur, Halsi, Barahiya anchals.

The value of second rank 0.15 hectare comes for Lakshmipur, Sikandra, Lakhisarai and Surajgarha anchals. The lowest value of 0.04, 0.06, 0.07 and 0.08 have been recorded for Jhajha, Munger, Jamalpur and Khaira anchals of the district where barren hilly and forest covered areas are not conducive for a greater dividend over net sown area.

Working Population of the Division

In almost all anchals of Munger about 20% to 30% population were male workers (2001). The highest male worker was 44.7% in Barahiya and the lowest 31.8% in Bariarpur. In other anchals, the condition ranges in between these two extremes (Table 4.4).

Among female workers the highest 17.9% in Halsi anchals of the division whereas in Munger and Jamalpur urban and industrial centres of the division, the percentage of female working force was very poor (2001). The main cause of such a distribution was due to poverty in rural areas which forces women folk to do some work but in urban areas female workers were not commonly found due to better economic conditions.

In the Munger division 37.18% people were workers. The lowest 24% of workers are found in the urban centres of Munger and hence the highest (74%) non-workers were found in these anchal. The highest 47.4 - 47.8% workers were found in Lakshmipur, Khaira and Sono anchals which are extremely rural pockets of the Munger division. In other anchals, the conditions were more or less the same *i.e.* 45.5% in Gidhaur, 44.8% in Jhajha, 42.9% in Beldaur, 42.3% in Chautham, 40.8% in Barhat and 40.4 in Alauli anchals (Table 4.4).

So far as the dependency ratio is concerned, the highest value (3.09) comes for anchals of Kharagpur, Dharhara, Barbigha and Gogri. The lowest dependency ratio of workers is 1.56 in Beldaur, 1.77 in Chautham, Alauli and Halsi anchals. In between these two extremes 3 dependency ratio comes for Munger and Jamalpur, whereas 2.04 comes for so many other anchals of the Munger division.

Table 4.4 : Working Population of the Munger Division (2001)

Name of Anchal	*% of Male Worker*	*% of Female Worker*	*% of Non-Worker*	*% of Worker*
Munger	33.3	3.6	76.0	24.0
Bariarpur	31.8	6.9	68.5	31.5
Jamalpur	32.2	3.8	76.1	23.9
Dharhara	32.0	7.0	69.3	30.7
Kharagpur	33.7	8.0	66.5	33.5
Asarganj	33.3	8.3	64.5	35.5
Tarapur	35.1	6.4	68.2	31.8
Tetiha Bambor	35.3	6.9	65.9	34.1
Sangrampur	35.8	8.3	65.8	34.2
Barahiya	39.8	8.6	64.6	35.4
Pipariya	44.7	10.1	65.3	34.7
Surajgarha	40.9	13.2	64.9	35.1
Lakhisarai	42.2	12.8	62.7	37.3
Ramgarh Chowk	46.0	14.8	63.9	36.1
Halsi	45.2	17.9	60.2	39.8
Barbigha	40.1	11.0	66.7	33.3
Shekhopur Sarai	43.4	12.9	63.6	36.4
Sheikhpura	41.7	13.0	65.0	35.0
Ghat Kusumbha	41.1	17.5	57.1	42.9
Chewara	45.2	18.5	61.8	38.2
Ariari	43.4	17.5	57.5	42.5
Islamnagar Aliganj	41.7	13.2	61.0	39.0
Sikandra	41.4	16.1	59.8	40.2
Jamui	39.1	10.8	65.4	34.6
Barhat	41.4	17.3	59.2	40.8
Lakshmipur	42.3	19.8	52.3	47.7
Jhajha	35.6	16.9	55.2	44.8
Gidhaur	39.0	13.3	54.5	45.5
Khaira	44.1	22.0	52.6	47.4
Sono	38.8	17.5	52.2	47.8
Chakai	35.8	10.3	60.3	39.7
Alauli	41.7	12.1	59.6	40.4
Khagaria	38.8	6.4	67.6	32.4
Mansi	36.8	5.3	67.4	32.6
Chautham	41.1	13.3	57.7	42.3
Beldaur	42.9	14.1	57.1	42.9
Gogri	40.8	8.2	66.0	34.0
Parbatta	41.3	9.3	65.3	34.7
District Average	39.44	11.92	62.82	37.18

Source : Census of India and District Census Handbook, Munger, 2001.

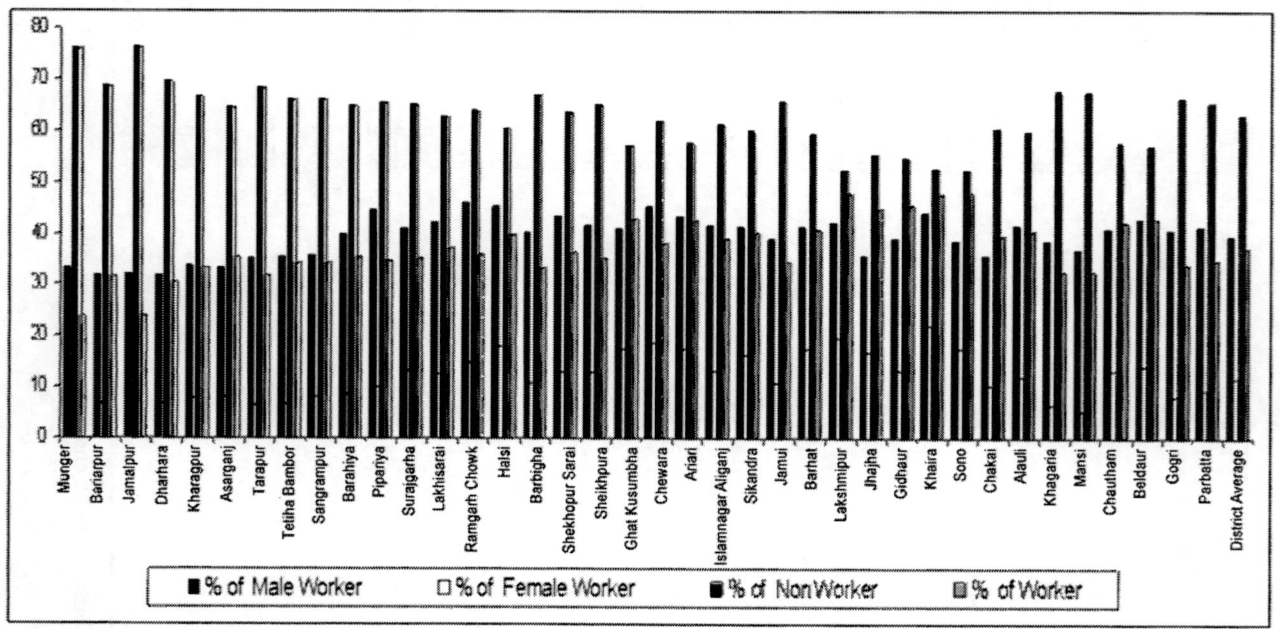

Fig. 4.3 : Munger Division : % Working Population (2001).

Table 4.5 : Occupational Structure, Scheduled Caste, Scheduled Tribe and Percentage of Literacy in the Munger Division (2001)

Name of Anchal	*% of Primary Workers*	*% of Secondary Workers*	*% of Tertiary Workers*	*% of Scheduled Caste*	*% of Scheduled Tribe*	*% of Literacy*
Munger	42	22	36	8.1	0.1	67.4
Bariarpur	78	10	12	14.7	0.7	52.6
Jamalpur	43	28	27	13.7	0.8	73.6
Dharhara	76	13	11	16.5	4.9	51.0
Kharagpur	82	9	9	16.2	5.4	50.4
Asarganj	81	10	9	12.9	0.0	51.6
Tarapur	86	4	13	11.9	0.0	54.4
Tetiha Bambor	87	12	11	14.3	0.6	50.3
Sangrampur	87	4	9	20.0	0.6	53.4
Barahiya	91	3	6	11.6	0.0	51.9
Pipariya	83	5	17	9.2	0.0	54.7
Surajgarha	82	3	10	14.8	2.0	51.1
Lakhisarai	80	6	14	16.0	0.5	46.0
Ramgarh Chowk	90	4	6	20.3	0.0	42.1
Halsi	92	3	5	22.0	0.0	42.6
Barbigha	81	6	13	21.1	0.1	56.1
Shekhopur Sarai	80	7	13	20.4	0.1	51.3
Sheikhpura	81	5	14	18.3	0.0	51.3
Ghat Kusumbha	85	7	8	17.9	0.0	39.0
Chewara	87	5	8	20.7	0.0	41.9
Ariari	91	4	5	20.2	0.0	40.3
Islamnagar Aliganj	90	4	6	18.5	0.3	44.8
Sikandra	89	5	6	21.4	0.6	42.7
Jamui	78	9	13	17.3	0.0	52.5
Barhat	88	5	7	24.4	5.0	45.5
Lakshmipur	68	16	16	21.6	7.2	38.4
Jhajha	58	26	16	11.7	5.1	43.2
Gidhaur	66	16	18	19.7	0.9	49.4
Khaira	91	4	5	17.6	2.9	42.7
Sono	77	18	5	14.1	3.3	34.0
Chakai	94	2	4	17.5	17.8	36.1
Alauli	93	3	4	25.4	0.0	31.4
Khagaria	85	3	12	15.7	0.0	45.4
Mansi	82	8	10	13.4	0.0	41.7
Chautham	90	3	7	17.5	0.1	36.5
Beldaur	94	2	4	15.5	0.0	32.0
Gogri	80	9	11	8.1	0.0	45.8
Parbatta	97	1	2	6.2	0.0	50.0

Source : Census Handbook, 2001.

In the Munger division higher percentage of people are engaged in primary services. As for example in Parbatta anchal 97% people are agriculturists, whereas in Munger 42% and in Jamalpur 43%. In other anchals 60% to 80% people are agriculturists and agricultural labourers. So far as the secondary and tertiary sectors are concerned, only Munger and Jamalpur have greater share of workers engaged in manufacturing and business activities (Table 4.5). In almost all other anchals the share of Secondary and tertiary sectors are very poor due to predominance of agricultural sector and lower level urbanization.

The share of Scheduled Caste and Scheduled Tribe population are quite unevenly distributed in the Munger division. The highest percentage of Scheduled Caste population is 25.4% in Alauli anchal and the lowest is 6.2% in Parbatta anchal. In between these two extremes all other anchals come under medium level population of Scheduled Caste and Scheduled Tribe (Table 4.5).

1,950,338 persons or 47.96 per cent population in the Munger division are able to read and write or have the higher educational level. The corresponding percentage for Bihar State is 47.0%. The division has, therefore, almost the same order of literacy as the State. So far as the anchal level pattern is concerned about 67.4% people are literate in Munger and 73.6% in Jamalpur anchals. The lowest literacy is 31.4% in Alauli anchal, 32% in Beldaur anchal. In other anchals the literacy value ranges between 34.0% to 56.1% (Table 4.5).

Migration

Migration is defined as leaving the place of birth by a person temporarily or permanently after breaking social and cultural ties at original place of residence. The migration which takes place outside of a place is called emigration and inside of the study area as immigration. The mass migration of people due to drought, famine and flood is known as mass migration.

The incidence of emigration is high in the Munger division. Emigration is most active from the southern part of the district, where the infertility of soil over a greater portion of Jamui District forces the people to find employment elsewhere, particularly in the coalfields of Chotanagpur and the tea-gardens of Assam.

Local level migration is also in vogue from north of the district of Munger every year in the months of January and February, large number of ranchers cross the Ganga to graze their flocks in Benua, Farkiya, Gogri and Jamalpur parganas, a tract of low-lying country, which is mostly flooded in the rainy season, but affords excellent pasturage in the drier months. Migration from village to village is rare, except along the banks of the Ganga and its affluents where the frequent shiftings of the river beds necessitate the movement of the villagers. Migration of a temporary nature from the neighbouring districts of Bhagalpur, Darbhanga and Muzaffarpur into pargana Pharkiya is also common after the rains. At this period a large number of cultivators come with their cattle and plough land at a low rate, and again they return home after the harvest of rabi crops.

In the Munger division, these born elsewhere in the division constituted 23.11% or less than one-fourth. Persons from other division of the State comprised 5.72%. Those coming from other States of India account for only 0.51% while foreigners constitute only 0.10% immigrants.

Table 4.6 shows the concentration of population in the Munger division on anchal basis. The optimum concentration is observed in Beldaur, Barahiya and Sikandra anchals of the division. The highest concentration is observed in Munger, Jamalpur, Tarapur and Barahiya anchals of the division where the values are 2.06, 1.92, 1.84 and 1.95 respectively. The areas where the population is sparsely located are Chakai (0.35), Lakshmipur (0.58), Sono (0.62) and Dharhara (0.63). In other anchals the concentrations of population vary between these two extremes.

Table 4.6 : Concentration of Population in the Munger Division (2001)

Name of Anchal	*Pi*	*Ai*	$\frac{Ai}{Ai} - Lc$
Munger	6.33	3.06	2.06
Bariarpur	3.54	2.06	1.72
Jamalpur	5.34	2.77	1.92
Dharhara	2.28	3.58	0.63
Kharagpur	4.33	4.79	0.90
Asarganj	3.04	1.65	1.84
Tarapur	2.48	1.79	1.39
Tetiha Bambor	2.56	1.83	1.38
Sangrampur	2.58	1.81	1.42
Barahiya	3.51	3.31	1.06
Pipariya	4.59	5.41	0.85
Surajgarha	4.98	5.30	0.93
Lakhisarai	5.11	4.38	1.16
Ramgarh Chowk	3.20	2.58	1.24
Halsi	3.06	3.48	0.87
Barbigha	3.67	1.88	1.95
Shekhopur Sarai	3.78	2.01	1.88
Sheikhpura	3.89	3.19	1.21
Ghat Kusumbha	3.68	3.45	1.07
Chewara	2.68	2.81	0.95
Ariari	2.49	2.59	0.96
Islamnagar Aliganj	4.52	4.44	1.02
Sikandra	4.62	4.50	1.02
Jamui	3.10	2.20	1.40
Barhat	4.04	5.89	0.69
Lakshmipur	4.06	6.91	0.58
Jhajha	4.03	5.42	0.74
Gidhaur	4.02	5.98	0.67
Khaira	3.50	5.43	0.64
Sono	3.13	4.97	0.62
Chakai	3.47	9.81	0.35
Alauli	4.16	3.48	1.19
Khagaria	5.01	3.32	1.50
Mansi	3.56	3.05	1.17
Chautham	3.31	2.96	1.11
Beldaur	2.85	2.83	1.00
Gogri	3.59	2.59	1.38
Parbatta	4.41	3.64	1.21

Source : Compiled by the author from Census Handbook, 2001.

Future Estimate of Population

Any attempt to make population projection is fraught with uncertainties, as there is no fixed law of population increment and various techniques of projection it has been decided here to use arithmetic and geometric techniques only. The estimation has been made from 2011 to 2031 A.D. for the Munger division. The estimated population is given below (Table 4.7).

Arithmetic progression is based on the basis of natural growth of population per year, while the geometric progression is calculated with the help of formula.

Table 4.7 : Population Projection of the Munger Division

(Rural In '000000 and Urban In '00000)

Year	Actual Population		Arithmetic Progression		Geometric Progression	
	Rural	Urban	Rural	Urban	Rural	Urban
1951	18.32	2.22				
1961	19.90	2.78				
1971	23.88	3.56				
1981	28.48	4.66				
1991	38.01	5.38				
2001	44.48	6.96				
2011			49.71	7.90	53.09	8.77
2021			54.95	8.86	63.39	11.02
2031			60.18	9.81	77.45	13.84

Source: Compiled by the author from Census Handbook 2001.

Calculation of Arithmetic Progression of Urban Population of the Munger Division

Population 2001 = 4,66,285
Population 1951 = 2,22,275
Difference = 2,44,010.
Growth in 30 years 2,44,010, i.e., 8,134 per annum
Estimated population for 1991 = 2,22,275 + 40 x 8134
Or, = 2,22,275 + 3,25,360
= 5,47,635.

Hence, estimated population of 2011 = 5,47,635.
Estimated population in 2021 = 2,22,275+50×8,134.
= 6,28,975.
Hence, estimated population in 2001 = 6,28,975.

Calculation of Geometric Progression

$Y=ar^{n-1}$
Where *Y* is the year chosen,
a is the first year,
r is the rate of growth, and
n is the interval of the year.
Population 1951 = 2,22,275.
Population 1981 = 4,66,285 (in the 31st year from 1951).
Thus *P* 1981 = P 1951 × r
Or. log r^{30} = 4,66,285

$$\log r = \frac{-1}{30}(\log 4{,}66{,}285 - \log 2{,}22{,}275)$$

$$\text{Or, } = \frac{1}{30}(5.6698 - 5.3491)$$

$$\text{Or, } = \frac{1}{30}(0.3206)$$

Or, = .001006 per annum
Therefore, *P* 1991 = P 1951× *r*
Or. log *P* 1991 = log 2,22,275+log *r* × 40 = 5.3492±.01006×40
= 5.7516
Antilog of 564400.
Hence, Estimated Population in 1991 = 5,64,400.

According to Table 4.7, it has been revealed that the arithmetic and geometric progressions give correct picture of the trend of population growth in the Munger division. The growth of urban population is linked with industrial and economic development and this leads to reduction in the pressure of population upon rural land, while rural population show substantial increase both in the arithmetic and geometric progression up to 2001 A.D.

Population Planning

Manpower planning is a strategy for the acquisition, utilization, improvement and preservation of human resources. Although we can distinguish three main stages in planning, i.e., evaluation of existing resources, an assessment of manpower requirement to achieve the objectives of the organisation, and measures to ensure that the necessary manpower resources are available as and when required, because all the three stages are complementary to one another. The result of each stage is to feedback into the other stage to ensure continuity. For example, if the labour supply is not available to meet a given target, then either the planned target is not feasible or the plan should make a provision for meeting the requirements of labour deficiency in close connection with population. The control of population is possible with the intensive use of family planning measures.

The district hospital Munger has a family planning centre. Maternity and Child Welfare centres are located at Maheshpur and Chautham, besides three mofassil subdivisional headquarters of Jamui, Lakhisarai and Khagaria. The state dispensaries and primary health centres are also supposed to propagate birth control measures and distribute contraceptives. In Table 4.8, the achievement of sterilization is shown which helps in checking the growth of population.

Table 4.8 shows that the sterilization figures are of fluctuating trends and the maximum sterilization has been done in 1995-2000. In spite of these measures the growth of population is going on unabated, this is because the superstitious people of the division do not favour birth control true to the sense.

Table 4.8 : Achievement of Family Planning Sterilization in the Munger Division (1994-2000)

Name of the Health Centres	*1994-95*	*1995-96*	*1996-97*	*1997-98*	*1998-79*	*1999-2000*
Munger	274	431	228	240	300	451
Jamui	1,222	860	45	43	37	239
Khagaria	133	310	107	20	52	14
Munger Block	8	253	14	19	40	110
Jamalpur	130	19	7	32	42	15
Dharahra	79	209	18	2	11	120
Kharagpur	49	211	58	19	24	139
Sangrampur	44	533	40	4	8	90
Tarapur	30	292	40	9	3	110
Ariari	8	39	59	3	1	181
Barbigha	44	438	54	30	42	245
Sheikhpura	11	83	39	2	18	181
Barahiya	50	119	34	39	29	110
Surajgarha	51	252	22	8	10	155
Halsi	22	486	32	4	19	100
Lakhisarai	9	41	40	2	34	201
Jamui	5	211	88	I	20	120
Jhajha	103	517	143	4	12	241
Sono	2	290	239	5	1	18
Lakshmipur	47	338	4	10	7	103
Sikandra	24	454	41	44	29	153
Khaira	50	40	21	5	4	149
Chakai	39	451	130	18	7	18
Khagaria Block	47	9	127	8	1	11
Gogri	50	330	33	15	33	118
Beldaur	45	733	100	5	5	58
Chautham	8	420	204	2	1	82
Alauli		822	182	'	21	114
Parbatta	37	422	85	13	13	210

Source : Munger Sadar Hospital, Munger, 2001.

SUSTAINABLE HUMAN DEVELOPMENT IN THE TWENTY-FIRST CENTURY : AN EVOLUTIONARY PERSPECTIVE

1. Introduction

> *"Order is not a pressure imposed upon society from without, but an equilibrium, which is set from within."*
>
> (J. Ortega y Gasset, 1927, quoted in Hayek, 1955)

Here is examined the state and nature of human development and identified factors that determine its enhancement for the twenty-first century. A general goal for human development is to enhance the quality of human life. However, the concept "quality of human life" is not well defined. It is determined by a set of interrelated factors that cut across many disciplines with varied perspectives and paradigms. These include the prevailing culture, health status, economic performance, political and social conditions, the building of human capacity and capabilities, and institutional development. For example, in an environment characterized by enhanced quality of human life, it is expected that people will be able to lead long and productive lives. They are also expected to enjoy good health, have access to knowledge and educational opportunities, and be treated by all with respect, in a socially equitable and dignified manner. In the sphere of political economy, they are expected to have the opportunity to participate in governance decisions that affect their lives and the community in which they live; and to have the potential to earn sufficient income to supply themselves with ample nutrition, shelter, and other material and aesthetic needs. Meanwhile, people are expected to maintain a sustainable environment and equitable social contracts across generations. In the present evolutionary perspective, a prevailing "culture" is characterized as a "weighted sum" of these context-specific factors.

However, these factors are not independent in their effects, nor do they act in harmony. For example, advances in medical science have greatly improved survival and health status in the developing countries. But they have also resulted in high rates of population growth, raising difficult challenges to development in many of these countries. Medical advances lead to extension in life expectancy at old age. However, excesses in such extension result in significant changes in the age structure and in the efficacy of related socio-economic and health institutions. In biotechnology, advances that enhance yield through genetic engineering have established this technology in major crop production around the world without careful examination of its net social benefit. Recent studies indicate serious unintended consequences to biodiversity, as the leading biotech corporations are discovering at present. The positive impacts of scientific advances on health, nutritional status, and life expectancy are qualified as net gains in measures of human development, but their negative externalities do not enter into these calculations.

These examples, among others, illustrate the complexity of measuring human development and achievement in the absence of a well-defined system of ranking. Indices of human development are not necessarily optimal, and their elements and weights are not constant over time; they should be assessed periodically. The social welfare consequences of some components of a human development index (HDI) may be nonlinear, or new knowledge about unintended negative consequences may be discovered. At best, HDIs are quantifiable approximations of a subjective and qualitative concept, the quality of life. For example, the United Nations Development Programme has published annually, since 1990, the Human Development Report (HDR), which includes rich information as well as a human development index (HDI). The HDI is based on three indicators: longevity, as measured by life expectancy at birth; educational attainment, as measured by a combination of adult literacy and the combined gross primary, secondary, and tertiary enrollment; and standard of living, as measured by real gross domestic product (GDP) per capita. Although a crude measure, it does serve the important function of focusing policy and academic attention on the wider aspects of human welfare not included in the standard GDP per capita. But the HDI should not substitute for careful analysis of the rich information provided in the HDR, and both undergo periodical assessment since the nature of human development is not static. It is continuously evolving.

Human development does not proceed independently of its environment. As humans alter their environment, the altered environment alters human destiny as well. It is for science to indicate the consequences and provide remedial and preventive measures. It is for a democratic system of social choice that includes a philosophy of human development to define its scope, assess progress, balancing competing concerns, and make allocative decisions.

2. Toward a Philosophy for Human Development for the Twenty-First Century

An attempt to examine the conditions and framework for a human development strategy requires a guiding philosophy that harmonizes and rationalizes the three universes of human culture: faith, science, and the arts. It should explore the synergistic nature of these basic elements of human culture, provide a ranking system of achievements in the various spheres of human actions, and provide a "balance" between an idealistic vision of human nature and social organization (Plato/Hegel), and one based on an empirical understanding of the dynamic interactions of humans with their external environment (Locke/Hume). The goal of such a philosophy is to provide balance and harmony in human affairs, leading to enhancement of the quality of life as well as to effective institutions and a sustainable environment.

Human intellect has advanced greatly, since the era of the great Greek philosophers, toward a consistent philosophy of human development. But there are many unsettled issues that persist, even as we enter the twenty-first century. To put the present state of human development into perspective, let us consider a brief historical review of how cultural evolution, with unprecedented versatility and adaptive capacity, substituted for evolutionary anatomical changes, and in turn is being replaced by a technophysio evolution.

A. From Biological Evolution to Cultural Adaptation

For hundreds of millions of years, living organisms and animals have been adapting to new environments or to new strategies for exploiting existing environments by the slow process of hereditary modifications of their body structure. Some animal species, within their genetic endowments, extend their inherited physical powers by using elementary tools or by co-operating in hunting or in defense activities. Humans, in the initial stages of cultural adaptation, used similar mechanisms to extend the power of their bodies to cope with an unaltered environment. But that early rudimentary human adaptive capacity, which was linked in the early stages of its development to evolutionary self selection, evolved independently from evolutionary anatomical change, into a complex and powerful social mechanism that is termed cultural adaptation or cultural evolution. "Cultural evolution" is a distinct human trait. It requires the use of intelligence and social organization. It allows humans to adapt more efficiently to their environments. It also empowers them to change these environments to make them more congenial to human needs and desires, thus achieving unprecedented control over their own destiny and the destinies of all other living species, as well as the physical environment itself.

Once this cultural versatility had appeared, evolution through adaptive radiation with respect to anatomical structures and physiological functions became a far less efficient strategy for dealing with environmental challenges than cultural amplification through inventions and technological progress.

Three basic qualities, acquired by humans as they interacted with their environment, were necessary to evolve in the direction of cultural adaptation. These are artisanship (the evolution of tool making into complex manufacturing and construction activities), conscious time binding (the ability to plan ahead and develop social institutions while benefiting from present and past experiences), and imaginal thinking (the ability to go beyond reality, essential for planned achievement). These three qualities are the foundations of the three principal realms of the present human knowledge. Natural science and engineering are the outgrowth of artisanship. Social sciences are fundamentally ways of directing social behaviour to avoid disaster and to improve the material state of humankind. The humanities are extensions of imaginal thinking. Societies that used these capacities more efficiently could acquire more food and defend themselves better against predators, and thus improve their chances for survival and reproduction.

There are two fundamental differences between cultural evolution or selection and biological natural selection that make the former far more flexible. The first is the speed with which cultural evolution can adjust to the environment. Biological natural selection proceeds more slowly and gradually. An alteration of a species that is viewed as sudden may actually take millions of years. Within less than 15,000 years, since humans started to acquire the attributes of "symbolic cultural evolution", the second phase of the cultural evolution discussed below, they were able to evolve into masters of their environment with a speed that has been greatly accelerating over time. It would probably have taken hundreds of millions of years to achieve these same results through the process of biological natural selection.

The second difference is the importance of education in the transmission of knowledge as opposed to gene transmission in the preservation of the evolutionary process. Education of human offspring and the preservation of the stock of knowledge, independently of genetic transmission through the biology of birth and death, are the key ingredients in the preservation and continuity of cultural evolution. Now it seems feasible, through the mechanism of cultural adaptation, to achieve dramatic changes in human development within one generation; an awesome responsibility for every human generation.

The ability of cultural evolution to develop, by non-genetic transmission, complex patterns of behaviour and social organization and institutions was made possible as a result of the human invention of symbolic language. The perfection of symbolism has lifted learning and scientific development to entirely new levels of complexity and continuity. That perfection led human evolution to take the decisive step away from the use of signs as the *modus operandi* for communication, which in turn led to the development of mathematical and logical structures essential for the development of science and human intellect on the global level. The experiences of Helen

Keller and Laura Bridgman, at the beginning of the twentieth century, illustrate that symbolism has made possible advances in human development and culture independent of the quality of sensory function. With feeble human bodies and a lack of essential sensory functions, being blind and deaf-mute, they were able to reach a high degree of mental development and intellect, once they had captured the symbolic use of words. Armed with symbolic skills, *Homo sapiens* became a formidable competitive species, as the demise of *Homo neanderthalensis* illustrates, but not necessarily a rational one. The evolutionary process is not necessarily linear. The *Homo sapiens* species, far from being the pinnacle of the hominid evolutionary tree, could be one more of its many terminal twigs. However, as the discussion indicates, *Homo sapiens* is developing a robust evolutionary niche that could sustain its presence.

B. Cultural Evolution, Symbolism, and Globalization

Without the invention of symbolism and its perfection, artisanship, conscious time binding, and imaginal thinking could not have evolved beyond rudimentary levels. The effect of cultural evolution and symbolism on human development has been enormous. By its very nature, symbolism includes the embryo of globalization. Symbolic thought is not attached to sensory data. Humans can relate through thoughtful debate based on logical structures: a more universal communication media, without the limitations of sign language and sense-based data. It took more than 3 billion years for many-celled animals to evolve on Earth, another 500 million years for apes to evolve, and an equal time for *Homo sapiens* to evolve: about 300,000 years ago. During that long span of living history, evolution, including the early development of non-symbolic cultural evolution, was basically opportunistic and epigenetic. It was not able to produce a pattern of recognized civilizations.

Symbolic cultural evolution, on the other hand, seems to have evolved as recently as 15,000 years ago. Within that short span, great civilizations and cultures flourished, starting with Sumeria, followed by Assyria, Babylon, Egypt, China, Greece, and Rome, up to the present "Western" civilization with its unprecedented progress in science and technology. Some 5,000 years ago in the Middle East, writing was invented and the first cities were established. Within 700 years of these momentous events, the Egyptian Pharaohs had built their famous pyramids and had established the earliest menageries and botanical gardens for pleasure, prestige, and to satisfy scientific curiosity. The evolution of these civilizations did not proceed in isolation. Although path dependent, the development of these civilizations was greatly influenced by mutual interactions through trade, migration, explorations, and conquest. Pre-Socratic philosophy, for example, which laid the foundation for Western civilization, evolved through interaction with and learning from previously established civilizations. It is well established that Western philosophy started in the sixth century B.C. at Miletus on the Ionian seaboard of Asia Minor. Ionia was the meeting place of East and West; it was also the land of Homer. The first Milesian philosophers, Thales, Anaximander, and Anaximenes, were open not only to oriental influences (Confucianism, 500 B.C.) and Homeric tradition (700 B.C.) but to the mathematics of Egypt and Babylon and to the ideas and information that flowed along the trade routes passing through Ionia from the far East. Thales of Miletus (624–546 B.C.), considered the founder of pre-Socratic philosophy, travelled to Egypt to learn astronomy, geometry, and practical skills to do with the measuring and management of land and water. It is remarkable that the five-pointed ancient Egyptian star, named "sba" by Egyptologists, is the same as that used in the American flag. It also has similar spiritual connotations in early Greek and later Masonic mythology. The history of civilizations indicates a continuity of learning processes among societies. Symbolism had a major influence in facilitating connectivity among civilizations, and in reconciling the anthropologists' diffusion hypothesis of the unity of civilizations with the empirical question of how isolated societies develop similar tools and modes of behaviour.

The historical development of symbolic cultural evolution was not smooth. There was both internal and external conflict. Internal conflict arose from a lack of harmony among the basic components of symbolic cultural evolution: natural sciences and technology (artisanship), social sciences (conscience time binding), and the humanities (imaginal thinking). Initially, harmony was sought through faith. Major religions—Judaism, Christianity, Islam, Buddhism, Hinduism, Taoism, and Confucianism—appeared during that period of symbolic cultural evolution to rationalize the role and place of humans in society and in the cosmic order, and to provide cultural stability through harmony among the main elements of symbolic cultural evolution. For example, in recent history, a self-regulating, invisible-hand paradigm was developed by Adam Smith, following Newton's view of a harmonious cosmic order. It contended that micro behaviour, when left to its own will, is guided by an inherent self-interest motive that converges atomistic behaviour into optimal macro order. While providing the foundation for scientific economic analysis, the paradigm built a bridge between scientific development, market behaviour, and established faith. However, advances in science and empirical knowledge, combined with "collective self-interest" that reduces individual mobility and motivation, have created tension in that synergy throughout the history of symbolic cultural evolution.

External conflict, partly a result of internal conflict, arose as a defense mechanism to safeguard as well as to diffuse what each culture viewed as superior elements of its way of life: its brand of socio-political organization and development path. Cultural diversity, instead of being viewed as offering human enrichment and progress, was viewed as a threat to the status quo and a call for socio-political alarm; a clash among civilizations seemed inevitable in this paradigm. However, the emergence and apparent inevitability of conflict seems to have been fueled more by changes in the internal dynamics of symbolic cultural evolution. Advances in science and technology led to the Industrial Revolution and the emergence of the present form of capitalism, in which material progress is the primary goal, progress that ultimately depends on the accumulation of resources. Global expansion became a necessity while opportunities were converging towards a zero-sum game status. Thus, although the past three centuries of symbolic cultural evolution have witnessed great advances in science, the arts, and material comfort, they have also witnessed great tension, wasteful wars, unemployed resources, and confused social priorities within and among nations and states.

As civilizations emerged, intellectual and religious leaders increased their efforts to inject human purposes and reason into the course of history, with varying degrees of success. There are optimists who believe that human societies are increasing their ability to chart and follow a purposeful course of change towards a better life for all. However, it seems that optimists are balanced by an equal number of pessimists. According to some authorities, "the inexorable laws of nature and evolution will eventually override purpose and cause the human species to decline and disappear, as other animal species have done in the past." In this pessimistic view, the future may not be secure, since as we begin to approach the limit of the earth's capacity we severely restrict our room to maneuver in response to change. It is true that many species, more than 90 per cent, have disappeared over the course of life history. However, as a "generalist" species, not dependent on narrow specific niches, *Homo sapiens* should be a robust one. The history of symbolic cultural evolution, its *modus operandi* and its continuous search for harmony among its three fundamental elements indicate that it is a powerful, robust, and adaptive engine that should be able to guide humanity into a better future. To examine this survival potential, we attempt first to understand the topography and course of that evolution. Every century seems to bequeath crises to the one that follows. What did the twentieth century inherit from the nineteenth century that affects human development and the quality of life? How did the twentieth century cope with its legacy, and what dilemma is it forwarding to the twenty-first?

(i) The Inherited Burden of the Twentieth Century

The nineteenth century bequeathed to the twentieth an age of doubt and disharmony in established cultures. Without harmony among the three universes of the symbolic cultural evolution, the foundations of human philosophies and, by extension, those of culture and human development were shaken and their conclusions questioned. For example, in isolation, theological learning was seen as "lifeless embalmment of knowledge." Scientific knowledge, deprived of conscious time binding (social sciences) and imaginal thinking (faith, philosophy, or the arts), seemed incapable of providing a value system to steer individuals safely through the complexities of life and the negative externalities of technological change. Overall, old ideas were being challenged. New ideas were being born. Meanwhile, the growth of industry, technological invention, and the expansion of colonial power were altering the face of nations economically, socially, politically, and environmentally.

The nineteenth century was a period of contradictions: an age of hope and doubt, of culture and anarchy, of freedom and slavery, of democracy and exploitation, of nationalism and colonialism, and of the birth of seminal revolutions in thought on the one hand and the death of established ideas on the other. It was the century that spawned the likes of Bentham, Darwin, Hegel, Mill, Marx, Nietzsche, Einstein, and Russell, building on the works of Galileo, Descartes, Ibn Khaldun, Hume, Newton, Rousseau, Helvetius, Kant, Malthus, and Adam Smith, among others. On the one hand, it was a century of significant scientific discoveries, such as John Dalton's proof (1808) that matter consists of atoms, or James P. Joule's finding (1851) that energy is indeed conserved, or Charles Darwin's *Origin of Species* (1859). There were also significant inventions such as the telegraph, the automobile, the telephone, and electrification, among others. On the one hand, it was a marvellous century. On the other hand, it was one of chronic malnutrition, illiteracy, and poverty for the majority of humankind.

At the end of the nineteenth century, there were serious doubts about the viability and moral justification of the national and international socio-political order, as well as concerns about the economic and social conditions of the working class. Two crises in particular had been responsible for the skepticism about the nature of cultural evolution and the future of humanity: a crisis of "science and faith" and of "empiricism and rationalism." There had been earlier signs, even before the dissemination of Nietzsche's controversial philosophy, of a widespread decline in the belief in a divine creator whose authority could decide baffling questions of faith and morals (although Nietzsche wrote, apparently despairingly, that people would rather have the void as purpose, than the void of purpose). Natural sciences, geology, biology, and physics had a hand in this crisis. Rational scrutiny and empirical documentation discredited many claims regarding the age of the universe or the origin of humanity that had been accepted as matters of faith.

Empiricism is the view that knowledge of the world is based upon and derived from sensory experience. It claims that whatever is in the mind must be first in the senses. Empiricism is clearly the opposite of rationalism. The latter maintains that reason alone can provide knowledge of the existence and nature of things; reality is a unified, coherent, and explicable system. By the end of the nineteenth century, empiricism, as a philosophy, slowly supplanted rationalism to become the prevailing point of view. The *modus operandi* of empiricism has been characterized by its reliance on sensory data as the only source of knowledge, its refusal to accept anything but material reality, its subjective and relative stance on moral and psychological matters, and its skeptical perspective on the most essential issues underlying human existence. The result is that while "empiricism" loosened many dogmas and made room for diverse attitudes, it deprived many of a secure basis on which their culture, belief, and action rested.

Not all students of human culture adhere to the empiricist view. Some continue to hold that ideas are the foundation and the essence of all things. In their view, knowledge is based on the ideas we ourselves hold, as well as those held all around us. It is through these ideas that we discover our identities, the societies we create, the political and cultural institutions we construct, even the direction in which we take history itself. Just as faith, economics, or the subconscious functions are the means through which other phenomena must be explained in the systems of Aquinas, Marx, or Freud respectively, it is "ideas" that perform that function for "idealism."

C. From Cultural Evolution to Technophysio Evolution

By any account, the twentieth century was pivotal. It affected, in fundamental ways, the *modus operandi* of human cultural evolution, and will probably have lasting effects on the direction of human development. It witnessed, as had previous centuries, major social and political experiments and great advances in the natural sciences and technology that had significant socio-economic and political impacts. But the speed, scope, and diffusion of advances in science, technology, and regulatory institutions that took place during the twentieth century are unprecedented, and the process is accelerating with no end in sight. There were, in the course of that century, major conflicts and setbacks at high cost to human lives and welfare, including the two World Wars, costly regional, ethnic and religious conflicts, and a prolonged world economic depression.

There were also great technological and scientific developments and inventions; these have the potential either to wipe out the whole of human existence and civilization, or to improve the chances of human survival, material comfort, and knowledge base. There were, in the twentieth century, significant advances in economic techniques and political theory that affected the organization of government and governance, in which economic efficiency emerged as a main policy goal. The century witnessed the continual ascendance of reason and the control of atomic and biological warfare. There were great advances in disease control and health status, in food production, and in material comfort in general, through advances in biotechnology, communication, and space technology, among other scientific developments. The twentieth century witnessed the birth of worldwide regulatory institutions (although short of creating a true world government with full implementing power). These have been designed to safeguard world health, labour rights, science, education and culture, world development and finance, and trade regulation, while other UN agencies and institutions were created to protect the world environment, human rights, and peace, among other purposes.

However, in spite of these scientific and organizational achievements, poverty persisted on a vast scale and the gap between the rich and poor, within and across countries, widened greatly at the close of the twentieth century. For example, a 1999 United Nations report indicated that of the 4.4 billion people in developing countries around the world, three-fifths live in communities lacking basic sanitation; one-third go without safe drinking water; one-quarter lack adequate housing; one-fifth are undernourished; and 1.3 billion live on less than US$1 a day. Nearly one-third of the people in the poorest countries, mostly in sub-Saharan Africa, can expect to die by the age of forty.

Accelerated advances in science and technology and in the knowledge base of the social sciences made classification, categorization, and specialization a necessity, thus reinforcing a propensity, inherited from Aristotle, for seeking definitions. Categorization of science tends to trade depth for understanding of synergies and to reduce interdisciplinary communication. This categorizing trend went beyond the sphere of science, especially in the first part of the twentieth century, to classifying people and societies into higher and lower orders depending on their colour, beliefs, traditions, or economic status: a tendency that paved the way for the bane of racism compounded by injustice.

The nature of cultural evolution also changed qualitatively and quantitatively during the twentieth century. Of the three basic elements of symbolic cultural evolution, the universe of science asserted itself, especially in the second half of the century, as the dominant force setting the pace for the evolutionary process of humanity, and providing the necessary harmony. In one perspective, science is viewed as a power that transcends social forces and is driven more by its own internal logic and the objective facts of nature and less by social forces; it is transforming society into a global "technical" civilization.

For others, advances in science and the direction they take are a reflection of the prevailing social system. Science is produced by scientists who "are neither saints nor devils but human beings sharing the common weakness of our species." In either perspective, human progress and development are being viewed increasingly as the self-generating outcomes of interactions between technological and biological development. This is a view that reduces the wider concept of symbolic cultural evolution into a technophysio evolution, a "form of human development that is biological but not genetic, rapid, culturally transmitted, and not necessarily stable."

Scientific development is moving the world closer to "one" human family. The concept of human races or groups differentiated by lags in anatomical evolutionary processes, which took hold at the end of the nineteenth century, was apparently buried as the twentieth century progressed. Around 1883 Francis Galton, a younger cousin of Darwin and noted mathematical biologist (biometry), coined the term "eugenics" upon reading the *Origin of Species*. The word served to describe the possibility of creating a perfect society, a "society in which human breeding would be taken as seriously as the breeding of domestic animals, a society in which the ideal of 'race improvement' would become the basis for a new ethics that would supplant Christianity." The eugenics movement stimulated programmes of research to establish its scientific basis; these were initiated by Karl Pearson and Ronald Fisher, both noted mathematical biologists. Fisher helped to establish a eugenics society at Cambridge University in 1911.

However, advances in anthropological research, the openness of Western education institutions to students from various social classes and races, and the emergence of a competitive global labour market, especially in the second part of the twentieth century, combined with advances in neuroscience, indicated that the premise of eugenics is unfounded. It is nurture rather than nature that accounts for most of the apparent "group" differentials in human achievement. Nietzsche's *Ubermensch* or "Superman" ("a new vast aristocracy based upon the most severe self-discipline, in which the will of philosophical men of power and artist-tyrants will be stamped upon for thousands of years", Russell, 1945, p. 764) is more universal and accessible to all *Homo sapiens*. In his acceptance speech for the prestigious Kyoto Prize, Mario R. Capecchi, the noted molecular biologist, reflecting on the struggles of his own life, tried to convey how genius springs from the most unlikely beginnings. Human society must find ways to recruit and nurture all its members regardless of social or economic background, since "unlikely beginnings can produce extraordinary lives." This is an essential perspective since, even as we enter the twenty-first century, there are some, especially in the policy field, who continue to adhere to the eugenics paradigm.

Nietzsche's *Ubermensch*, however, is not necessarily the ultimate goal of the evolutionary process, although, in some of his writings, Darwin seems to imply it: "all corporal and mental endowments will tend to progress towards perfection." But Darwin's view of "progress", like that of Malthus, was based on "the uncertain outcome of daily struggle, not the predetermined unfolding of a progressive tendency." There are other views of the evolution of life. For example, there are authorities such as Berlin, Popper, or Gould who doubt that progress defines the historical evolution of life processes. In their essentially atomistic view, history is not directional. It is rather a set of random processes, a "random walk" shaped by exogenous boundaries. Others believe in the directionality of history, a purposeful and predetermined chain of events that culminates in the whole of human society approaching a perfect organism, partly a result of underlying Hegelian processes. These are basic divergences. They reflect fundamental developments in the history of the philosophy of science. It was Aristotle who classified "causes" into four categories: material, formal, efficient, and the unmoved mover (or final cause, which played a pivotal role in his system). Aristotle believed in "the scientific importance of final causes, and this implies that purpose governs the course of development in the universe." That view has a "teleological" as opposed to an atomistic or mechanistic perspective, a perspective that evidently supports the thesis of the directionality of history mentioned above. But the term "teleological" has been applied indiscriminately to highly diverse phenomena ranging from the biological to inanimate phenomena, raising unproductive controversies in the development of a consistent philosophy of human development that persisted through the twentieth century. Bertrand Russell (1945) summarizes the controversy as follows :

> When we ask "why?" concerning an event, we may mean either of two things. We may mean: "What purpose did this event serve?" or we may mean: "What earlier circumstances caused this event?" The answer to the former question is a teleological explanation, or an explanation by final cause; the answer to the latter question is a mechanistic explanation. I do not see how it could have been known in advance which of these two questions science ought to ask, or whether it ought to ask both. But experience has shown that the mechanistic question leads to scientific knowledge, while the teleological question does not. The atomists asked the mechanistic question, and gave a mechanistic answer. Their successors, until the Renaissance, were more interested in the teleological question, and thus led science up a blind alley.

Recent analysis has indicated that the controversy is partly a result of inadequate scientific knowledge of biological processes, and partly a lack of clarity about the concept of teleology. The application of teleology to biological phenomena must be distinguished from its application to inanimate phenomena. Aristotle's teleological explanations relate more to biological phenomena. Similarly, Kant, although a strict mechanist with respect to inanimate nature, adopted a teleological explanation for living processes, a result of inadequate biological knowledge at the time, as Ernst Mayr indicated.

There is also a middle ground that views "human development" as the outcome of a process of maximizing survival probabilities subject to environmental and resource constraints. Survival probabilities lack certainty and predictive power, and do not necessarily reflect progress in human development. Scientific advances provide better understanding of the physical structure and thinking machinery of the brain, but not necessarily of the motivational forces of the mind that *run* its programmes. However, there are signs of progress in understanding the brain-mind association. Evidently, the field of "nature-nurture" research has gone beyond biological determinism or the debate of biological potentiality versus biological determinism, towards genetic determination. But, given the enormous number of brain neurons and synapses that provide contacts between neurons, combined with the complexity of subjective and qualitative psychological phenomena, a scientific view of how the brain-mind works is difficult to attain, especially if approached through reductionism. There are limits to reduction; even in mathematics, the whole is always greater than the parts. Proceeding with "partial constructs" is more productive when dealing with complex systems. Many scientists accept parallelism in psychophysiological analysis as a workable assumption: the belief that there is one-to-one correlation (equivalence) between one's mental states and brain states. An assumption lacking empirical verification, at present, it (parallelism) is no more than an unjustifiable assumption of current science rather than a probable conclusion consciously based on empirical evidence. The gap between present scientific knowledge about the physical world (the brain) and the mental phenomena (the mind) is fundamental and dynamic since, as Gödel observed, "the mind, in its use, is not static, but constantly developing."

Others believe that advances in modern neuroscience could benefit from intertheoretic reductionism where the function of the brain and that of the mind could be better understood in the context of a "global" theoretical framework that encompasses the domains of both the brain and the mind. The two main problems facing those who seek the biological basis for the *conscious* mind, namely that the mind is observable only to its owner and, second, the conflict between observer and observed, are not unsolvable.

Biology in general and neuroscience in particular have been so remarkably successful at unraveling a great many of life's secrets, the current description of neurobiological phenomena is far from being complete. Consequently, it cannot be declared that the conscious-mind problem is insoluble because the brain has been comprehensively studied without finding the mind. Neither neurobiology nor its related physics have been fully studied. Some neurologists are confident that the gap between the mental states and the brain states will be bridged in the coming decades. The biological processes now presumed to correspond to mind processes *in fact are* mind processes and will be seen to be so when understood in sufficient detail.

Advances are forthcoming, however. Controlled experiments on the mammalian brain seem to be moving genetic engineering beyond plant and animal breeding towards genetic transmission of selective health and intelligence attributes in human development. These advances are certain to accelerate as findings from the Human Genome Project unfold. These findings are expected to open a new era of research opportunities in neuroscience with significant philosophical and policy consequences to human development. They open the door for an unending, built-in catalytic growth of interdisciplinary scientific knowledge. These findings also provide information waiting to be abused by unscrupulous insurers, employers, eugenicists, or social Darwinists.

There is also evidence from recent research on neural networks that the unprecedented power of present computers could be increased further through the development of biocomputers controlled by hybrid chips. These hybrid computers could approach the versatility of the human mind, while retaining the speed and discipline of electronic mechanism. Hybrid chips combine living nerve cells (e.g. neurons from leaches, lampreys, or spiny lobsters) with silicon circuits. Computing with neurons or DNA could be commercial reality in the first decade of the twenty-first century. The twenty-first century may witness advances in biotechnology that focus not only on disease control but also on learning processes and personality traits. The potential for scientific progress seems boundless. But the stability of the system and its long-term consequences for human welfare are uncertain. As Einstein, one of the great scientists of the twentieth century, reminds us, the rational calculus of techniques "may not be enough to solve the problems of our social life. The intellect has a sharp eye for the methods and tools but is blind to the ends and values." Over the course of history, there has been a conflict between science and ethics. Science treats people as material objects, and its rules are the physical processes that cause behaviour through natural selection and neurophysiology. Ethics, on the other hand, treats people as equivalent, sentient, rational, free-willed agents, and its rules are the calculus that assigns moral value to behaviour through the behaviour's inherent nature or its consequences. History indicates that the dominance of the natural sciences over faith and imaginal thinking created human suffering as well as their own contradictions (the Marxist revolutions of the twentieth century).

If as a consequence of the emerging technophysio evolution, the sphere of science ends up encompassing conscious time binding and imaginal thinking, then faith and the arts, for example, have to adjust to scientific advances and discoveries. Such adjustment is illustrated by the apparently continuing tension between science and theism in the United States. In such a state of human affairs, "either scientists must be prepared to fudge their data or all of us must be prepared to give up our values." However, "to give up" implies a thesis of discontinuity and radical heterogeneity of human culture. Such a thesis is not necessary to understand the historical course of symbolic cultural evolution, a course, as we emphasized earlier, that was never free from internal or external conflicts. Development itself, including advances in science, implies changes in established values (value endogeneity) in contrast to value heterogeneity (a socially tolerated variance around an established value system). Value endogeneity may not proceed without resistance, especially since scientific change has been proceeding rapidly, while change in values proceeds slowly, since the latter safeguard the established socio-political order and scheme of things in society.

The evolution towards a global scientific technical civilization would have been predicted *a priori* as an outgrowth of the very nature of symbolic cultural evolution. Symbolism provided the foundation for cumulative advances in the basic domains of cultural evolution, especially in the case of science and technology. It provided a universal medium for science and technology to advance, and to disseminate its accomplishments. Advances in the sciences, especially in communication and information technology, combined with changes in the international power structure, have accelerated the "globalization" process in the last part of the twentieth century.

The century started with doubt and skepticism, but ended with increasing certainty or belief in the power of science: in its ability to provide not only new and exciting opportunities for human development and enjoyment but also workable solutions that minimize negative externalities. But, in the absence of advances in the other domain of symbolic cultural evolution, certainty in the power of science is a lop-sided belief that is bound to destroy the internal equilibrating mechanism of cultural evolution whose internal multi-universe dynamics provides purpose and richness to human destiny. It is probably a major socio-political challenge for the twenty-first century, to develop a universal democratic system based on the principle of the rule of law (isonomy) that is able to evolve and be operational in the context of the emerging globalization environment with its economic and political power structure.

3. Human Progress and Prospects at the End of the Twentieth Century

Our review of the nature of symbolic cultural evolution, especially as it adapts to a technophysio evolution, illustrates that there have been, in the twentieth century, fundamental changes in its structural dynamics, with significant consequences for human development. Advances in science, especially in information, communication and biotechnology, and its emergence as the major force in symbolic cultural evolution, have influenced the main forces determining human development prospects in the twenty-first century. Main determinants include the evolution and diversity of cultures, demographic change and socio-economic development, education and learning innovations, the globalization of labour and financial markets, equity, gender, and sustainable environment. These seven sets of factors are examined briefly in the remainder of the paper, using the evolutionary perspective adopted in the present discussion.

A. Culture : Evolution and Diversity

In the present discussion, we make a distinction between "symbolic cultural evolution" and "culture." Symbolic cultural evolution is a dynamic process that continually attempts to maximize the survival probabilities of humankind using the fundamental domains of artisanship, conscious time binding, and imaginal thinking. Culture, in contrast, is a transient concept defined in space and time. It is the outcome of the complex and dynamic processes of the cultural evolution that, although uniform in general form, have produced different patterns of cultures depending on the constraints of artisan development and environmental endowment. The distinction between "culture" and the "symbolic cultural evolution" is essential, especially when attempts to infer moral conduct from observed cultural experiences are divorced from historical context. For example, for millennia, African societies developed a system of gender roles essential for survival: men specialized as game hunters for food and as warriors for defense; women, on the other hand, specialized as home keepers and as subsistence cultivators of small plots of land. The system evolved as optimal given the prevailing technology and resource endowments. However, when these societies were colonized, the colonial power needed labour to work in gold mines or in cash-crop cultivation (cocoa or cotton). The authorities prohibited hunting and introduced policies that made defense at the family and tribal levels obsolete. Men's time-honoured vocations were dismantled, while women continued their traditional vocations with their roles becoming more visible and vital for family survival. It is tempting for post-colonial observers, unaware of the historical and anthropological background, to conclude that African men are exploiting women; the system represents unjust gender roles that are culturally set. Nothing could be further from the truth. What is being observed is a snapshot of a "culture in transition," of a society attempting to maximize its survival chances while adjusting to a changing set of exogenous shocks.

Analogous to Pinker's computational theory of the mind, symbolic cultural evolution provides humans with a powerful and flexible "software" programme that, combined with their developed anatomical endowments that include sufficient redundancy (i.e. enough unutilized cells that can be engaged in additional activities), is able to provide varied solutions to complex survival challenges. There have been hunting/gathering, nomadic, agrarian, early and late industrial societies, with varied cultures. Cultures are not static. They are the outcome of the dynamics of symbolic cultural evolution and the unique and changing environment of each society, and as societies interact with each other. The plurality of "culture" is a natural outcome of human evolution. Plurality is embedded in the UNESCO definition of culture as "the whole complex of *distinctive* spiritual, material, intellectual, and emotional features that characterize a society."

Furthermore, like genetic diversity, plurality is an important factor for the vitality and sustainability of the global symbolic cultural evolution. As Arnold Toynbee indicated, civilizations in decline are consistently characterized by a tendency toward standardization and uniformity. Conversely, during the growth stage of civilization, the "tendency is toward differentiation and diversity." The vitality of the global symbolic cultural evolution and its potential contribution to human welfare may be enhanced when diverse cultures are allowed, not only to coexist, but also to interact freely and peacefully, discovering common values and heritage. The altruistic behaviour that is apparent in many cultures illustrates that cultures are not necessarily based on competitive behaviour. They include co-operation as a basic element. However, not all altruistic behaviour is derived from a system of ethics based on reasoning; most is based on instinctive behaviour based on inclusive fitness. As Darwin emphasized, the social instincts never extend to all the individuals of the same species. The inclusive nature of co-operative behaviour presents a challenge as to how an ethical system based on cooperation could evolve globally, especially in the present environment of global market competition. It is true that the shift from an instinctive altruism based on inclusive fitness to an ethics based on decision-making was perhaps the most important step in humanization. It is equally true that if we jointly prefer a co-operative approach to a competitive one, we have the ability to modify our society for the good of all. But can culture be engineered? Is such engineering desirable?

There is a revival of interest in the role of culture as a pillar and vehicle for human development. However, culture should not be viewed as a commodity, similar to other marketed commodities. The production and marketing of cultures go beyond the ancient Chinese debate in which Xunzi (298–230 B.C.) believed that goodness is not an innate faculty as Confucianism asserts. Humans had to be taught goodness, a theme followed by some social scientists more than two millennia later. Evidently, although Xunzi emphasized the social benefit of teaching (producing) goodness, he did not assign a market value for such learning. Attempts to engineer cultures have been costly. Witness the experience of the Chinese Cultural Revolution in the second half of the twentieth century. The commercial or authoritative view of culture seems to contradict the nature and origin of its dynamic evolution, while introducing difficult policy challenges in the arena of human development. It is one thing to understand the factors that promote freedom of thought and expression in the arts, faith, and the sciences and "invest" in the enhancement of that freedom. It is a different matter to use such knowledge to restrict imaginal thinking and enforce corporate "truth." Education and creative thinking should go hand in hand. However, the present globalization era, with its pervasive technical civilization, presents new challenges to the role of culture in human development. At present, according to social observers, the future of human development and its sustainability is being shaped by the conflicting trends of globalization and identity. The information technology revolution, and the restructuring of capitalism, have induced a new form of society: the network society. This new form of social organization, in its pervasive globality, is diffusing throughout the world, as industrial capitalism and its twin enemy, industrial statism, did in the twentieth century, shaking institutions, transforming cultures, creating wealth and inducing poverty, spurring greed, innovation, and hope, while simultaneously imposing hardship and instilling despair.

However, this is not necessarily a predestined road. We should not underestimate the power of "identity" in human evolution. But the idea of "networking" and "identity" has an origin in the analysis of class development and class struggle, a controversial subject in sociology. It has been argued that Marxist class theory failed because it did not anticipate the institutionalization of class conflict in industrial societies, which tended to reduce the incidence of conflict. Marxist class struggle lost its worst sting in the twentieth century. It has been converted into a legitimate tension between power factors, which balance each other. Capital and labour continue to struggle with each other. But they come to compromises, negotiate solutions, and thereby determine wage levels, working hours, and other conditions of work. How will the culture of conflict resolution, developed in the industrial era, hold or evolve in the information and globalization age? This is a critical question for the twenty-first century.

B. *Global Demographic Transition : Socio-Economic Consequences and Potential*

Demographic change is an integral part of the adaptive mechanism of the cultural evolution that has been accelerating in its latter symbolic and technophysio phases. The historical evidence suggests that throughout the history of *Homo sapiens*, the long-run birth rate was kept as low as possible consistent with survival: as low, that is, as the death rate. Why? The answer seems to be that the hominids were exploiting a unique evolutionary niche by relying on culture, learning, and social organization as their mode of adaptation. The twentieth century witnessed basic demographic transitions that affected population growth, structure, and spatial distribution. The demographic transition is simply a shift of the birth and death schedules (reproduction and health behaviour), from high levels at the early stage of the transition to lower levels at the later stage. However, the timing and speed of the decline of the two schedules are dissimilar, with fertility decline lagging behind that of mortality. The transition implies that demographic behaviour and the environment are interactive. The speed and pattern of demographic behaviour during the transition will vary depending on the specific socio-economic, cultural, political, technological, medical, and public health status, and the environmental and resource endowment characteristics that prevail in society. In turn, these characteristics will change as a result of demographic change.

(i) Demographic Change : Alarm About Depopulation and Stagnation?

In the industrialized countries of Europe and North America both mortality and fertility, although initially not uniform across social groups, were declining early in the twentieth century, with fertility falling below replacement levels in some societies. In some European societies, fertility started a secular decline as early as the beginning of the nineteenth century, for example in France. These apparent secular declines raised policy and academic concerns about potential depopulation coupled with population aging leading to labour shortage, undesirable immigration, insolvency of social security systems, and economic stagnation.

In the second half of the twentieth century, negative population growth became evident in some industrialized countries. For example, in the mid-1980s, levels of fertility in West Germany implied a negative intrinsic annual rate of growth of –1.7 per cent. If this negative rate were to be maintained for 200 years, it would shrink a population to one-thirtieth of its original size. As the realization of such a prospect sinks in, countermeasures will be put to work. In many developed societies, measures to increase the birth rate were introduced. However, some reviews of the causes and dynamics of below replacement fertility, and of the performance of pronatalist policies in industrial countries, concluded that some of these strategies amount to destroying the family in order to save it. Such concerns are probably an echo of those of Rousseau, Montesquieu, and Hume about the effect of modernity and its civil institutions on the robustness of natural institutions, including that of the family. There is, however, recent evidence indicating that marriage and childbearing are being viewed as social capital, the demand for which is not declining with modernity.

The concept of "social capital" had its origin as early as 1916. However, it was only recently that it was given its wider use in the sociological and economic literature. Put simply, social capital may be defined as a set of informal values shared by members of a group that permits mutual co-operation and organizational efficiency with minimum transaction cost. Although apparently simple, the definition cuts across many disciplines. For example, a group could vary from a family, a social group, a country, to the global society, while co-operation may entail the enforcement of social contracts or rules for economic exchange. There are recent extensive surveys and critical reviews of the concept of social capital and its use in sociological and economic analysis. Some economists are concerned that the concept of social capital encompasses too much to be defined adequately for economic measurement and analysis. Some examined in depth the concept of "trust", an important component of social capital, and concluded that although important for market performance, it is difficult to interpret its evolution and stability in the context of economic thought and analysis; they questioned the economic investigation of trust: given that for the economist comparative advantage does not lie in discerning duty and morality—in seeing right or wrong—are there subjects that should be closed to us, subjects that simply do not allow an economic and game-theoretic analysis, are poisoned by it, and poison us in turn?

(ii) Demographic Change : A Window of Development Opportunities?

In contrast, in less-developed countries, where most of the world population resides, the pattern of demographic transition was different. Both fertility and mortality remained at high levels through the mid-twentieth century. Mortality started a relatively fast decline in the late 1950s, but fertility did not. As a result, population growth accelerated at high rates in the developing countries, raising concerns about imbalances between human needs and resource endowments. Many developing countries adopted stringent population policies to contain population growth, some with spectacular success, for example China and other countries in East Asia. By the mid-1980s, fertility started a sustained decline that accelerated in the 1990s, in almost all the developing countries. The nature of fertility decline varied across countries and social groups, and was affected by patterns of institutions and social organizations with different social implications. In some societies, the decline was led by delayed marriage. In others, it was led by contraception techniques. These demographic patterns are shaping age structures across the developing countries well into the twenty-first century.

There are potential economic benefits for countries passing through the phase of delayed fertility decline. It presents a "one-time demographic gift": as the burden of dependency declines, potential saving increases and general health status improves, based on the biology of increased infant and maternal survival as a result of better birth spacing and reduced teen marriage. If these demographic benefits were to be combined with sound internal reform, the countries in question might enjoy a more educated labour force and the creation of productive employment to absorb the growing labour : a "quantity-quality" transition. The "quantity-quality" transition implies, over the longer term, a slower growth of the labour force with higher levels of human capital per worker.

It also implies a growing demand for skills. For these potential benefits to materialize, internal reform should include economic policies that pay closer attention to the efficiency and effectiveness of public organization and to the role of information in decision-making, and should direct savings to productive investments. Internal reform should also include socio-political policies that introduce transparency in governance, improve educational quality and relevance, and equalize opportunities.

There are other longer-term consequences of the demographic transition. The initial increase in the size of working age cohorts will translate, within a generation, into an increase in the elderly population, an increase that requires a larger transfer from a relatively declining pool of working cohorts. In the twenty-first century, population aging will be a worldwide phenomenon not confined to the developed countries. A failure by the developing countries to implement sound policies during this vital phase of the transition could have negative effects: since the population of working age will grow at high rates for at least a generation, lack of quality education will translate into non-competitiveness and inter-cohort conflict. Rapid fertility decline, combined with accelerated mobility and migration and the integration of societies into the global market of production, finance, and ideas, has significant consequences for family and community structure and behaviour. Furthermore, advances in contraception technology have separated reproduction from sexual behaviour with repercussions on family formation and stability, and the stability of social norms. The stability of established social contracts and intergeneration transfers is questioned, and, accordingly, the efficacy of national social policies. Meanwhile, the internal sovereignty of the state to implement social policies that reduce potential socio-political tensions is being diminished in the emerging globalization environment.

Most developing countries are not capturing the full potential of the demographic window of opportunity. Although some countries, especially in East and Southeast Asia, were able to take advantage of the demographic window of opportunity, many others have not done so. One reason is the lack of adequate internal reform, especially in education and in market and democratic institutions. The second reason is the volatility of the international finance capital markets in the emerging global environment, in which international regulatory institutions are in the process of experimental reform.

(iii) Migration and Human Development

Population movement is an integral part of symbolic cultural evolution processes. It is an outcome and cause of these processes. This section reviews briefly the various patterns of population movements. There may be exogenous changes in the environment, such as droughts, or humans may alter their environment as a consequence of negative externalities of their economic and social behaviour. As a result, the carrying capacity changes relative to reproduction and some may leave their localities, sometimes on a mass scale, seeking better fortune. Conflict often arises as a result of human movements and settlements in new localities or countries, with adverse effects on both the environment and human development and welfare. Conflict generates additional movements. As mentioned earlier, the historical cultural evolution was not smooth. There were internal and external conflicts because of lack of harmony among its basic components. Major population movements occurred in the twentieth century as a result of these age-old disharmonies, or of attempts by society or social groups to homogenize their own cultures or escape political or religious persecutions. There are also purposeful and positive population movements.

Part of the internal dynamics of symbolic cultural evolution is the emergence of the human drive for discovery and change. Humans seek new knowledge, attempt to extend their reach and improve their socio-economic status and welfare, with mobility and movement as a main mechanism. In this respect, voluntary human migration is viewed as investment in human capital and as a mechanism that provides for increased macro efficiency and welfare. The speed and extent of voluntary migration accelerated as a result of technological advances and the growth of industry and trade relative to agriculture. Rural-urban migration within countries accelerated, as did international migration, a result of the unequal pace of industrialization and development in different countries. At the close of the twentieth century, as the new globalization era evolved, there seems to be a tendency to reduce the aggregate scale of international population movements while changing its pattern towards skill selectivity, a return to the "brain drain." For example, in the United States, the potential for immigration of skilled workers was high, since ambitious, skilled young people from allover the world are frustrated by backward and inflexible economic and social systems. Immigrants are attracted by a dynamic economy and the possibility of upward mobility.

This tendency for skill migration is also partly a consequence of the enhanced and more secure movement of finance capital and trade, and partly a result of accelerated population aging, especially in the developed countries of Europe and North America. The full implications of these dynamics are yet to unfold, however.

C. Education, Learning Innovations and Motivation

Symbolic cultural evolution depends on education and the accumulation and preservation of knowledge, rather than on genetic transmission, for its continuity. As Kenneth Boulding succinctly observed, it is knowledge that is evolving; humans are only agents in the evolutionary process. High levels of human capital are required to assure scientific progress and growth, maintain established knowledge across generations, and promote demand for the outputs of the technological establishment by generating skilled consumers able to purchase sophisticated products and services. The evolutionary perspective views the accumulation of human capital as necessary for the continuity of the scientific revolution. Education and human capital accumulation are also necessary for empowering the less-developed countries to catch up in the development process. However, as mentioned above, these are not sufficient conditions, especially in the context of the emerging global market. Two dimensions of human capital formation require attention. The first is the quantitative dimension of the quantity and quality of education. The second is more qualitative and psychological. It refers to the motivational aspects of human development.

(i) Education Quantity and Quality

The twentieth century has witnessed increased awareness of the socio-economic importance of education. A skilled labour force is viewed as necessary for integrating the economies of the developing countries into an increasingly competitive global market. There have been purposeful attempts towards increased investment in education and training, and the development of quality assessment measures, in the developed and developing countries. School enrollments have increased in all the developing countries. However, there is a wide variance in enrollments and number of years of education. Although early studies indicated sizable returns to the quality of the education process, the emphasis on education planning and assessment of performance has been based on input measures: enrollment, years of education, teachers' qualification, curriculum, or educational technology, and not on performance. Recently, there has been a systematic effort to measure quality of education in terms of outcome, with UNESCO providing technical support to conduct such studies in various countries. The recent case of Kuwait is illustrative. The Kuwait Society for the Advancement of Arab Children (KSAAC), with technical support from UNESCO, conducted a major study to measure the performance of children in basic education. A special characteristic of the study is its attempt to go beyond the standard focus on science and mathematics, to include the social sciences, language, and culture in its coverage, an important focus on the wider role of education in human development. The methodology was comparable and consistent with international standards. The findings of the study indicated that although expenditure per student in basic education is one of the highest in the world, its returns in quality are lagging. This finding, among others, supports the general conclusion that there is significant variance in the quality of education within and across countries, especially in mathematics and the sciences, and in logical reasoning. Furthermore, other studies indicate that access to educational opportunities is unevenly distributed within and across countries by gender and by social groups. Attempts to reduce the gender gap in education seem not to give adequate attention to quality or content, or adopt an approach that integrates education and development.

On the other hand, advances in science and technology have lengthened the gestation period of basic and higher education, and on-the-job training. These changes have, in the context of a human capital framework, increased the opportunity cost to parents of providing quality education to their children, the opportunity cost of higher education to students, and the public cost of education and vocational training. Furthermore, lack of adequate knowledge about costs and returns to education leads to inefficient micro and macro allocative decisions, especially in less-developed settings. In some areas, where fertility started a secular decline and the supply of labour has been growing, exerting downward pressure on returns to investment in human capital, the expected quantity-quality trade off has been faltering. The combination of increased cost, lower returns, and inadequate information raises a policy dilemma since the social benefit of education is increasing during the present phase of symbolic cultural evolution. On the supply side of human resources, public investment in quality basic education that focuses on mathematics and the sciences, and on innovative learning technologies, is necessary for economic growth, especially in the context of the demographic transition discussed above. On the demand side, public investments that increase the efficiency and effectiveness of labour markets, reduce social imperfections, and enhance the institutional framework to attract national and international investment are equally necessary. Since these investments enhance the demand for labour and accordingly increase the returns to education, they are necessary to increase parental commitment to quality versus quantity in childbearing and child-rearing behaviour.

(ii) Learning Innovations

The scientific and technology establishment is becoming increasingly complex, expensive, and globally competitive. Scientific development is mainly a result of investment in human and physical capital with education, training, and research and development (R&D) playing a pivotal role. Education is being increasingly viewed as input for technological and scientific progress. Innovations in learning are developed to enhance the efficiency and effectiveness of education and training systems in order to contribute to economic growth. Progress in learning technologies through advances in computer and information sciences and interactive software provided worldwide opportunities for learning and certification through the Internet. These learning innovations have enhanced the effectiveness of education and training around the world. However, access and utilization of learning innovations are highly uneven, and limited by the high cost of the necessary implements and required background. Students in poor countries or from poor families are at a disadvantage. For example, according to the United Nations Development Report 1999, the cost of transport and communication has declined dramatically since the 1960s. The cost of telephone calls fell from US$46 for three minutes in 1960 to US$3 in 1990 (in 1990 dollars), and the price index of computers fell from 12,500 in 1960 to 100 in 1990 (1990 = 100). However, the gap between the rich and poor in acquisition of new technologies and in their use has been growing. In 1996, for example, there were 204 personal computers per 1,000 people in countries ranked "high" on the human development index (mainly high-income countries), 7 per 1,000 in medium rank countries, and close to zero in the low rank. The case of Internet hosts indicates the same skewed pattern in 1998. There were, however, positive experiences in the last part of the twentieth century. Some countries in East and Southeast Asia—for example Singapore, Korea, China, Malaysia, and Thailand—achieved high rates of economic growth by investing in quality basic education, and by taking advantage of advanced learning technologies and of the demographic window of opportunity. Their investment in quality basic education was supported by a strong higher education system and effective R&D, and accordingly attracted domestic and international demand for their skilled human resources.

There are concerns about the effect of learning innovations other than their potential negative effect on equal opportunities. Some educators view the learning innovation process as an efficient and mechanistic process that is leading learning institutions towards structured curricula and narrower vocational focus, and increased privatization and commercialization of the education establishment, but not necessarily better quality or relevance to the wider concept of education. This process of learning innovation has been labeled the age of digital diploma mills. The wider role of education, as a promoter of imaginal thinking and cultural development, is being subjected to the calculus of market returns, although its social value may not be revealed in the marketplace. That wider role of education is essential for the attainment of the fuller dimension of human development as well as the very survival of symbolic cultural evolution. The challenge for human development policies in the twenty-first century is how to combine education and creative thinking in an environment driven by symbolic cultural evolution in its technophysio phase.

(iii) Motivation and Morals

Attempts to link motivation to human development have been part of the development literature for decades. Achievement motivation, the propensity to strive for success in situations involving an evaluation of one's performance in relation to some standard of excellence, vision of opportunities, discipline, work ethic, commitment, and self-esteem are essential qualities of human development that define a motivated and progressive society, qualities that have been recently labeled spiritual assets. These, however, are qualities that are not transmitted via school curricula alone. They require time-intensive investment by parents, schools, and community, as well as the larger systems of government and governance to assure their continuity across generations. The process of moral development raises some of the most hotly debated questions in philosophy and the social sciences. Is there a set of universal values that guide moral development everywhere? Are there levels or stages of moral judgment, for example from self-interest, to social approval, to abstract ideals? The questions are complex and have no clear answers. A review of psychological studies of children's moral development indicates that "moral identity—the key source of moral commitment throughout life—is fostered by multiple social influences that guide a child in the same general direction. Children must hear the message enough for it to stick. The challenge for pluralistic societies will be to find enough common ground to communicate the shared standards that the young need."

These conclusions support the view that transparency and accountability in governance and isonomy in the rule of law are prerequisites for the spread of moral conduct. (The ideal of "isonomy" refers to the certainty of being governed legally in accordance with known rules, as described by Hayek in 1955.) It is an essential foundation for democratic practice and raises challenging questions for sustainable human development, especially in the context of the emerging globalization environment. Some of these issues are examined briefly in Section 3.4.) For example, the Secretary of the Singapore Development of Planning made the point in his address to the Oman 2020-Development Vision Conference (May 1995) that equal opportunity is a basic ideal of Singapore development. Promotion of students or employees or the conduct of any other state affair is based on merit and merit alone. There are heavy civil penalties for lack of compliance with that system. What is more important is the high social penalty. Social penalties, however, are part of the moral code and can only develop when "the message is heard enough and implemented fairly with consistency." Although parents are the original source of moral guidance for most children, the education system should be able to supplement the inputs of parents and communities by providing, in the formative years of education, the basis for imaginal and independent thinking in its educational approach, in and outside the classroom. Transparency in governance, isonomy in the rule of law, and other democratic practices and institutions provide the necessary support and enforcement beyond the formative years.

D. Globalization and Human Development

(The influence of writers such as Hayek, Schumpeter, and Russell on the discussion in this section should be evident.)

Symbolism implies universality in scientific communication. It has been the nucleus of the present globalization process, a process based on the global spread of technology, finance capital, and a global labour market, rather than on military conquest. These processes are evidently leading human society towards a new global moral system, a global technical civilization, one in which efficiency is supreme. Adam Smith's "division and specialization of labour" and Marx's "workers of the world unite" are both operative, in the "new global environment", with unexpected consequences. The globalization processes are varied, complex, and evolving. They are affecting not only production and exchange relations but also established human values, roles, and relations. In this section we focus briefly on two issues: government and the rule of law in a global environment, and the equilibrating mechanism of a global market.

The "new global environment" differs significantly from that of the past. The past environment, excluding military conquest and occupation, was one of "interdependence" among states, regulated by international institutions such as the International Monetary Fund (IMF) or the General

Agreement on Tariffs and Trade (GATT) that attempt to set rules for orderly co-operation as inter-dependence increases among sovereign states. Government sovereignty, especially internal, was not challenged by these regulations, taken mainly as safeguards, although by putting limits on levels of tariffs or by exchange rate manipulation, governments' external sovereignty was compromised. The effect of the "new globalization environment" on state sovereignty is subtler. It influences both the external and internal sovereignty of the state, while reducing the interest of national elites in local affairs. In the global environment, global corporate networks challenge a state's internal sovereignty by altering the relationship between the private and public sectors. By inducing corporations to fuse national markets, globalization creates an economic geography that subsumes multiple-geography. A government no longer has a monopoly of legitimate power over the territory within which corporations operate, as the rising incidence of regulation and arbitrage attests. This process does not imply that private sector actors are deliberately undermining internal sovereignty. Rather they follow a different organizational logic from that of states, whose legitimacy derives from their ability to maintain boundaries. Markets, however, do not depend on the presence of boundaries. While globalization integrates markets, it fragments politics.

The implications of these dynamics for human development are better understood by examining two questions. The first is related to insecurities introduced in labour markets as a consequence of the internal dynamics of the present globalization environment. The second is the influence of globalization on democratic practice and the rule of law.

As discussed earlier, the present globalization processes are not new. They are the inevitable consequence of the *modus operandi* of symbolic cultural evolution, which elevated the role of artisanship relative to that of conscious time binding and imaginal thinking. However, it is not intuitively evident whether the present phase of the globalization process is structurally different from past phases or differences are merely qualitative. Three major factors characterize the present globalization environment: innovation in communication and information technology, the dominant role of finance capital, and the emergence of global regulatory institutions. However, these characteristics have been present in various forms for many centuries. For example, there were major technological breakthroughs in the late nineteenth century and in the 1920s that had significant socio-economic impact. There were also regulatory institutions, some acting through colonial powers that had extensive global reach combined with strict enforcement power. The significant role of global finance capital is not new. History indicates that following each major innovation, finance capital follows good economic principles but eventually it gives way to speculative ventures, economic facts give way to psychological fantasy, and share prices become independent of performance, leading to eventual stock market collapse and prolonged economic recessions. The result has been the recurrence of large economic and social losses.

The present globalization phase seems to have more than qualitative differences from the previous phases. The *modus operandi* of the present globalization environment is being fueled by the significant and accelerated pace of innovations in information and communication technology. These innovations facilitated the global spread of finance capital, while international regulatory institutions are being redesigned to provide the necessary security for capital movement. However, as in the past, the psychological element of finance capital takes over, creating greater worldwide volatility, in part because of the efficiency of the information technology itself. Workers' security and human development could suffer greatly as a result of investors' psychology and not necessarily because of weakness in economic facts. Indeed, economic reality takes hold in the final analysis since irrational increases in equity prices cannot be sustained for long without the support of real earnings. Similarly, irrational capital flight should eventually reverse its course once the economic facts override psychological and political forces. However, the loss to human development could be sizable and *selective* in the interim (and the interim could be of long duration), while, as discussed earlier, the social role of governments is being compromised in the global environment.

There is also historical evidence that periods of massive industrial consolidation and dramatic technological innovation have been followed by periods of political, social, and institutional reform. One eventually creates the need for the other, as the economic changes produce social conditions that come into conflict with democratic ideals. What does the globalization environment offer as a supplement or substitute, when the synergistic relations between "democracy", the "rule of law", and the "role of government and governance" erode as a consequence of the rules set by its *modus operandi*? Historically, the struggle for freedom has been a struggle between the power of government with arbitrary authority and the rule of law; according to Hume, it is the evolution from a "government of will to a government of law." A "government of law", "equality before the law", or "rule of law" are all terms that refer to the ancient Greek concept of "isonomy" which identifies a society of human freedom as opposed to an arbitrary government of tyrants. In ancient Athens, where isonomy was first established, Athenians were given not so much control of public policy, as the certainty of being governed legally in accordance with known rules.

At that time, the ideal of isonomy was used and practiced as a justification for the ideal of democracy, people's control over public policy. It was feared that democracy without the rule of law would eventually erode the ideal of isonomy from its true substance, and in turn, democracy. Democracy was viewed as the child of isonomy, not the other way round. In this view, the law should be obeyed so people could be *free* from arbitrary rules set for some and not for others. However, although the rule of law provided for individual freedom and for creating and maintaining "trust" in the laws, it erected boundaries and set limits on the freedom of both government and subjects. As early as 1690, Locke defended this ideal in order to limit the power and moderate the dominion of every part and member of society. However, although the ideal of isonomy gave ample power to governments to deal with a wide range of actions to enhance human development, from safeguarding values and promoting equality to enhancing individual freedom, it has been on the decline for centuries in all countries of the world, including England. This has been the case although it has been established that the absence of isonomy erodes trust between governments and the governed, and accordingly reduces true democratic participation and practice, probably as a result of the power of interest groups. These processes lead to the decline of social capital at the various levels of society and lead to the erosion of institutions of conflict resolution that evolved during the Industrial Revolution to reduce class conflict within and outside industry, as discussed earlier.

The concern about the impact of globalization on government authority, discussed above, is well founded but these consequences should be evaluated as to their effect on governments of states as well as on the evolving global government. For the former, the impact on sustainable human development could be negative not only in authoritarian governments but equally in democratic ones, especially those lacking in the ideal of isonomy. For the latter, the adoption of the ideal of isonomy and the development of a global political citizenship seems to be of great urgency. As early as 1930, in his evaluation of the *Report of a Committee on Ministers' Powers*, Ivor Jennings argued that "this rule of law is either common to all nations or does not exist." Jennings' statement seems to be equally valid in today's globalization environment, although it requires additional qualification: global rules should be set in the context of democratic processes. In the absence of such processes, global rules may not be sustainable. Rules designed to benefit the few without compensating the losers will be resisted. The recent experience of the World Trade Organization (WTO) and the IMF is illustrative. The demonstrations that took place during their meetings in Seattle (November 1999) to protest against their mandates indicate weaknesses in the present international regulatory system. The focus on individual or state-level governments' behaviour rather than on the behaviour of the emerging global government has probably left a gap in analyses of human development. There are optimists who believe that the processes of globalization will eventually lead to a spontaneous rise of co-operative norms that enhance and reinforce the accumulation of social capital, without the support of an authoritarian hierarchy. Such optimistic views have not been substantiated as yet by theory or experience. There are many structural factors that tend to inhibit the realization of such hopes. These include size, boundaries, repeated interaction, established cultures, and lack of isonomy, justice, and transparency in national and global governance; all these make the spontaneous emergence of systems of cooperative norms unlikely without the helping hand of "rational hierarchical authority, in the form of government and formal law."

The internal logic of the "new globalization environment" includes a self-regulating mechanism that affects human development in various ways. Competition from foreign labour reduces labour power and acts as a check on wage hikes in the advanced economies,while built-in high rates of labour growth keep wages in line in the developing part of the global economy. If relative wages increase or productivity falters, given the present phase of the demographic transition, financial and investment flight provide for market discipline. This will guarantee the control of inflation through unemployment and unemployed capacity, while government capacity to provide social safety nets or enforce trade union rights is being constrained. A main policy theme that has emerged in the globalization environment, supported by conceptual modelling and empirical findings, and adopted by the international regulatory agencies, is that the reduction of trade and finance barriers leads to higher growth and welfare levels. The robustness of the supporting evidence has been questioned recently. The relations between openness, growth, and welfare are far from simple. Even if trade policies that restrict international trade were to reduce economic growth "it does not follow that they would necessarily reduce the level of welfare."

The globalization process, on the one hand, seems to present broader opportunities for economic growth to many countries, while seeming to deepen inequalities between social groups and within and between states, on the other hand. The high mobility of international finance capital and its volatility, combined with the accelerated speed of technological development, introduced uncertainties and insecurities into labour markets and to the processes of human development in general. In an evolutionary perspective, the present globalization process may be presented schematically as follows :

> "Symbolic cultural evolution (SCE) → increased advances in science and technology → increased complexity and longer and more costly learning requirements → increased inequalities in acquired capabilities → more government and less isonomy → global interdependence in trade, finance capital, and information flows → development of global rules to govern trade and finance (e.g. WTO) → more collaboration in global production processes and R&D → higher rates of economic growth accompanied with increased income and wealth inequalities → increased economic shocks as a result of speculative finance capital flight → emergence of unemployment, poverty, inequality, and political unrest → more international rules and regulation without global government adhering to the ideal of isonomy → strict enforcement of market efficiency rules as designed, for example, by WTO, IMF, or by the disciplinary rules of international finance capital, and less by social objectives → more political and social unrest → new domestic and international revisionist arrangements."

The emerging globalization environment may not lead to the vicious circle implied above. It may, with enhanced democratic participation, and the evolution of effective global governance and regulatory agencies that are more sensitive to human and environmental needs, lead to higher and equitable global welfare. However, that path is neither necessarily a natural outcome of the emerging technophysio evolution nor a policy option without cost to the major economic and political powers in the present global environment. This is evidently the case since, as Binmore (1998) has pointed out, it is only in "a *well-ordered* society, [that] each citizen honors the social contract because it is in his own self-interest to do so, *provided* that enough of his fellow citizens do the same." These conditions are not as yet satisfied in the present globalization environment, an environment lacking in both self-order and isonomy. The struggle for a decent, democratic, and humane society has little to do with scientific developments or the increase in the volume of international trade. It exists, according to some observers, in an entirely separate realm—the realm of democratic citizenship—that is now being undermined by trade. For pessimists, that realm is influenced by market behaviour, a market that is the final step in a process that first leaches out the moral content of a culture and then erodes the autonomy of its citizens by shaping their personal preferences, while the evolving market cannot be corrupted because its institutions have been corrupted already. Whether the constituents in the emerging global society develop a collective long-term horizon based on "initial positions" of moral justice will depend to a large extent on the course of symbolic cultural evolution, whether its present technophysio *modus operandi* is balanced by conscious time binding and imaginal thinking.

E. *The Income and Capability Gap : Unequal Opportunities and Poverty*

It is not evident *a priori* whether the scientific revolution of the twentieth century combined with the *modus operandi* of symbolic cultural evolution is leading to more equality in opportunities and in returns to investment in human capital, regardless of colour, gender, or location. Indicators at the close of the century, referred to earlier, are not optimistic. Such pessimism is enforced by uncertainties introduced by the emerging global environment. But, as mentioned earlier, the potential for narrowing the gap is evidently great. Human development in the age of symbolic cultural evolution is based on advances in the sciences and technology that continuously extend humans' outreach beyond their limited anatomical capabilities. Extension of human capabilities, as mentioned above, is a function of innovative learning techniques, advances in computer and information sciences, and in neuroscience. Present knowledge in neuroscience indicates that the anatomical structure of the brain is identical, in general form, in all members of *Homo sapiens*. They all have, given the normal distribution of populations by natural abilities, the same anatomical potential for learning. Even before the present advances in neuroscience, many educators of the nineteenth century were convinced that it is nurture, rather than nature, that is responsible for human development. It is differentials in health conditions, and in educational quality and opportunities, that account for most of the differentials in achievements. In his advice to his son, who had been showing signs of genius at a very young age and was heading for his first independent trip to Europe at the age of fourteen, James Mill told him :

> "John : until this moment, mindful of the fact that over-estimation of one's own merits is a grievous defect, I have carefully concealed from you the extent to which your intellectual attainments surpass those of most boys of your age. Some may even be so thoughtless as to pay you compliments, and suggest to your mind the erroneous belief that you possess exceptional abilities. In fact, whatever you know more than others cannot be ascribed on any merit in you, but to the very unusual advantage, which has fallen to your lot, of having a father able to teach you, and willing to give the necessary trouble and time. That you know more than less fortunate boys is no matter of praise; it would be a disgrace if you did not."
>
> (Quoted in Russell, 1934, p. 96)

James Mill's model of educating his son, John Stuart Mill, was influenced by the educational philosophy of Helvetius (1715–1771); he believed in the power of knowledge and education, and especially the impact of governance and customs : "The principal instructors of adolescence are the form of government and the consequent manners and customs. Men are born ignorant, not stupid; they are made stupid by education." James Mill's statement to his son, quoted above, indicates the potential for both equality and inequality in human development. The statement implies that, on average, the potential for learning is the same throughout *Homo sapiens*; controlling for the random dispersion of *native/natural* talents, John was most probably an exceptional child by nature, located on the extreme part of the *natural* ability curve.

Mill's statement also implies that learning depends on the quantity and quality of inputs including health and nutrition. If these are not distributed equally, then equality of socio-economic opportunities in the marketplace will not be highly correlated with earnings since the ability curves will diverge from their random nature. Health and nutritional status have improved across all developed and developing countries, but with significant differentials within and between countries in both access to and quality of services. There is evidence that health status and health utilization vary by education and socio-economic factors, while health utilization depends on health and nutritional status and socio-economic factors. Inequality in education and health status is both a cause and an effect of poverty and economic inequality. Furthermore, because of the role of parents as providers of critical learning, health, and nutritional inputs to their children, poverty and low levels of abilities are transmitted across generations. The nurture doctrine of Helvetius and James Mill becomes clearly operative.

On the macro level, there have been several attempts to link income inequality to economic growth. The best known is that of Kuznets and Myrdal, known as the Kuznets hypothesis. In this hypothesis the processes of industrialization and urbanization lead to a worsening of income distribution in developing countries because in the early stages growth is concentrated in the modern sectors. It is only at relatively high levels of income that technological progress affects the bulk of the economy and that redistribution through income transfers becomes significant.

The validity of the hypothesis has been questioned. Recent empirical studies and a survey of the theoretical evidence on the relation between inequality and economic growth find no systematic relation between inequality and the stage of economic development. An interesting observation that emerges from the survey is the importance of organizational change : "inequality heavily depends on the type of flexibility chosen by firms in the management of human resources." Firms in Germany and Japan, for example, choose to promote workers from the lower end of the occupational or skill structure to higher levels, whereas firms in the United States and United Kingdom tend to rely on external flexibility by hiring from outside and firing their unskilled workers. Wage inequality did not increase in the two former countries, whereas it increased sharply in the US and UK. The emerging globalization process seems to follow the external flexibility strategy and tends to produce results with similar patterns to those expected from the Kuznets and Myrdal hypothesis, although its dynamics and allocative mechanism are different from those implied in the latter framework. Initially, global finance capital is attracted to the skilled and efficient segment of the labour force in developing countries, while the unskilled are left with low-paid jobs and higher levels of unemployment. It is only when skills are dispersed more widely that inequality declines. That decline, as mentioned earlier, is conditional on successful internal reform, patterns of organization, and better understanding of the behaviour of external economic and political forces. On the other hand, given a constant demand, a wider dispersion of skills will increase the supply of skills relative to that demand. Equality may improve at the cost of a decline in wages of the skilled and in returns to investment in higher education. These are complex dynamics, however. They require careful modelling and empirical verification.

More recently, convergence was introduced to growth models : the lower the initial levels of development indicators, other things being equal, the higher the speed of catching-up. The hypothesis provides hope for less-developed countries and poor communities in general. However, it does not illustrate how the catching-up process functions in the context of the globalization process, a context that, although promoting economic growth, does not necessarily provide for equality or poverty alleviation even in the advanced economies. For example, in the United States, poverty seems to persist in the face of a decade of unprecedented economic growth combined with innovative social policies. Furthermore, globalization "helps push down US wages [while] trade accounts for roughly one-quarter of the rise in US income inequality since 1970s, studies show." In the newly industrialized economies of East and Southeast Asia, the economic and financial crises that occurred in the late 1990s, partly as a result of volatility in finance capital and partly through inadequate diagnosis, had severe socio-economic and political repercussions with high costs to human development. The crises set back these economies, although their basic "internal" structural parameters, responsible for their previous spectacular success, did not change. There are many uncertainties and much volatility in the emerging global labour markets, which present challenges to the conventional wisdom of the determinants of poverty and socio-economic inequalities. These factors have an impact on human resource development and require innovative thought and policy analysis. However, the East and Southeast Asian crises seem to suggest that economies with robust human development systems, that is, competitive education quality, flexible labour markets, and isonomy in the rule of law, can withstand market shocks better, although the interim losses can be severe.

In a recent contribution with a wider perspective, Landes (1999) attempted to explain why some nations are so rich while others are so poor. The perspective is partly based on Ibn Khaldun's theory of the role of geography and organizations in the rise and fall of civilizations, partly on a social Darwinist framework in which the adaptive capacity of social organizations is pivotal, and partly on a neoclassical framework of market behaviour. In this multidimensional framework, the evolution of human development seems unidirectional, with Western civilization, and in particular that of Western Europe, epitomized as the ideal. Joining the Western wagon, so to speak, is not open to all societies, a result of misfortune of geography or having a stagnant culture or religion. However, the role of the present phase of globalization in facilitating inter-society learning and the role of science in rehabilitating unfavourable geography are not adequately incorporated in this framework.

F. Gender

The previous discussion leads to the important issue of gender bias in human development, unequal opportunities, and role and status differentiation. There have been significant advances in the understanding of gender differentiation by opportunities, status, and life rewards, especially in the second half of the twentieth century. There have also been significant policy, technological, and behavioural changes that have not only the effect of bridging the "gender gap" but also the potential to change the patterns of long-established familial and social relations and organizations. Probably, changes in gender roles and status will be one of the most significant forces shaping human development in the twenty-first century.

It was the anthropologist Margaret Mead who first made the distinction between gender and sex. Sex is the biological category, whereas gender is the culturally shaped expression of sexual difference : the masculine way in which men should behave, and the feminine way in which women should behave. It is the "central aim of much feminist thought to uncover concealed asymmetries of power in differences of gender, and to work for a society in which the polarization of gender is abolished."

In the contemporary feminist ethical debate there are some who argue that biologically based gender differences exist in reasoning; women tend to value community, caring, and bonding more than men do. It is not evident that these apparent differences reflect innate faculties rather than the way in which men and women have been conditioned to emphasize different aspirations and ideals. There is some evidence that differences in behaviour are outcomes of the combined effects of nature and nurture, although the former is based on a "social Darwinism" paradigm. The issue of gender roles is both complex and socially fundamental. It requires a perspective that integrates developmental, genetic, evolutionary, and cultural approaches. The result, according to Waal (1997), could be the "power of breaking down of old barriers between disciplines. Most likely what will happen in the next millennium is that evolutionary approaches to human behaviour will become more and more sophisticated by explicitly taking cultural flexibility into account." Culture may cease to be viewed as the antithesis of nature.

Meanwhile, from a policy perspective, the argument for nature is not fully supported by scientific evidence, while that for nurture is subject to a reform in educational and social policies, thus focusing attention on the role of social organizations and market behaviour as determinants of the gender gap. The power of the nurture paradigm is demonstrated by the significant change in women's roles and status in society, achieved in the second half of the twentieth century. Recent assessments indicate that women as a whole are gaining more control over their lives. They are able to control reproductive behaviour as a result of more effective and accessible contraception technology. Furthermore, female labour-force participation has increased, in more occupations, and with higher earnings. Also, there are more women entering higher education than men in many developed and developing countries; women are becoming more active in politics, and more of them vote. These trends are fundamental. They are bound to affect the future structure of employment and labour markets, and the pattern of government and public dialogue around the world. Although these trends seem universal, there are significant variations; generalization is necessarily suspect.

There are significant differentials in the extent of the gender gap, and the rate at which it is being closed, within and across societies. A recent extensive review of gender and jobs documented the prevalence of high levels of occupational segregation by sex in today's world. In that review by Richard Anker, occupational segregation by gender was shown to be extensive in every region, at all economic development levels, in all political systems, and diverse religious, social, and cultural environments. In short, occupational segregation by sex is an important worldwide phenomenon. Furthermore, not all women have full control of reproductive behaviour. Contraceptive use is not universal, especially in eastern and western Africa. In the 1990s, total fertility rates (TFR) in these two regions continued to exceed six children per woman while less than 13–17 per cent of the women have access to modern contraception compared to over 40 per cent in the rest of the world. The gender gap in education was not completely closed by the end of the twentieth century. Furthermore, there is a large female disadvantage in education in the countries of western and central Africa, North Africa, and South Asia. Wealth gaps are even more widespread in the developing countries and the interaction of gender and wealth results in large gaps in educational outcomes in these countries. In the socio-political domain, women's socio-economic status and political participation in many developing countries, although improving, are relatively low as indicated in the recent UN Human Development Report. Although the overall trends are positive, significant disparities persist, especially in Africa and rural South Asia.

G. *Environment*

As symbolic cultural evolution proceeded and accelerated its outreach to alter the environment, three questions about the role and place of humans in society and the cosmic order have become increasingly pertinent :

> "Is the earth, which is obviously a fit environment for [human] and other organic life, a purposefully made creation? Have its climate, its relief, the configuration of continents influenced the moral and social nature of individuals, and have they had an influence in modelling the character and nature of human culture? In his long

tenure of the earth, in what manner [have humans] changed it from its hypothetical pristine condition?"

(Glacken, C. J., quoted in Goudie, 1993, p.1. Human and humans substituted for man and men)

The answer to the first question has been based on faith while the last two questions have not been fully answered. The symbolic evolution provided *Homo sapiens* with increasingly powerful tools to alter the environment but not to assess the long-term implications of such actions, whose consequences become more serious since, in their conscious or unconscious attempts to change their environment, human values and cultures are being equally changed. It seems that the "spiritual" notion that humans are set above and against nature has been formulated into a philosophy of science and progress.

The combination of accelerated technology, consumption, and population growth has increased awareness of the global nature of environmental sustainability. The first conference on environment and development—the United Nations Conference on the Human Environment—was held in Stockholm in 1972. The Conference focused the international community on environmental protection, especially as it relates to economic development. It was followed by the seminal Report of the World Commission on Environment and Development (1987). In 1992, the World Bank published a special issue of the World Development Report dealing with development and the environment that, in the same year, paved the way for the United Nations "Earth Summit", which recommended several actions and programmes to protect the human environment. The World Development Report, 1992 provided a methodological framework for action. It outlined environmental priorities for development (water, air pollution, solid and hazardous wastes, land and habitat, and atmospheric changes) and the necessary data, institutional and information frameworks for policy analyses and action.

It became increasingly evident that the future of human development is contingent on a balanced coexistence between humans and their environment. Balanced coexistence depends to a large degree on understanding the synergistic nature of environment, technology, and population dynamics, and on the development of preventive and curative measures to preserve the integrity of the environment. That synergy is complex and there is a growing theoretical and empirical literature on the subject. These synergies are not examined in the present discussion. For details, the reader is referred to the World Commission on Environment and Development (1987) and the World Bank (1992, 1995), among others. Rather, this section attempts, in the context of the present evolutionary approach, to differentiate between two levels of environmental concerns, a dichotomy that has not been given adequate attention in environmental discussion. The first may be labelled internal environmental sustainability, internal that is, to human physiology and biochemistry, and the second, mentioned above, external environmental sustainability. As will become evident, these two levels of environmental concerns constitute a complex synergistic system.

One of the most fundamental developments in the survival of *Homo sapiens* is the evolution of its immune system against parasites, viruses, toxins, and other hazards that assert themselves by interacting with human biochemistry. This immune system seems to have developed much earlier than the estimated 300,000 years of the existence of the species in its present form, and probably evolved some hundred million years ago when the earliest vertebrates evolved from their invertebrate ancestors. This evolution was not a random process. Recent findings indicate that this endogenous evolution of the immune system has always served the sole purpose of defense against infection. The system that evolved in the genetic and anatomical structure of *Homo sapiens* is called here the "internal environmental mechanism."

This ongoing struggle between host and pathogen has been to a large extent accommodating in nature. But recent experience indicates that the mechanism is losing its preventive power, largely as a result of medical and social developments. The future may not be as comfortable or as certain. Changes in some key aspects of humanity's conditions of existence have created uncertainties. For perhaps a century, since the advent of the technophysio evolution—a trivial amount of time from an evolutionary perspective—a significant proportion of the human population has lived in an artificial environment of its own construction, largely freed from parasites and many pathogens to which the immune system used to respond. The human defense system is becoming under-employed, while the microbes are not resting on their laurels, as illustrated for example by the global outbreak of human immunodeficiency virus (HIV) and, more recently, the Nipah virus outbreak in Malaysia in the spring of 1999. In their attempt to contain the spread of the Nipah virus, which killed more than 120 people in less than three weeks, the Malaysian authorities destroyed more than 900,000 hogs without being able to reach a full understanding of the nature of the virus. "The immune system won a battle but then lost the war. The virus next time may be even worse."

Even in well-understood host-pathogen systems, such as the malaria parasite, the battle for malaria eradication is not won. After decades of defense against the mosquito host and the parasite itself, in the late 1990s, malaria caused an estimated million annual deaths aside from the sizable morbidity incidence and productivity loss, mainly in sub-Sahara Africa, according to the 1999 World Health Report of the World Health Organization (WHO). Measures such as the use of screens, draining wetlands, sewage systems, paved streets, "tight" housing (i.e. housing protected from the elements), and keeping animals away from human accommodation succeeded in banishing the host mosquitoes and virtually eliminated malaria in the well-to-do countries. This has not been the case where poverty and the environment are favourable for mosquito and parasite breeding, mainly in the region south of the Sahara in Africa, and in parts of India. Attempts to tackle the parasite itself through the use of drugs have not been effective in these regions, while attempts to develop effective vaccines are still in the experimental stage. It continues to be an active area of research in molecular biology and immunology. The reason for the elusive hunt is that the malaria parasite is a unicellular protozoan with great adaptive and maneuverable capacity, requiring researchers to work with dozens of new proteins to induce the complex mix of antibodies needed for the development of an effective malaria vaccine. The WHO is restarting a major anti-malaria campaign in those regions where malaria is resurgent, to halve its death toll, at an annual cost approaching one billion dollars.

Adding to the risk of an under-employed immune system are the growth of world population, the emergence of mega-cities and the widespread use of air travel. These developments, on the one hand, increase the ease with which people can become exposed to agents of disease, while adding to the complexity of tracing the causal host-pathogen chain, on the other hand. The case of *Pfiesteria* outbreaks, which have been killing millions of fish while posing direct and indirect health hazards for humans, is illustrative. It is being traced to human activities in animal husbandry, and residential location. Many of these factors are a result of accelerated developments in symbolic cultural evolution and its technophysio phase that increased both protection, which reduces natural immunity, and human mobility, which increases the degree of exposure. Part of the advantage of the technophysio evolution is its success in the development of countermeasures. Vaccination, for example, is a key development in the struggle against microbes. The basic theory of vaccination is simple. Stimulate the "under-employed" immune system by giving the body something that looks like an invading microbe. The process will put the immune system on guard again whenever an actual microbe invades. Experiments with HIV vaccines, although promising, have not been a total success. Part of the difficulty is the ability of HIV to evade the immune system. Furthermore, the cost of HIV vaccine developed thus far has been prohibitive, especially for populations that ©*Encyclopedia of Life Support Systems* (EOLSS) have the highest rate of infection, for example those in sub-Sahara Africa. Pharmaceutical for-profit companies hesitate to take the risk of developing vaccines where the technical outcome is uncertain and economic returns are not positive. It is not evident, however, that losing the natural capacity for internal defense that took millions of years to acquire, while the risk of exposure has been increasing, is an optimal strategy.

The synergistic effects of technology, population dynamics, and affluence (consumption levels) on the external environment have been examined and debated extensively in the literature, with many unsettled questions, including the choice of an optimal strategy and policy to safeguard the environment : whether to use direct or indirect control policy approaches. In a 1991 analysis of the causes of rising pollution in the United States it was concluded by Commoner that the dominant contribution to the rising pollution levels during the 1950s and 1960s came from technology change rather than population growth or affluence. Others give a larger share of the blame to population growth, especially in the developing countries, while others blame consumption patterns and levels, especially in the developed countries, for the lion's share.

Perhaps, the words of John Stuart Mill (1848 [1965]), the young genius of James Mill, indicate a hopeful direction for symbolic cultural evolution to take in a harmonious coexistence with the environment :

> "It is scarcely necessary to remark that a stationary condition of capital and population implies no stationary state of human improvement. There would be as much scope as ever for all kinds of mental culture, and moral and social progress; as much room for improving the Art of Living and much more likelihood of its being improved, when minds ceased to be engrossed by the art of getting on. Even the industrial arts might be as earnestly and as successfully cultivated, with this self sole difference, that instead of serving no purpose but the increase of wealth, industrial improvements would produce their legitimate effect, that of abridging labour. . . . Only when, in addition to just institutions, the increase of mankind shall be under the deliberate guidance of judicious foresight, can the conquests made from the powers of nature by the intellect and energy of scientific discoverers, become the common property of the species, and the means of improving it and elevating the universal lot."

J.S. Mill's statement illustrates the importance of collective judgment in the development of environmental policies. Such judgments depend on existing knowledge, the power of interest groups, and the presence of democratic institutions that allow participation in the decision processes. These are essential prerequisites especially since there are extreme views in the field. At one extreme, there are optimists who believe that the human-environment system is self-regulating; *laissez-faire* policy is optimal. At the other extreme, there are pessimists who believe that the continuation of present trends in population growth and resource use will lead the world to environmental collapse; active intervention is necessary. Both perspectives lack convincing analysis. Solid knowledge is required. But the knowledge base is not fixed. Institutions need to be adaptive to new knowledge as well as sensitive to the needs of human development with diverse levels of well-being.

4. Concluding Remarks : Opportunities and Challenges

The present phase of human development is unprecedented in mechanism and speed, with uncertain local and universal outcomes. The evolutionary approach of the present essay attempts to place the present phase of human development in historical perspective. For many millennia, the evolution of *Homo sapiens* has been based on extending human reach through the accelerated processes of symbolic cultural evolution rather than the slow processes of anatomical evolutionary change. The *modus operandi* of symbolic cultural evolution, in its present technophysio phase of scientific materialism, is knowledge-based; it helps to maintain and promote scientific and technological developments, their technical application and utilization, and the intergenerational preservation of such knowledge through advances in educational methods, content, and quality. These evolutionary developments open a new window of nurture-nature evolutionary processes that present opportunities and challenges to human development in the twenty-first century.

At present, human destiny is to know, if only because societies with knowledge dominate societies that lack it and there is, once reached, a threshold level, a built-in catalytic growth of learning. Scientific development, especially in biology and neuroscience, illustrates that the evolution of the anatomy of the human brain has rendered similar basic structures that, given a random distribution of disabilities, allows for equal *potential* of learning for all members of the family of *Homo sapiens*. The DNA of an individual is made up of about equal contributions from all ancestors that extend for hundreds of thousands of generations, of which present parents contribute an insignificant part. Diversity of the gene pool should be a cardinal value for the enrichment of symbolic cultural evolution. Furthermore, advances in communication and information technologies have facilitated and accelerated the speed of the flow of new knowledge and reduced the global cost of its acquisition. Apparently, the main constraints on acquiring and developing scientific knowledge are relative cost and benefit, the absence of necessary institutions, and the lack of unified goals at local and state levels. These are not simple hurdles for the majority of the world population to overcome. This is the case, given the persisting tendency of *Homo sapiens* for categorization. This tendency is illustrated by a recent discussion of the technophysio evolution : "Technophysio evolution implies that human beings now have so great a degree of control over their environment that they are *set apart* not only from all other species, but also *from all previous generations*." Such an assertion could pave the way for a new round of injustice since it does not take into account the fact that large segments of present generations may not have the full privilege of the technophysio evolution, although not lacking in potential to contribute to its advancement. These are not the only challenges.

Symbolic cultural evolution implies a continuous search for a healthy synergy between scientific development (artisanship), social sciences (conscious time binding), and the humanities (imaginal thinking) that attempts to produce a humane and harmonious balance for the evolutionary process. However, acceleration in scientific development has left science, on the one hand, and the social sciences and humanities, on the other, heading towards separate worlds of discourse. The scientific bases of the social sciences and humanities have not developed at the same speed as those of science and technology. The present methodologies of the social sciences and humanities seem to be more attached to older scientific and technological environments. However, efforts to develop strategies for human development continue to depend heavily on the methodologies and perspectives of the social sciences and humanities. Innovative social theories such as those of Marx, Adam Smith, Spencer, or Toynbee did not have their expected outcomes, partly because their underlying premises of human nature, as discussed earlier, had no scientific basis in biology or neuroscience. The invisible hand remained invisible because the sum of the actions of millions of poorly understood individual human beings, their anatomical and motivational behaviour not fully determined in science, cannot be aggregated and computed with scientific authority.

The stated purpose of governments everywhere is to achieve levels of human fulfillment that are higher than those required for animal survival. The purpose is for their societies to join actively in the evolutionary process of symbolic cultural evolution, to be producers of science and technology and other knowledge, and not merely passive recipients. Without necessary knowledge, education, and health, humans are reduced to a sense-based survival status, a pre-symbolic status that is deprived of the rich emotional potential of cultural development, and one that is unfortunately prevalent among a sizable segment of the world population. Most programmes for human development are based on the premise that education in science and technology is the key element in the goal of human fulfillment. But such a singular focus inevitably leads to the basic dilemma of the present phase of symbolic cultural evolution : the tension and widening gap between the sphere of the social sciences and humanities and that of science. Education should bridge this gap by probing deeper into the scientific foundations of the latter and the cultural consequences of the former. It should also be able to develop an environment that promotes understanding and mutual respect for cultural diversity, by being closely linked to the development of democratic institutions, isonomy in the rule of law, and transparency in governance.

State authorities, in their attempts to implement programmes for human development, are being increasingly constrained by the emerging globalization process, an outgrowth of symbolic cultural evolution, especially in its present phase of scientific materialism. The emerging global environment, although based on innovative scientific developments with unlimited potential for human welfare, seems to be governed more by a competitive ethic of an obsolete Darwinian anatomical evolutionary process, a presymbolic cultural evolution. There are attempts to reorganize the global society away from these selfish trajectories, apparently constrained by the ancient genetic rules of human nature. The institution of transparency in world governance, isonomy in the rule of international law, and the development of binding socio-economic and legal international institutions are necessary conditions for the evolution of a global human development system with higher and more equitable levels of welfare. Similarly, advances in analyses, methodologies and measurement techniques in economics, social accounting, management, demography, and epidemiology, among others, provided for better understanding of outcomes, mechanisms, and determinants of human development. However, these necessary conditions are not sufficient without national governments implementing internally oriented sustainable reform (IOSR), in which the education systems bridge the gap between the three spheres of symbolic cultural evolution. Such systems could allow future generations to combine scientific development with imaginal thinking, breeding generations that rebel against outdated social and political constraints. As Dyson (1995) expressed it : "We should try to introduce our children to science today as a rebellion against poverty and ugliness and militarism and economic injustice."

Symbolic cultural evolution cannot be sustained for long if its outreach destroys its fundamental sustenance. Short-term gains in material benefits tend to produce interest groups that divert attention from negative environmental consequences that threaten the very foundation of human development. The synergy between the dynamics of built-in population growth, technology, and affluence, especially if associated with poverty and unequal development, inevitably leads to serious conflicts and detrimental environmental effects. International democratic institutions that develop and implement rules of conduct between production technology and social behaviour on the one hand and the environment on the other are essential for sustained human development. The presence of such institutions requires development in both science and ethics.

There are two other environmental consequences of the present technophysio phase of symbolic cultural evolution that are "internal" to human development. The first relates to the effects of medical advances and public health measures, combined with urbanization and the increasing volume of travel across national and regional boundaries. These reduce the efficacy of natural immunity on the one hand, while increasing the risk of exposure to disease on the other. The result is that *Homo sapiens* must continually be on guard, providing new measures and countermeasures against emerging and old microbes, a process that is costly and not necessarily efficient from an evolutionary perspective. The second internal consequence is the aging of the global human population, as a result of the development and diffusion of efficient fertility control technologies that accelerated the final phase of the demographic transition, and of new medical advances that increase life expectancy at old ages. The inevitable result of the global aging process is the development of new patterns of intergeneration transfers and new local and global social and political arrangements, with significant distribution effects and patterns of investment that have impacts on both human development and our external environment in the twenty-first century.

The present phase of symbolic cultural evolution presents enormous opportunities for advances in human development and welfare. It also presents as many challenges. In formulating any satisfactory modern ethic of human relationships, it is valuable to remember Bertrand Russell's remark (1995) that it is essential to recognize the necessary limitations of humans' power over the non-human environment, and the desirable limitations of their power over each other. In the twentieth century, humans have been extending their reach at an accelerating speed, which seems to be leading the human symbolic cultural evolution into a new phase in uncharted territories. Humans are no longer adjusting to environmental conditions. They are changing their own capabilities and outreach in ways that require adjustments in both their environmental and their human conditions, adjustments that may lag behind the scope and tempo of the induced changes. Human development is becoming a continuum of self-generating processes and adjustments. These processes, however, require reflection and careful analysis as to their control, destination, and long-term consequences for the future of human development and welfare.

For example, advances in neuroscience combined with innovative electronics present curative and preventive opportunities in the medical field. They also extend to anatomical possibilities that have the potential to expand thought processes, as illustrated by recent experiments that modify the function of memory in small mammals. It is important that the latter possibilities be limited by existing knowledge, not only about the anatomy of the brain, but more essentially by how the "human mind" works. The development of the transistor towards miniaturization and complexity presents an illustrative example. Recently, Keyes, a pioneer in the physics of information processing systems, indicated that the future of transistors requires expanded knowledge of solid state physics since as chips grow more complex they require more fabrication steps, and each step can influence the next. For instance, when doping atoms are introduced into a crystal, they tend to attract, repel, or otherwise affect the motion of other dopants. Such effects of dopants on other dopants are not well understood; further experiments and theoretical investigations are therefore needed. The anatomy of transistors, which are at the heart of modern computers and the revolution in information technology, is not complicated by the presence of the enormous complexity of the "human mind" or by the expanded horizon and synergies presented by the evolving globalization environment. It is true that decoding the human genome has the potential of revolutionizing science and human health, but the road is neither simple nor straightforward. Computational biology indicates that the human genome comprises 3 billion bits of data, with each "bit" being one of four different nucleotide bases; not all are genes, the rest have supporting roles. It is not easy to identify genes. The process of identifying them is further complicated since they are not necessarily continuous, with portions scattered throughout the genome. Innovative computational and statistical techniques are applied for finding gene. However, locating genes is the easier task; it is a long way from understanding their functions.

As the world moves towards a "global village" environment, Casset's statement (in the opening quote of Section 1) becomes almost a tautology. In a global society, equilibrium is inevitably set from within. However, the present global environment is in an early evolutionary state of its institutions and global social values, a stage controlled by technology and finance capital whose allegiance is to shareholders of capital and not necessarily to equitable human development. The challenge is how to master the synergies of human nature and the global environment. To master the synergy of human nature, its anatomy and motivation, requires a highly complex and interdisciplinary knowledge base, while our knowledge about the global socio-economic and environmental consequences of human actions is still evolving. It is not sufficient to believe that human destiny is independent of past history as Berlin or Popper implied. Nor is it comfortable to believe that human destiny is a relentless progress towards an idealized social organism. The first paradigm implies a loss of conscious time binding in the process of symbolic cultural evolution, a loss that leads to regression in human development. The second may lead to instability and uncertainty about the meaning of human progress. More important, humans have developed a highly complex and artificial environment that, although meant as a defense against hostile environmental elements, has introduced a potentially unstable prospect. An experiment with highly developed social species from the insect world may provide a disquieting parable. The displacement of the "army ants", one of the most developed social organisms, from their environment in Central America to an artificial environment in an exhibit in a gallery in New York, led to their death, all 2 million of them, in one day. There was no adequate explanation for the episode, especially since ants are known to be highly efficient and adaptive. The ants were located in a huge square bin, walled by high plastic sides and supplied with adequate food and sand, similar to that of their native environment. Thomas's (1974) description of the episode is apt :

> "It is a melancholy parable. I am unsure of the meaning, but I do think it has something to do with all that plastic—that, and the distance from the earth. It is a long, long way from the earth of a Central American jungle to the ground floor of a gallery, especially when you consider that Manhattan itself is suspended on a kind of concrete platform, propped up by a meshwork of wires, pipes, and water mains. . . . I do not believe you can suspend army ants away from the earth, on plastic, for any length of time. They will lose touch, run out of energy, and die for lack of current." [Emphasis added]

The prospect for *Homo sapiens*, all 6 billion and counting, to survive as one large social organism could be equally uncertain, especially given the increasingly artificial environment they have created as their own habitat. What has been created mainly as defense against hostile environmental elements could end up presenting its own dangers. Although it might be true, that we, as *Homo sapiens*, have no adequate sense of where we are heading, the lessons of symbolic cultural evolution remind us that a good moral compass, combined with an intelligent use of scientific achievements, might keep us going, even prospering, for a long time. There is a need for more knowledge about these complex synergies combined with moral direction, especially in the present technophysio phase of symbolic cultural evolution.

Present knowledge about the determinants and consequences of human behaviour is still at a far distance from the goal set in 1983 by Gell-Mann for the research programme of the think tank of the Santa Fe Institute : to tackle the great, emerging syntheses in science : ones that involve many disciplines. That task was coined as the science of "complexity." The world and human evolution are neither linear nor orderly. This view of the world follows Heraclitus's (*c.* 500 B.C.) view that this world-order "always was and is and shall be : an ever-living fire, kindling in measure and going out in measure" (see Russell, 1945, pp. 38–47; Collinson, 1987, pp. 10–12; Waldrop, 1992, p. 335). Complexity does not view the world as being in a basic state of equilibrium in which the actors' job is to maintain its "natural" equilibrium status, whenever untidy forces push the system away from that Heavenly Newtonian equilibrium. Rather, it views the world as a continuous process of flow and change in which the exact patterns are not repeatable. The shift to a metaphor of "complexity" is fundamental. It is a different metaphor in which a simple system could give rise to immensely complicated and unpredictable consequences. Instead of relying on the Newtonian metaphor of clockwork predictability, complexity seems to be based on metaphors more akin to biological processes; messiness and the liveliness in the economy can grow out of incredibly simple, even elegant theory.

The rules and synergies of the new globalization system are not that simple and are growing in complexity, making it difficult to project future events. Attempts to forecast the future path of globalization, even by established students of world system analysis, have proved to be elusive and mostly off the mark. Cost-benefit analysis techniques that assume perfect system knowledge are clearly inadequate to guide human development policies in the globalization era. Observing and understanding the evolution and emergence of institutions that govern and shape the direction of the globalization process is a prerequisite for the development of policies for sustainable human development in an environment in which North-South or East-West, or human-nature dualities are vanishing. The future of human development in the age of global technophysio evolution in the twenty-first century is full of promises and uncertainties.

Gandhian Policy for Sustainable Human Development

"It will give you talisman. Whenever you are in doubt or when the self becomes too much with you try the following experiment : Recall the face of the poorest and the most helpless man whom you may have seen and ask yourself, if the step you contemplate is going to be of any use to him, will he be to gain anything by it? Will it restore him to control over his own life and destiny? In other words, will it lead to Swaraj or self-rule for the hungry and also spiritually starved millions of our country? Then you will find your doubts and self-melting away."—*M.K. Gandhi*

Many feel that the policies pursued in the field of holistic rural development have not been very successful. One of the reasons has been lack of accountability and non-involvement of stakeholders. It seems that at times of the policy prescribed for improvement had the negative impact. Instead of empowernment it made them further dependent. Policies and technologies have definitely not improved the capacity and capability to be competitive. There is nothing wrong with the masses as far as desire, urge, motivation to improve their total lot is concerned. The real problem is policies and implementation. One can expect that this joint effort would definitely help in correcting those weaknesses. He also said that various civilisation have come and gone but Indian civilization has continued to flow mainly because of "inherent capacity of purging themselves of error."

In this process of purging, cleaning, opening up, re-examining and reprioritization the lessons of positioning of decentralised sector will have to be reassessed. Many social, economic, cultural activities which were not attended to have suddenly become relevant and timely.

While looking at the challenges separately the consultant has tried to codify some of his experiences based upon Wardha childhood experiences, then Army experiences, and subsequently, working in state like U.P. and followed by Khadi and village industry days. It was here real opening of activities measuring problems, lessons, hopes, aspirations and 'despondency reasons' were identified and studied. As a part of study some of the VRAT or VOWS of the Father of the Nation which I understood include the following :

"*Satya* (Truth) : Truth is God. Truth in a much wider sense means *Truth in thought, Truth in speech and Truth in action*. Where there is Truth, there is true knowledge, and where there is true knowledge, there is always bliss (*Ananda*).

"*Ahimsa* (Non-Violence) : We may not give up the quest for truth, which along being God himself, *Ahimsa* is the means, truth is the ———*Ahimsa* it is not possible to seek and find truth.

"Patience, sacrifice, humour : These are the time tested parameters for learning to live with satisfaction and tolerance while enjoying.

"*Aswad* (Control of the palate) : Control of the palate is very closely connected with the observance of *Brahmacharya*. What we eat should only be to sustain the body and not for self-indulgence.

"*Asteya* (Not-stealing) : It is theft to take anything belonging to another without his permission. However, to take or collect something from others for which we have no real need, also amounts to theft.

"*Aparigrah* (Non-possession) : Civilisation, in the real sense of the term, consists not in the multiplication of wants but in their deliberate and voluntary reduction. This along promotes real happiness and contentment and increases the capacity for service.

"*Abhaya* (Fearlessness) : Fearlessness is indispensable to attain ineffable peace and connotes freedom from all external fears or fear of disease, bodily injury and death, of dispossession, of losing one's nearest and dearest, of losing reputation or giving offence and so on.

"*Manavata—Radical* humanism : This is a part of spiritual—religious effort to open the mind of individual and societies and understand the unity of all creations by the Creater—It is essentially developing consciousness to awaken conscience for sustainable holisitic development.

"*Sharir-Shram* (Bread Labour) : Everyone, whether rich or poor, must perform his daily work, which should assume the productive form i.e. Bread Labour. How can a man, who does not do physical labour, have the right to eat?

"*Sarva Dharma Samabhav* (Tolerance for other Religions) : Tolerance for other faiths imparts to us a truer understanding of our own. True knowledge of religion breaks down the barriers between faith and faith. Reverence for other faiths need not bind us to their faults. We must be keenly alive to the defects of our own faith also, and try to overcome these defects.

"*Swadeshi* (Service of the immediate neighbour) : A votary of *swadeshi* dedicates himself to the service of his immediate neighbours. One who allows himself to be lured by the 'distant scene' but falls in his duty towards his neighbours, violates the principle of *swadeshi*."

Based upon the above gist the consultant tried to link up the economic focus issues along with the cardinal principles of Gandhian approach to rural sustainable development. In this connection, Shri Y.A. Pandit Rao, has mentioned the following *cardinal principles* to Gandhian approach : "(i) Total and integrated development of an area from the viewpoint of self-sufficiency in basic need namely, food, clothing, shelter, education, medical care etc. by using local resources, local man-power and local consumption to the maximum extent. Area could be a village or group of villages or a block or a district, depending upon situation, (ii) total development of man, (iii) voluntary effort and voluntary action, (iv) self-help and minimum dependence on government, (v) primary, secondary and tertiary sectors of the economy working in close cooperation, (vi) maximum use of small machine, (vii) first concern for the poorest among the poor, (viii) moral considerations in selecting economic activities, (ix) non-political work i.e. above party politics. It would be clear that although

Gandhiji believed in simplicity of life but all the same he believed in improving the quality of life."

It seems that because of the various reasons of *our time and neglect* of the *local wisdom, well documented lessons and practiced traditional practices by the family systems have somehow been missed in the speed of development.* Time has come to be clear about the lessons of the developmental processes pursued so far allover the world. In this connection, any development process must have some integrative and sustainable inseparable so that the question of empowerment of self-employment, organisership could all be examined with reference to various Gandhian institutions which were started for social, economic, cultural upliftment etc. activities before the independence. The 50th year of independence in 1997 could be the culmination of these studies for ensuring sustainability not *only of values of our time but also the lessons learnt and not learnt* despite their known availability. The fact is that the documentation, picturisation and computerisation coupled with fax and TV has made our job much easier and convenient. Maybe by dovetailing of Parliamentarians' Forum and UN Systems efforts, the next two years may see a radical sustainable shift in the planning action process. *Anyway, Time alone would be the best Judge.*

Gandhiji—Trusteeship—Corporatc Sector and Rural India

It is normally understood that Corporate sector and rural India have hardly any areas of mutuality. Maybe this view is perpetuated by exploiters and propagators of *status quo*. Traditionally, Corporate sector using big machines and capital intensive nature of manufacturing and aggressive marketing through their lobbying and influencing, manipulative powers managed to escape from their responsibilities of helping the unorganised. Efforts by few have resulted in better publicity, image building, tax concessions and overall unreplicable development models. This point was discussed on several occasions with Shri K.C. Sharoff, Chairman of Excel Industries and also Shri M.L. Dantwala. The gist is that a time has come when through transparency, viability and understanding through cost-effective systems development led by economic compulsions, it is possible to work better for *win win win situation for all.*

Maybe the Parliamentarians' Forum and UN systems convergence along with World Bank's policy and funding prescriptions could help us in developing replicable sustainable those models which would without destroying environment, people's will to work and need to be self-sufficient and self-dependent with dignity and self-respect would be emerging. Whether we like it or not there are clear indications and patterns in the horizons which compel us to examine whether innovations of technology, enterprise development need to attend to mass aspirations and to develop transparent models based upon cost-effectiveness and redefined viabilities. There are forces both of time and experience which make it essential to happen.

REFERENCES

Agosin, M.R. and Bloom, D.E., "Global Integration in Pursuit of Sustainable Human Development : Analytical Perspectives".

Choudhary, P.C.R., Bihar District Gazetteers, Munger, Patna, 1960, p. 74.

Dayal, P., "The. Bihar Plain—Regional Study", Transactions of the Council of Indian Geographers, Vol. 5, 1968, p. 20.

Isard, W., "Methods of Regional Analysis", An Introduction to Regional Science, New York, 1960.

Mandal, R.B., 'Nature of Population Geography', Dimensions in Geography, (Eds. Sinha and Mandal), Associated Book Agency, Patna, 1979, p. 6.

Hasegawa, Sukehiro, "Development Cooperation : Global Issues and the United Nations", 2001.

Sirageldin, I., "Sustainable Human Development in the twenty-first century : an evolutionary perspective, in *Sustainable Human Development*", [ed. Ismail Sirageldin], in *Encyclopedia of Life Support Systems (EOLSS)*, Developed under the Auspices of the UNESCO, Eolss Publishers, Oxford, UK, 2003 [http://www.eolss.net]

Taori, Kamal, "People's participation in Sustainable Human development" (A Unified Search), Concept Publishing Company Pvt. Ltd., New Delhi, 1998.

5

Forest Resources and Sustainable Development in New Millennium

Plants sustain life. The existence and expansion of forests constitute one of the primary determinants of a healthy ecology for man. Our ancestors intuitively understood the importance of forests in maintaining the optimum ecological balance. Such attitude towards tree can certainly ensure the durability of our species.

Forest is the greatest renewable natural resource and one of the most striking features of the landscape. The importance of the use of forest resource and its conservation has been admitted all through the ages and men have made extensive use of it for their comfort and luxury. It has shaped the economy of the Munger Division and increased human consumption by providing raw materials which are needed most for our industries. No one can deny the importance of forest in the regional ecology as well as in the economy. A vast wealth is hidden in the thick forest because it provides men with essential commodities like fuel, medicinal herb, fodder, timber and different kinds of fruits. Forest plays a vital role in arresting the Monsoon rains. Where there is forest there is a plenty of rains and thus it helps the land to grow more food for mankind. In the modern age it has been considered to be the laboratory of modern workshop and development of fodder resources. Various types of natural resources are in abundance in the forest region.

History

The history of forest resource use is as old as civilization in India. One millennium ago there was a time when the broad picture of India was that of a sea of forest with scattered islands of cultivation. The economic and cultural life of the country centred largely around forests. *Rigveda* has a striking hymn to the Goddess of forests. *Manusamhita* (the ancient Hindu law code), regards the destruction of trees as a serious offence and prescribes a heavy penalty for it.

In Kautilya's *Arthashastra,* there is a mention of forests and wild-lives and gave much attention to afforestation and protection of wildlife. Forest superintendents were appointed who had a number of assistance. Offenders against forest laws were severely punished. The forests at that time were classified as under :

1. Those set apart for religious seclusion,
2. Reserved forests for supplying of forest produce, and
3. Forest set apart for :

   (i) Grazing royal elephants,
   (ii) Hunting grounds of the royal people, and
   (iii) Hunting grounds for the public.

During his reign Ashoka laid much importance on the planting of trees along the road and on camping sites. Medicinal plant farming was also encouraged.

After the Muslim invasion in India, the clearance of forests began with sincere efforts, while the old Hindu tradition held forests to be sacred and discouraged cutting down of trees. On the other hand, the new invaders did not have the same religious feeling of sentimental scruples and planted many noble avenues around their seats of Government. They pursued a deliberate policy of encouraging agriculture at the expense of forests because the former yielded more revenue.

Sustainable Forest Resource

The United States Government has no clear goal, and only disjointed commitments, to suggest that a sustainable forest resource is important to the nation.

Does it matter? ......If so, what would it look like to have such a policy? ......What would it accomplish? ......How do we make something happen?

Do a search of the United States federal statutes looking for the words "forest" and "policy" and you'll find statements embedded in statute many years ago by such legislative efforts as :

— The Multiple-Use Sustained Yield Act of 1960;
— The Forest and Rangeland Renewable Resources Planning Act of 1974;
— The National Forest Management Act of 1976;
— The Federal Land Policy and Management Act of 1976;
— The Forest and Rangeland Renewable Resources Research Act of 1978;
— The Cooperative Forestry Assistance Act of 1978.

All of these were important in their time.

There have been other laws passed over the years that are now clearly outdated. For example, it's written in federal statute to permit...

> "...citizens of Malheur County, Oregon to cut timber in the State of Idaho..." And also, "The President is authorized to employ so much of the land and naval forces of the United States as may be necessary effectually to prevent the felling, cutting down, or other destruction of the timber of the United States in Florida, and to prevent the transportation or carrying away any such timber as may be already felled or cut down; and to take such other and further measures as may be deemed advisable for the preservation of the timber of the United States in Florida."

There are also more recent additions such as *The National Forest Dependent Rural Communities Economic Diversification Act of 1990*, and *The International Forestry Cooperation Act of 1990*, or *The Healthy Forests Restoration Act of 2003*. Some of these continue to impact forestry concerns, others...such as the 1991 directive to form a President's "*Commission on State and Private Forests*"...... are already residual artifacts.

But nowhere are these volumes of legal requirements and Congressional findings stitched together by a stated, common goal to provide present and future generations of the United States with a sustainable forest resource. Instead, forestry professionals must persistently try to address broad-scale, highly interconnected challenges as if they were confined within discrete boundaries.

These boundaries may be useful to define ownership, or geography, or authority. But allowing ourselves to be placed within boxes of limited vision ensures that we will never be successful at delivering that which society deserves, and no doubt expects. Looking at forest resources, the threats they face and the inadequacy of our current answers to these threats...... clearly we have no coordinated, unified goal to achieve a sustainable forest resource in this country. Some examples :

— Despite growing XX per cent more wood in this country than we harvest each year, wood imports have grown by over XX per cent since 19XX. The US has, arguably, the highest standards of forest practice in the world, but increasingly looks to other countries for wood regardless of the comparability of their environmental and social accountability.

— The Southern Forest Resource Assessment, published by a consortium of federal agencies in 200X, says that conversion to residential and other developed uses is the largest single threat to one of the largest wood producing regions in the world, the Southeastern United States.

— Fragmentation has caused the average size of private forest holdings in the US to shrink from XX acres to XX acres in the past XX years.

— Amendments to the Internal Revenue Code in the 1970s, that were virtually un-noticed by the professional forestry community, have spurred massive changes in the ownership of industrial forest properties, leaving only a handful of fully integrated, publicly held forest products firms still in business.

— In the US today there are 190(??) million acres of forest at risk of catastrophic fire due to the lack of proper management.

— In the US today there are XX million acres of predominately dead standing trees, due to insect and disease, with no foreseeable treatment.

— Hardwood forests in the Eastern United States suffer a host of additional maladies, from the lack of regeneration due to deer overpopulation, to the impacts of acid rain, to projected over-harvest in the Southeast, to decades of high-grade degradation carried out under the guise of "selective" harvest and poorly conceived diameter limit harvests.

— What were once strong, forest-based rural economies in all corners of the country.... Georgia, Maine, Minnesota, Oregon, New Mexico..... now have similar concerns about long-term community health and viability due to the loss of forest industry.

— Federal budgets show a long-term trend of disinvestment in federally and privately owned forest lands across the full range of values.... recreation, wilderness, access, wildlife, water and timber.

— Wildlife dependent on contiguous forest, such as interior nesting neotropical migratory birds show serious declines.

— In many parts of the country it is not uncommon for timberlands, historically owned for a range of forest values, to have a real estate market value that dwarfs the net present value of sustainably managed forest products.

— In many parts of the country the loss of logging and milling infrastructure is making it increasingly more difficult for non-industrial private landowners to realize expected timber values from their properties.

These issues are not unconnected. Yet there are no policies to acknowledge that a reduction in harvests on federal lands coupled with the forest practice standards in Brazil have an impact on the continued viability of a paper mill in northern Wisconsin.....that the existence of that paper mill can determine whether a small private landowner keeps his or her land in forest, or sells it to a developer.....that whether or not the land stays in forest impacts the successful nesting of neotropical migratory birds ....that if the paper mill goes away and the land is parceled to second home developers, family wage jobs in manufacturing, forestry, transportation and service are all lost to the community....and that when these incomes exit the community then schools, roads and all other forms of essential societal needs begin to degrade.

It does matter that the United States has no unified goal to provide citizens with a sustainable forest resource. And it clearly matters that we have no coordinated set of policies and actions to help realize such a goal.

What Would a Sustainable Forest Resource Policy Look Like?

In 1987, the Bruntland Report, more formally known as *Our Common Future*, published by the UN World Commission on Environment and Development, broadly advanced the notion that sustainable development must meet the needs of the present generation without compromising those of future generations. The report went on to say that doing so required the integration of economic, social and environmental values.

Following on that concept, the 1992 UN Conference on Environment and Development produced "*A non-legally binding authoritative statement of principles for a global consensus on the management, conservation and sustainable development of all types of forests.*" Article 2(b) of what are now called "The Forest Principles" states :

> "Forest resources and forest lands should be sustainably managed to meet the social, economic, ecological, cultural and spiritual human needs of present and future generations. These needs are for forest products and services, such as wood and wood products, water, food, fodder, medicine, fuel, shelter, employment, recreation,

habitats for wildlife, landscape diversity, and other forest products. Appropriate measures should be taken to protect forests against harmful effects of pollution, including air-borne pollution, fires, pests and diseases in order to maintain their full multiple value."

These Principles were adopted by consensus on the part of nearly 180 countries in attendance at this conference, most commonly known as the 1992 Rio Earth Summit.

Using these Principles and relevant portions of *"Agenda 21"*, also agreed to in Rio, the Montreal Process subsequently developed *"Criteria and Indicators for the Conservation and Sustainable Management of Temperate and Boreal Forests."* Endorsed through the 1995 "Santiago Declaration" by a dozen countries, including the United States, who represent 90% of the world's temperate and boreal forests, this document defines in more specific detail the kinds of social, economic and environmental values we expect from forests in terms of what we should strive to measure as we track progress towards sustainable forests.

In 1998, the United States Roundtable on Sustainable Forests was initiated to coordinate efforts among federal agencies as they began development of the country's *"First Approximation Report"*, an attempt to produce the nation's first assessment under the concept of criteria and indicators for sustainable forest management. Since then the Roundtable has become a broadly-based forum to advance the concepts of sustainable forestry, two national-level criteria and indicator assessments have been produced and the next "RPA Assessment" produced by the US Forest Service, due in 200X, will be based on the criteria and indicators framework.

Clearly embodied in this last nearly twenty years of work is a concept of sustainable forest resources that is globally endorsed and that represents a solid foundation for the development of a national policy.

Building on this concept, a **National Policy would say** that :

— The management and conservation of forest resources in the United States should be guided by a mandate to meet the forest related needs of the present generation without compromising the ability of future generations to meet their needs.

— Doing so requires that economic, social and environmental values from forests be provided within a framework where these values are mutually supportive. That is, we recognize that economic returns from forests are what pays for the provision of environmental and social values; that by protecting environmental values we are ensuring that these resources can continue to return economic and social benefits; and that by addressing social needs and concerns we maintain the social license to manage these lands to most effectively deliver economic and environmental goods and services.

— And to borrow a notion from Aldo Leopold.... Laws and programmes that promote this vision of sustainable forests and the interconnectedness of environmental, social and economic values are acceptable expressions of federal policy. Government functions that do otherwise are not.

What Would A Policy Accomplish?

To be effective the policy must articulate a clear requirement for implementation. Implementation could start with an assessment of federal laws and regulations to evaluate their alignment with the achievement of the goal of a sustainable forest resource as defined by this policy. Given an honest assessment (which is an imposing "given") the next step would be a wholesale rewrite of federal statute to create alignment. The hope would be to create a more effective, and potentially radically different, model of federal government involvement in the promotion of sustainable forests. To be more effective the model needs to recognize the existence of geographic boundaries and ownership differences, but bring forward solutions to the task of integrating social, environmental and economic values by looking across all forest resources simultaneously.

Sustainable Forest Management

Sustainable forest management (SFM) is the management of forests according to the principles of sustainable development. Sustainable forest management uses very broad social, economic and environmental goals. A range of forestry institutions now practice various forms of sustainable forest management and a broad range of methods and tools are available that have been tested over time.

The "Forest Principles" adopted at The United Nations Conference on Environment and Development (UNCED) in Rio de Janeiro in 1992 captured the general international understanding of sustainable forest management at that time. A number of sets of criteria and indicators have since been developed to evaluate the achievement of SFM at both the country and management unit level. These were all attempts to codify and provide for independent assessment of the degree to which the broader objectives of sustainable forest management are being achieved in practice. In 2007, the United Nations General Assembly adopted the Non-Legally Binding Instrument on All Types of Forests. The instrument is the first of its kind, and reflects the strong international commitment to promote implementation of sustainable forest management through a new approach that brings all stakeholders together.

A definition of SFM known as sustainable forestry to this present day is understanding the forest management that was developed by the Ministerial Conference on the Protection of Forests in Europe (MCPFE), and has since been adopted by the Food and Agriculture Organisation (FAO). It defines sustainable forest management as :

The stewardship and use of forests and forest lands in a way, and at a rate, that maintains their biodiversity, productivity, regeneration capacity, vitality and their potential to fulfill, now and in the future, relevant ecological, economic and social functions, at local, national, and global levels, and that does not cause damage to other ecosystems.

In simpler terms, the concept can be described as the attainment of balance—balance between society's increasing demands for forest products and benefits, and the preservation of forest health and diversity. This balance is critical to the survival of forests, and to the prosperity of forest-dependent communities.

For forest managers, sustainably managing a particular forest tract means determining, in a tangible way, how to use it today to ensure similar benefits, health and productivity in the future. Forest managers must assess and integrate a wide array of sometimes conflicting factors—commercial and non-commercial values, environmental considerations, community needs, even global impact—to produce sound forest plans. In most cases, forest managers develop their forest plans in consultation with citizens, businesses, organizations and other interested parties in and around the forest tract being managed. The tools and visualization have been recently evolving for better management practices.

Because forests and societies are in constant flux, the desired outcome of sustainable forest management is not a fixed one. What constitutes a sustainably managed forest will change over time as values held by the public change.

Criteria and Indicators

Criteria and indicators are tools which can be used to conceptualise, evaluate and implement sustainable forest management. Criteria define and characterize the essential elements, as well as a set of conditions or processes, by which sustainable forest management may be assessed. Periodically measured indicators reveal the direction of change with respect to each criterion.

Criteria and indicators of sustainable forest management are widely used and many countries produce national reports that assess their progress toward sustainable forest management. There are nine international and regional criteria and indicators initiatives, which collectively involve more than 150 countries. Three of the more advanced initiatives are those of the Working Group on Criteria and Indicators for the Conservation and Sustainable Management of Temperate and

Boreal Forests (also called the Montreal Process), the Ministerial Conference for the Protection of Forests in Europe, and the International Tropical Timber Organization. Countries who are members of the same initiative usually agree to produce reports at the same time and using the same indicators. Within countries, at the management unit level, efforts have also been directed at developing local level criteria and indicators of sustainable forest management. The Center for International Forestry Research, the International Model Forest Network and researchers at the University of British Columbia have developed a number of tools and techniques to help forest-dependent communities develop their own local level criteria and indicators. Criteria and Indicators also form the basis of the Canadian Standards Association certification standard for sustainable forest management.

There appears to be growing international consensus on the key elements of sustainable forest management. Seven common thematic areas of sustainable forest management have emerged based on the criteria of the nine ongoing regional and international criteria and indicators initiatives. The seven thematic areas are :

- Extent of forest resources
- Biological diversity
- Forest health and vitality
- Productive functions and forest resources
- Protective functions of forest resources
- Socio-economic functions
- Legal, policy and institutional framework.

This consensus on common thematic areas (or criteria) effectively provides a common, implicit definition of sustainable forest management. The seven thematic areas were acknowledged by the international forest community at the fourth session of the United Nations Forum on Forests and the 16th session of the Committee on Forestry. These thematic areas have since been enshrined in the Non-Legally Binding Instrument on All Types of Forests as a reference framework for sustainable forest management to help achieve the purpose of the instrument.

Ecosystem Approach

The Ecosystem Approach has been prominent on the agenda of the Convention on Biological Diversity (CBD) since 1995. The CBD definition of the Ecosystem Approach and a set of principles for its application were developed at an expert meeting in Malawi in 1995, known as the Malawi Principles. The definition, 12 principles and 5 points of "operational guidance" were adopted by the Fifth Conference of Parties (COP5) in 2000. The CBD definition is as follows :

> "The ecosystem approach is a strategy for the integrated management of land, water and living resources that promotes conservation and sustainable use in an equitable way. Application of the ecosystem approach will help to reach a balance of the three objectives of the Convention. An ecosystem approach is based on the application of appropriate scientific methodologies focused on levels of biological organization, which encompasses the essential structures, processes, functions and interactions among organisms and their environment. It recognizes that humans, with their cultural diversity, are an integral component of many ecosystems."

Sustainable forest management was recognized by parties to the Convention on Biological Diversity in 2004 (Decision VII/11 of COP7) to be a concrete means of applying the Ecosystem Approach to forest ecosystems. The two concepts, sustainable forest management and the ecosystem approach, aim at promoting conservation and management practices which are environmentally, socially and economically sustainable, and which generate and maintain benefits for both present and future generations. In Europe, the MCPFE and the Council for the Pan-European Biological and Landscape Diversity Strategy (PEBLDS) jointly recognized sustainable forest management to be consistent with the Ecosystem Approach in 2006.

Independent Certification

Growing environmental awareness and consumer demand for more socially responsible businesses helped third-party forest certification emerge in the 1990s as a credible tool for communicating the environmental and social performance of forest operations.

There are many potential users of certification, including : forest managers, investors, environmental advocates, business consumers of wood and paper, and individuals.

With forest certification, an independent organization develops standards of good forest management, and independent auditors issue certificates to forest operations that comply with those standards. This certification verifies that forests are well-managed—as defined by a particular standard—and ensures that certain wood and paper products come from responsibly managed forests.

This rise of certification led to the emergence of several different systems throughout the world. As a result, there is no single accepted forest management standard worldwide, and each system takes a somewhat different approach in defining standards for sustainable forest management.

Third-party forest certification is an important tool for those seeking to ensure that the paper and wood products they purchase and use come from forests that are well-managed and legally harvested. Incorporating third-party certification into forest product procurement practices can be a centerpiece for comprehensive wood and paper policies that include factors such as the protection of sensitive forest values, thoughtful material selection and efficient use of products.

There are more than 50 certification standards worldwide. Some common certification standards are :

- Canada's National Sustainable Forest Management Standard (CSA)
- Forest Stewardship Council (FSC)
- Sustainable Forestry Initiative (SFI)

- Programme for the Endorsement of Forest Certification schemes (PEFC)

The area of forest certified worldwide is growing rapidly. As of December 2006, there were over 2,440,000 square kilometres of forest certified under the CSA, FSC or SFI standards, with over 1,237,000 square kilometres certified in Canada alone.

While certification is intended as a tool to enhance forest management practices throughout the world, to date most certified forestry operations are located in Europe and North America. A significant barrier for many forest managers in developing countries is that they lack the capacity to undergo a certification audit and maintain operations to a certification standard.

Sustainable Development and Resource of Forest in Munger Division

The area of Munger division is 7,920 square kilometres with only 132,770.2 hectares or about 16.5% area under forest. Forest lands are found in the Division of Sono, Chakai, Jhajha, Kharagpur, Bariarpur, Jamalpur, Lakshmipur, Barhat, Gidhaur, Dharhara and other anchals. Large expanse of forest is found in Kharagpur hill block which has been subdivided into four management blocks such as Kharagpur, Dharhara, Malleypur, Lakshmipur, Jamui, Jhajha and Chakai corresponding to seven forest rangers in which these fall. They are of 19,762.03, 14,613.37, 18,408.4, 15,918.36, 23,994.32 and 1,939.1 hectares in area respectively. Other forest areas are Gidheshwar, Amkola and Maheshwari.

Prior to 1946 forests in the Munger division were privately owned. The principal owners being the Maharaja of Darbhanga, Baneli Raj, Gidhaur and Khaira estates later on with the vesting of these estates to the State of Bihar under Land Reform Act, government has become the proprietor of these forests.

Gidhaur Estate

All provisions of the Indian Forest Act are applicable to the reserved forest. There were 36,579 hectares of forest in Gidhaur Estate.

Other Private Forests

The management and control of private forests of Munger Division were taken over by the state government under Bihar Private Forest Act, 1946 and later on the Bihar Private Forest Act, 1947.

Due to the absence of any system of forest conservation in the past, these forests were very much maltreated and subjected to reckless cutting.

But after taking over of the management by the State Government, the forests are being managed purely on scientific lines.

In 1952, the Government of India enunciated a new forest policy which comprised (i) adaptation of a national festival of tree plantation (Van Mahotsava), (ii) wildlife conservation, and (iii) soil conservation on all India basis. The Munger Division also gave considerable importance to the preservation of existing forest areas and afforestation through planting various species on new suitable lands, by increasing the area under forests.

Objectives of Study

As per requirements of man, forests are the greatest renewable natural resource that we possess. By virtue of their biological characteristics they are able to produce while conserving and conserve while producing and thus they can be of use in perpetuity if wisely exploited. Ours is a capital scarce-economic region and we can achieve a fast rate of economic growth only if we exploit our natural resources efficiently and rationally.

Hence, this study has been designed to throw light on the forest resource in the Munger Division.

Problems

In the Munger Division a smaller proportion of land is under forest. Hence our strategy should be such that we maximise returns from the inputs and have an optimal mixture of trees that have a short gestation period as also of slow growing trees.

Distribution of Forests

Distribution of forest in the Munger Division is quite uneven. The northern plain is devoid of forest. On the other hand, the southern part is covered with dense forest especially anchals of Dharhara, Khaira, Kharagpur, Chakai, Sono, Lakshmipur, Barhat, Gidhaur and others. The sparse and scrub type vegetation are found in anchals of Jamui, Kharagpur, Bariarpur, Sangrampur, Tetiha Bambor, Jamalpur, Barhat, Gidhaur and Lakhisarai. There are anchals like Sikandra, Islamnagar Aliganj, Lakshmipur, Barhat, Gidhaur, Pipariya and Surajgarha which have sparse forest under private ownership.

The percentage of forest in Khagaria anchals is 0.2% of their total land whereas in Sangrampur and Tetiha Bambor 1%, and Tetiha Bambor 2%, Jamalpur 3%, Pipariya 5%, Lakhisarai 6%, Sikandra 12%, Islamnagar Aliganj 13%, Lakshmipur 17%, Gidhaur 18%, Barhat 18%, Surajgarha 18%, Bariarpur 23 %, Sono 30%, Kharagpur 32%, Chakai 33%, Khaira 36%, Jhajha 39% and Dharhara 50%. Thus we find that the anchals of Dharhara, Jhajha, Khaira, Chakai and Kharagpur have a respectable proportion of their total land under forests (Figure 5.1 and Table 5.1). Anchals which have no forests are as follows : Munger, Asarganj, Tarapur, Sheikhpura, Barahiya, Ariari, Barbigha, Ramgarh Chowk, Halsi, Shekhpura, Shekhopur Sarai, Ghat Kusumba, Chewara, Ariari, Jamui, Alauli Mansi, Chautham, Gogri, Beldaur and Parbatta.

Table 5.1 : Distribution of Forest in the Munger Division (2001)

(Area in Square Kilometres)

Name of anchals	*Total Geographical area*	*Area under forest*	*Percentage of land under forest*
Dharhara	282.85	141.4	50
Jhajha	427.39	166.7	39
Khaira	418.71	150.7	36
Chakai	774.04	255.4	33
Kharagpur	306.79	98.2	32
Sono	392.27	117.7	30
Bariarpur	160.52	36.9	23
Surajgarha	389.66	70.1	18
Barhat	232.16	39.5	18
Gidhaur	71.11	12.8	18
Lakshmipur	251.77	42.8	17
Islamnagar Aliganj	172.89	22.5	13
Sikandra	184.01	22.0	12
Lakhisarai	283.30	17.0	6
Pipariya	58.39	2.9	5
Jamalpur	85.45	2.6	3
Tetiha Bambor	101.38	1.01	1
Sangrampur	85.94	0.9	1
Khagaria	262.36	0.3	0.1

Source : Working Plan, Divisional Forest Office, Monghyr, 1981.

Forest Type

Forest type is fundamentally controlled by soil characteristics, temperature, sunlight, rainfall, and evaporation. All these elements are themselves interrelated and interdependent. The relation between rainfall and evaporation greatly affects soil conditions and the nature of vegetation growth. In the Munger Division growth of forest resource is variously affected by the physical characteristics of land and climatic conditions. The average annual rainfall for the last 16 years was 102 centimetres and has been received by the onset of South-West monsoon. Most of the rains to three months—July, August and September—while winter rains contribute a little.

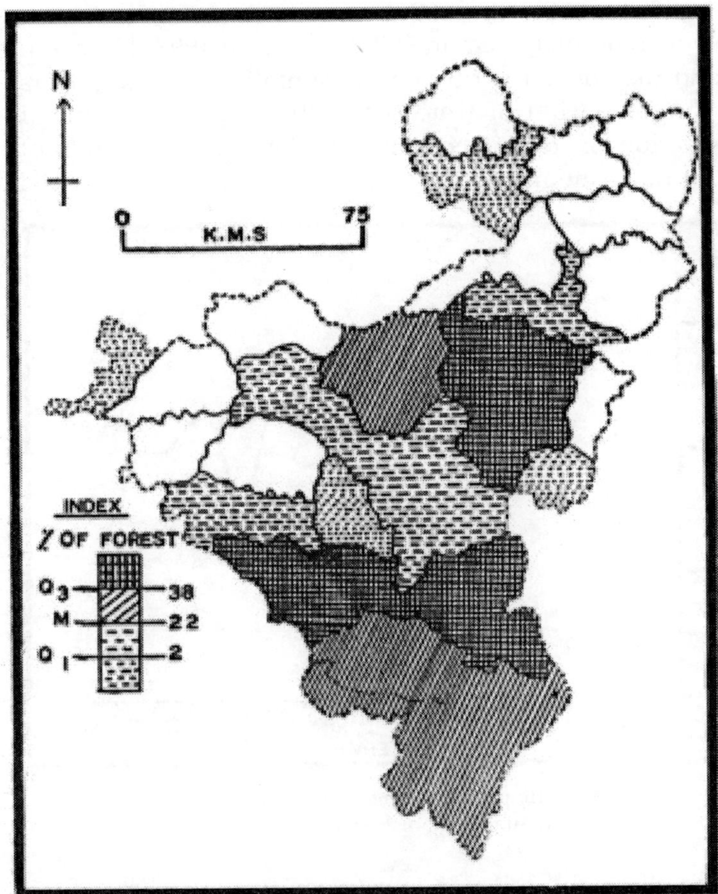

Fig. 5.1 : Munger Division : Distribution of Forest.

In Figure 5.2, X axis represents the years interval while others show the regime of rainfall on Y axis in the southern part of Munger where forest covers are widely distributed over the mountain slopes. This region shows variations of the distributions of the average rainfall in different years from 1988 to 2003. It has been found that 1989, 1990, 1995, 2002 are the years when the rainfall made troughs in the graph as the synclines whereas the years 1993, 1997 and 2000 showed high amount of rainfall as the anticlines of the graph. The high rainfall 152 cm has been recorded in 1998 which is followed by the lower trend of 142 cm in 1993, 122 cm in 1997, 120 cm in 2000 and the above distribution of rainfall for varying years for a longer period of time shows the higher amount of precipitation of forest economy on the marginal lands of agricultural economy.

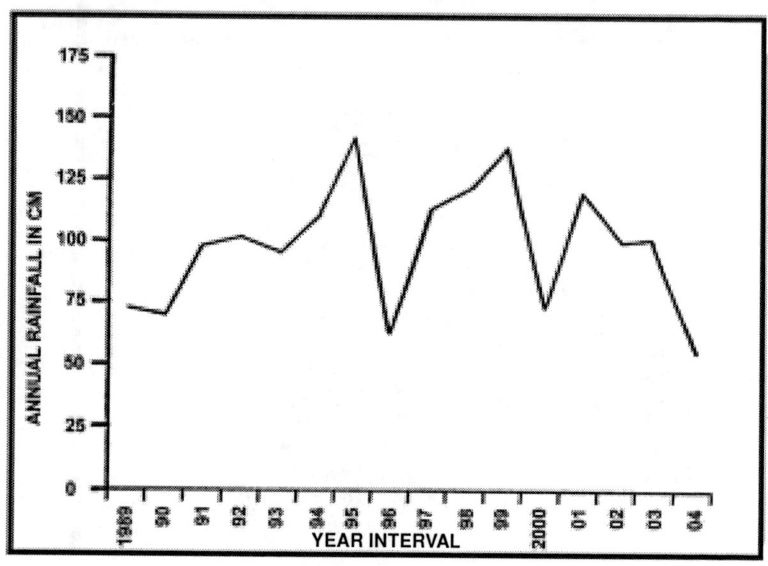

Fig. 5.2 : Munger Division : Annual Variation of Rainfall in Forest Area (1989-2004).

The climate of the area is between parching heat of the west and moist of the east in North Indian Plains. During the hot weather, westerly wind comes from the western part of North Indian Plains, causes high temperature and 'Loo'.

So far as the rainfall is concerned the winter months of January, February, March, November and December are devoid of rainfall while the months of June, July, August, September and October show a good amount of rainfall, more than 25 centimetres in August. This shows that rainfall is a general condition with the incoming of monsoon which is beneficial for the Kharif and the paddy crops and it is a good sign for better prospect of maize and rice cultivation as these two crops are widely grown in the Munger division. But in practice with the accumulation of water by the Ganga, the

Burhi-Gandak, the Kamla-Balan and the Kosi river the situation aggravates, each and every village in North Munger where the less or high amount of rainfall matters little due to flood condition, especially in the rainy season.

So far as the temperature is concerned January and December are marked by 11°C, while the months of June, July, August and September show a high temperature (42°C) followed by high amount of rainfall (1000 mm) and humidity (85%). The humidity is low especially in March due to less rainfall in winter and the beginning of summer season. The season is good for the ripenning of wheat and other summer crops. The temperature rises up to 16°C and this is quite often followed by storms and summer shower especially in the month of April. The summer season is marked by high temperature, scorching heat, dreary looking environment and patches of summer crops which are raised with the help of input of chemical fertilizers. On the other hand, the rainy season is characterised by heavy precipitation, dry weather, high humidity, high temperature, waterlogging conditions and a rigging of flood in most of the cases.

In areas of high rainfall due to less evaporation the soil is suitable for plant growth. In such areas forests are dense with trees 40 metres high and where there is comparatively less rainfall, evaporation is high, forests are unevenly distributed with trees having just 20 to 25 metres high.

Table 5.2 : Forest Type

The Munger division is characterized by three major forest types :

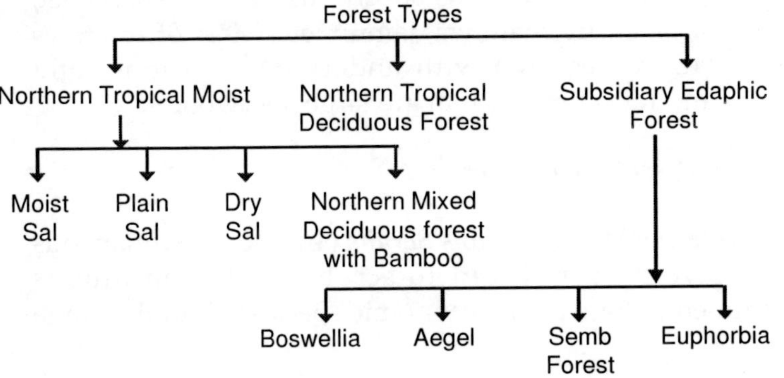

The following local type of forests can be distinguished :
(A) *Sal Forest*
Sal Forest is classified under following sub-groups :

(i) Moist Sal, Plain Sal, and
(ii) Dry Sal.

(B) *Mixed Miscellaneous Scrub*

(i) Inferior Miscellaneous Scrub, and
(ii) Boswellia.

(A) Sal Forest

(i) Moist Sal : This local type of forest occur in moist valleys and by the side of streams where deep soil is found. The height of the tree has been recorded up to 30 metres and girth up to 1.5 metres in Rameshwar kund and Madhuban, Baghail, etc. Some evergreen species are also intermixed with undergrowth of climbers, canes, bamboos, sal, lendi, asna, teak, shisham, jamun and semal.

(ii) Plain Sal : This is the type which is found on the Chakai plateau and Dullumpur Beats. Most of them are of young age represented by poles and saplings in varying degrees of the maltreatment.

(iii) Dry Sal Forest : This type of forest is found on northern slopes of hills where the slope is moderate to steep. Sal tree forms 80 per cent to 90 per cent of the forest, together with a few dry miscellaneous species like palas, khair, bet, babool, acacia, shrub mahuwa, jamun, etc. Most of the trees are 15 to 18 metres high with undergrowth of grass and bamboos and few climbers that are large and woody.

(B) Mixed Miscellaneous Scrub

(i) Inferior Miscellaneous Scrubs : Inferior miscellaneous forests have been reduced to scrub due to continuous maltreatment. There are no economic species worth the name and these forests are not of much use even from the point of view of soil conservation.

***(ii) Boswellia Forest (Salai)* :** All drier sites like hill tops and ridges have considerable Boswellia surrata which occurs in varying proportions, i.e., 6 to 32 trees per hectare.

***(iii) Aegle Forests* :** This type of forests occur progressively in dessicted soil where edaphic retrogression is found. Sometimes edaphic retrogression occurs in sal bearing forests.

***(iv) Euphorbia Forests* :** This type occurs in areas dominated by maltreatment and retrogression of soil since long. Even the Sal forests have been degraded to semi-desert state by man as in Jojo forest of Kharagpur.

Grass Land

There are no grass lands in the study area but wherever the canopy is open on hilltops and slopes grass covers the ground.

Forest Products

***Timber* :** In the forest of Munger division various types of wood are found in abundance. It is, therefore, easy to get any variety of timber. But all species found here are not utilized by paper and wood based industries. Nowadays, the demand is increasing with the abundant supply of good timber like Sal, Sisam, Jamun, Semal, etc.

***Firewood* :** Firewood acquires an important place in the forest use. A very large proportion of wood is utilized in this way. In the district of Munger more than 94 per cent of population is rural which depends on firewood for heat and generation.

The average annual revenue from timber and firewood is approximately Rs. 40 lakhs.

***Bamboo* :** Bamboo is an important type of forest produce. The average annual revenue from bamboo is approximately Rs. 25 lakhs. The entire bamboo working circle has been leased to M/s. Titagarh Paper Mills.

Kendu Leaves **:** Kendu leaves are plucked throughout the study area annually. General plucking is done in the months of April and May. Large number of persons are engaged in forests for plucking of Kendu leaves.

Sabai Grass **:** The other main product of economic value is Sabai grass. The annual revenue from this source is Rs. 45,000 only.

Table 5.3 : Forest Products in the Munger Division

Forest Products	*(1992-93)*		*(2000-01)*	
Timber	2,85,189	Cft.	99,099	Cft.
Firewood	2,48,098	Cft.	10,490	Cft.
Bamboos	31,705	Cft.	5,623	Quintal
Kendu Leaves	25,510	Quintal	23,680	Quintal
Sabai grass	875	Quintal	1,034	Quintal

Source **:** Working Plan, Divisional Forest Office, Munger, 2001.

In Figure 5.3, X axis represents different kinds of forest products which includes timber, firewood, bamboos, kendu leaves and sabai grass and Y axis represents production of timber and firewood in '000 quintals. This graph shows that the production of timber and firewood get prominence in the Munger Division whereas the production of bamboos, Kendu leaves and Sabai grass are very meagre in 1992-93 and 2000-2001.

Figure 5.3 also shows the change in production in 1992-93 and 2000-01. This change is also marked especially under that of timber and firewood but in case of bamboo and Sabai grass the production is static in both the abovementioned periods.

This also shows that in coming years the prospect of timber and firewood productions are better because the rise in production within ten years is more than 100% and the annual growth rate of production is 10%.

In Figure 5.4, X axis represents the age group of Sal, Sabai, Khair, Dhaura and Asan trees respectively. While Y axis is represented by trees whose diameter ranges from 10 to 30 cm.

in most cases (Table 5.4). In this graph it has been shown that the trees with lower age group are generally smaller in diameter, while large trees reaches the age of up to 30, 50 and 60 years. The diameter increases from 7 to 30 cm and above. This shows that the forest cutting or the depletion of forest economy in lower age group for trees of economic value must be prohibited because the wood having higher diameter certainly be a matter of economic value particularly for hill and valley Sal.

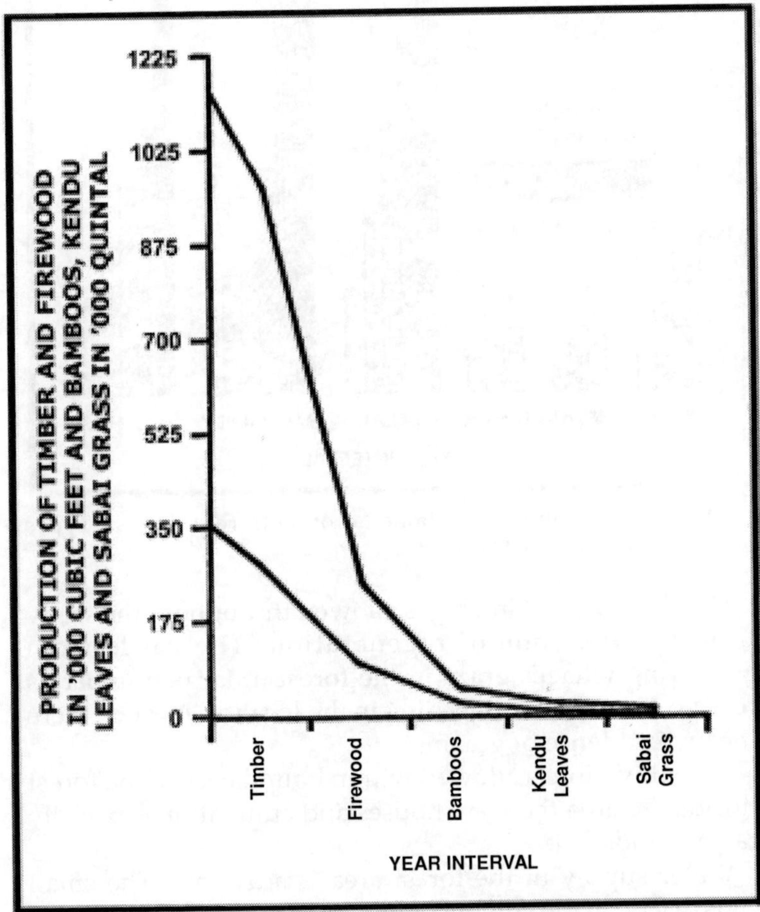

Fig. 5.3 : Munger Division : Change in Forest Products (1992-93 to 2000-01).

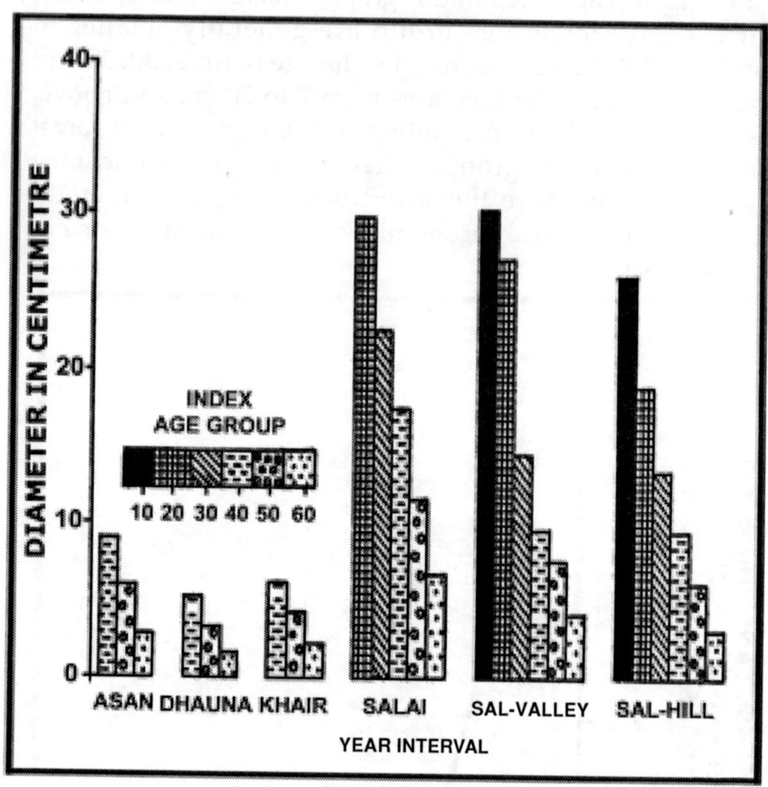

Fig. 5.4 : Munger Division : Diameter of Different Forest Types Age-Group 10-60 Years (2000).

Cattle grazing : Grazing is allowed throughout the study area except the zone of regeneration. The cattle from neighbouring villages graze in the forest in large number as there is no system of stall feeding in the locality. As such there is a heavy incidence of grazing.

No cultivation is allowed within boundaries of the forest but inside the area there are houses and cultivation especially on raiyati lands.

Water supply in the forest area is adequate. The small streams dry up in the winter. After winter ridges and hill slopes are completely devoid of water. The river Man and its tributaries, the Anjan and the Morway are perennial having a constant source of water supply. Besides these rivers and

Table 5.4 : Diameter of Different Forest Trees (2000)

Species	*Age*	*Diameter (in centimetre)*
Sal (Hill)	10	3.00
	20	6.30
	30	9.50
	40	13.50
	50	19.30
	60	26.00
Sal (Valley)	10	3.90
	20	7.50
	30	12.30
	40	17.00
	50	28.50
	60	30.90
Salai	10	6.90
	20	11.90
	30	17.60
	40	22.60
	50	24.90
Khair	10	2.00
	20	4.00
	30	6.00
Dhaura	10	1.30
	20	3.00
	30	5.30
Asan	10	2.80
	20	5.60
	30	8.90

Source : Working Plan of the District Forest, Munger, 2001.

streams, Kharagpur lake on the river Man including, Morway Dam also help in continuous water supply in the lower slopes.

Wildlife **:** No systematic survey of biotic resources has been carried out in the area under study. A survey of tiger was done in May, 1972.

Tigers are no longer commonly found in the area. In the Census of 1972 no tiger was located. However, during the rains of the same year a tiger was seen by Beat Officer, Dharhara

Table 5.5 : Wildlife Resources in the Munger Division (1999-2000)

Name of the wild lives	*1999*	*2000*	*2001*
Wild boar	347	1,023	1,091
Nilgai	111	181	250
Kotra	90	204	227
Peacock	150	204	227
Sambhar	34	136	182
Bear	43	68	113
Hyena	13	45	22

Source : Working Plan of the Divisional Forest Office, Munger, 2001.

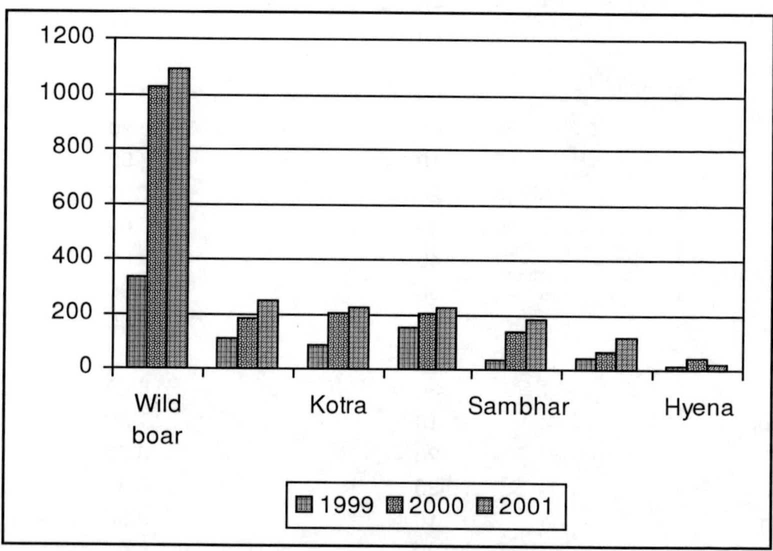

Fig 5.5 : Munger Division : Growth and Conservation of Wildlife.

near Paisra forest. Leopards are commonly seen in the area, though in the recent past a heavy toll of leopard population was taken by the poachers. Sambhar is found in a fairly good numbers inside the area. Chital is also found but a few in number. Nilgai is commonly found in the plains. Bear is found in all hills and forest which live upon white ants, plum and other fruits. Besides these, various types of snakes are found allover the forests.

Figure 5.5 and Table 5.5 show the presence and potentialities of biotic resources in the Munger Division. This map has been prepared under the heading, "Growth and

Conservation of Wild Life Resources for the Census Years, 1998, 1999, 2000". X axis represents different types of wildlives, i.e., Wild Boar, Kotra, Peacock, Sambhar, Bear, Hyena, Leopard and Y axis represents the number of wildlives in the district. In the Munger Division more than 300 Wild Boar were found in 1978 whereas in 1979-80 the number rose to more than 1000. Similarly, in almost all cases, in 1978 the number of wildlives were lower, while it has increased in 1980 except Hyena. This shows that the effect of conservation of wildlife resource has gained momentum in the Munger Division and the people are now more conscious towards saving the life of wildlives in order to save these precious resources from extinction and making the 'Wild Life Week' success which generally falls during 7th November to 16th November in India.

Utilisation of the Forest Produce **:** The inhabitants of Munger Division are mostly rural and depending mostly on agriculture. As the people are mostly agriculturists their demand for wood is considerably less though of course firewood is required by all for cooking food. Besides firewood there is a local demand for *GOL.* The villagers cut a thicker pole and are outbask and sap and prepare *GOL* out of it. These *GOLs* are very commonly used here. Besides they prepare beams and ridges—pieces out of poles of 6.3 centimetre and up in the girth. They require two types of plough as well. Small plough is used for ploughing Kharif lands and bigger one is for Rabi cultivation. They prepare bigger and smaller ploughs out of 9.6 metres girth poles. They use bamboo and grass for roofing their houses.

Timber and firewood are almost consumed locally. Negligible amount of firewood is sent outside the district. A considerable amount of natural tooth brush (DATWANS) are sent to Gaya and Lakhisarai. But this does not give any revenue to the forest department as local villagers take these out without paying for them. Fruits and flowers of Mahua, Kendu, Pear, etc. are consumed locally. Considerable amount of timber and poles are being supplied outside as well. This district is supplying electric and transmission poles to Bihar State Electricity Board. Besides above bamboo, Salai and Sabai grasses are being supplied to paper mills also. Some of the Salai goes outside for packing as well. In furniture there is a great scope and market for them if proper effort is made by the forest department. Kendu leaves have also very good market for manufacture of *Biri.*

The bamboo extracted from this division are mostly consumed by Titagarh Paper Mills. A considerable amount of bamboo is used in the preparation of *lathis* which are supplied to various parts of the State of Bihar and ouside as well. But a small amount of revenue is realised out of it from the local villagers who extract and supply them. By local sale and by auction about 20,378 quintals of bamboos are removed annually.

At present M/s. Bengal Paper Mills have taken lease of Salai along with other hardwood species in Jamui, Jhajha and Chakai. The balance Salai areas are auctioned every year with other hardwood species. The sale of Salai on the whole is not satisfactory.

The Salai grass is fully consumed locally by the villagers for string making. Figure 5.6 shows revenue income and

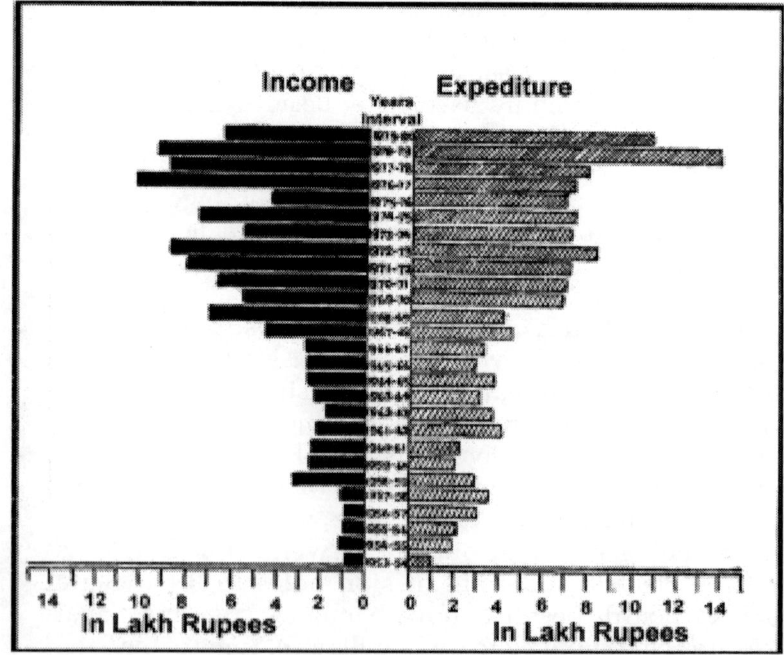

Fig. 5.6 : Munger Division : Revenue Income and Expenditure of Forest Resources (1953–1979-80)

expenditure on forest resources in the Munger division. For this income and expenditure pyramid has been prepared in which the left hand side horizontal bars show revenue expenditure from 1946-47 to 1972-73 while the right hand side horizontal bars show the expenditure for the same period in lakh rupees. It has been found that both income and expenditure are simultaneously increased from 1946-47 towards 1972-73. From 1969 to 1973, the graph shows that the income is lesser than expenditure on forest resources. This is due to the increase in staff expenditure, conservation of forest soil from erosion. This decreasing income and increasing expenditure is really alarming, for the growth of forest resource in the Munger division, because this may disturb our ecological balance in between natural vegetation, wildlife and man.

The main line of Eastern Railway runs from north-west to south-west of the forest division between Kharagpur and Gidheshwar hills. The main station of export of forest materials is Jamui from where the bundle of bamboo and wooden logs are exported to various consignee. Fuel, Salai and Datwan are also exported from Jamui and Jhajha. Materials of export are timber poles and fencing materials which are received from the Gangta forest depot. Other stations on the main lines are Kiul, Mananpur, Gidhaur, Jhajha and Simultala. The stations concerned on this loop line are from west to east Kajra, Dharhara, Jamalpur and Bariarpur. These stations export bamboo and *lathi* made of bamboo in plenty.

The roads emerging from Munger are Munger-Kharagpur-Jamui-Sikandra-Pakri Barwa-Nawadah, road which connects Patna, Ranchi and Gaya also. Almost all the important timber, firewood and bamboo depots lie on this road, i.e., Shampur, Lakshmipur, Kharagpur, Gangta, Melleypur, Jamui and Sikandra. The road which runs from Bariarpur to Bhagalpur also import and export forest products. The old Patna road extends from Munger to Barahiya. It is a very important road for the forest area. The Santhal Parganas road connects Jamui with Santhal Parganas. It passes via Jhajha and Chakai and other important centres of forest consumption. There are other cross-roads which are known as important forest export centre.

Forest Conservation

The saving of forest from destruction is known as conservation of forest. Over a vast portion of land forests help in conserving irrigation water, soil, regulation of stream flow, provides places for public recreation and offer a suitable environment for wild-life. Apart from this if properly managed they provide employment and security for millions of people. Hence, forest itself is one of the greatest conservation of other natural resources which need preservation and care for the welfare of people.

The history of modern forest conservation in the Munger division began in the 18th century when forest areas in Jamui, Jhajha, Chakai, Dharhara and in other blocks were abundantly found.

The management and control of all other private forest of this division were taken under Bihar Private Forest Act, 1946 and the replaced Act of Bihar Private Forest Act, 1947. Due to the absence of any system of forest conservation in the past, huge timber losses resulted from improper cutting methods, insects, disease, fire, overgrazing and other destructive practices. The magnitude of forest destruction indicates that a combination of better prevention, control and utilisation of this loss would go to shortfall in contributing our future timber requirements.

Problems of Forest Areas

The forests have deteriorated greatly not only in the quality but also in types. One of the greatest cause of forest fires are Mahua pickers, who carelessly burn the undergrowth of Mahua trees to facilitate Mahua picking.

The following are the evil effects of fire :

(i) Humous is destroyed which contains nutrients thereby rendering the soil poor.

(ii) Loss of moisture makes the soil hard, thereby reducing percolation and increasing run-off, result of which is erosion.

(iii) Grazing is also harming the forest because the continued grazing of large number of animals on the forest floor makes it hard and compact. This also does great damage to the regeneration coming up.

(iv) Man is by far the worst enemy of the forest. Tremendous irregular cutting and unregulated grazing have brought the best sal forests of past to this stage. Man is responsible for the vast damage done to the forest in the past.

Even though the forest department has tightened the control, the history is being repeated again and again. This has got to be checked and its demerits explained to the people. The semi-desert condition that has reached in various areas in these forests is nothing but irresponsible act of man in the past.

1. Besides this, the other problems of forest resources are to increase productivity of forest;
2. To establish a linkage between the development of forests and forest based industries; and
3. To gear up the development of forest to the growth of rural economy.

Remedies

Forest management needs artificial afforestation through planting because it is a means of getting barren land back into production. Besides, the idea of social forestry should be adopted.

Social forest is a new concept of forest creation, management and judicious utilization of goods and services. It aims at improving the land, labour and water resources to optimise production of farm manure, firewood, food, small constructional timber and to stabilise soil. The main constituents of social forestry are : farm, rural and urban forestry. Hence we may say that social forestry programmes should be launched for forest development.

Although return from forest products is low, however, there is good scope for the establishment of cardboard industry based on firewood and saw mill wastes.

Besides above, the district has some 200,000 Mahua trees which have potentiality to produce about 300 tonnes Mahua kernels per annum. Another 200 tonnes of seeds of Neem, Karanj, Palas, etc. are estimated to be produced per annum. Thus these non-edible oilseeds have a production potential of about 5,000 tonnes per annum. Small scale oil mills can also be started on the basis of these raw materials.

It is estimated that some 1,000 tonnes of saw dust will be available from the saw mills of Munger division as well as those of the adjoining districts of Giridih and Santal Parganas. An oxalic acid scheme based on raw material can be started in the district.

It is also estimated that about 200 tonnes of tamarind leaves and fruit pulp will be available in Munger division as well as in the adjoining division of Santal Parganas. It will be feasible to start Tartaric acid scheme in Munger based on this raw material.

On the basis of above raw materials the following forest based industries are feasible to be located in the Munger division :

(A) Medium industries based on wood are :

(i) Particle board plant at Munger,
(ii) Small scale industries based on timber,
(iii) Mechanised boat building plant,
(iv) Pencil making plant,

(B) Small scale industry based on minor forest produce including saw dust are :

(i) Plant for the manufacture of oil based on non-edible oil seeds at Munger,
(ii) Oxalic acid plant at Munger, and
(iii) Tartaric acid plant at Munger.

In the Munger Division deciduous type of forest is found with abundance of Sal, Mahua, Semal, etc. The southern part of Munger having dense forest leading to the congregation of floral and faunal resources which are the healthy sign of resource development and the maintenance of the ecological balance in the district. In spite of forest and wildlife conservation schemes the burning down of forest and the prey of wild beasts are continuing in the southern hilly tracts of Munger. But the industrialists are getting raw materials for paper industry, cardboard factory and several others which are changing the scene fast by the introduction of forest based industries especially in urban areas.

REFERENCES

Ahmad, Enayat, "Bihar : A Physical, Economic and Regional Geography", Ranchi University, 1965, pp. 77-78.

Criteria and Indicators for Sustainable Forest Management : A Compendium. Paper compiled by Froylán Castañeda, Christel Palmberg-Lerche and Petteri Vuorinen, May 2001. Forest Management Working Papers, Working Paper 5. Forest Resources Development Service, Forest Resources Division. FAO, Rome (unpublished).

Evans, K., De Jong, W., and Cronkleton, P. (2008) "Future Scenarios as a Tool for Collaboration in Forest Communities". *S.A.P.I.E.N.S.* 1(2).

Ghosh, Shankar, op. cit., p. 10.

Ghosh, Shankar, "The Planning of Our Forest Resources", in Yojana, Planning Commission, 1976, Vol. XX/16, p. 10.

Guidelines for Developing, Testing and Selecting Criteria and Indicators for Sustainable Forest Management, Ravi Prabhu, Carol J. P. Colfer and Richard G. Dudley. 1999. CIFOR. The Criteria and Indicators Toolbox Series.

Gupta, A. K., "Sustainable Development of Indian Agriculture : Green Revolution Revisited", Indian Institute of Management, Ahmedabad, W.P. No. 896, 1990.

Khan, M.A.W., "Ecology and Potentiality of the Bamboo Forests of Bilaspur Division, Madhya Pradesh", Indian Forester, 1960, p. 10.

Mamoria, C.B., "Forest Resource and Forestry Development in Agricultural Problem of India", 17th Edition, Kitab Mahal, Allahabad, 1973, p. 123.

Mozgeris, G. (2008) "The continuous field view of representing forest geographically: from cartographic representation towards improved management planning". *S.A.P.I.E.N.S.* 1 (2).

Quintal, O., "Agriculture and Sustainable Development".

Roy Choudhary, P.C., "Forest and Wild Life in Bihar", Indian Geographer, 1959.

Roy Chaudhary, P.C., ''Bihar District Gazetteers Munger", Secretariat's Press, Patna, 1960.

Roy, M.P., "Yield Regulation in the Forest of West Bengal", Indian Forester, 1961, p. 2.

Sharma, R.P., "Forestry in Rajasthan, Indian Forester, 1955, p. 1

Sahay, S.S., "Working Plan of the Divisional Forest Office", Munger, 1976, p. 12.

Srivastava, B.P. and Pant, M.M., "Social Forestry in India", Yojana, Planning Commission, 1979, Vol. XXIII/I2, pp. 18-23, Ministerial Conference on the Protection of Forests in Europe.

United Nations Forum on Forests (2004).

Wilson, Art and Tyrchniewicz, Allen, "Agriculture and sustainable development : policy analysis on the Great Plains", International Institute for Sustainable Development, Winnipeg, Manitoba, Canada, 1995.

6

Agricultural Resources and Sustainable Development in New Millennium

Agriculture is the backbone of Indian economy. Economic historians generally concur that there are no cases of successful development of a region in which a rise in agricultural productivity did not accompany industrial development. The Rostow's stages-theory of growth has observed that agriculture plays a multiple and converging role in the transitional process of the take-off into self-sustained growth. The operation of the planned development in India over the last 20 years bears witness to this fact. The performance of the last three Five Year Plans and 3 One Year Plans have clearly shown that the stimulation of a high but sustained rate of increase in per capita output in India (from Rs. 249.6 in 1948 to Rs. 323.0 in 1968-69 at 1948-49 prices, and from Rs. 249.6 to Rs. 542.3, at current prices) is difficult to have without a strong development base in her agriculture.

It may thus be observed that agriculture occupies a major place in the dualistic set-up of the national economy. Its development operation has snow ball effects over the whole of the economy. It constitutes the most important constriction the process of growth of the economy by supplying food for population; by supplying some basic raw materials for the expansion of certain consumer goods industries, by enlarging the demand for industrial output through an increase in agriculture income and through extending the capacity to absorb the monetary flow of industrial investments by employing a large pool of labour force, and by raising foreign exchange earnings through exports. It is unfortunate that Indian Agriculture continues to be backward having multi-directional facets although among all the enterprises 'agriculture is strictly the most fundamental enterprise.'

Agriculture includes land devoted to the production of crops and rearing of livestock. However, some writer restrict the term to growing of crops alone. According to J.L. Buck 'Land utilisation is the satisfaction, from which the farm population derived the type of agriculture developed, the provision for future production and contribution to national needs.' As such the scope of agricultural economy includes regional agriculture analysis, food and commercial crops, agricultural problems and planning, and food supply in relation to population. Considering the modern technological advancement and commercial economics, agricultural land use seems to be more complex than that in a simple hand labour and animal energy economy. The great importance of agriculture in determining man's economic and cultural progress is due largely to the diversification, relative scarcity and localization of agricultural resources on a particular piece of land, controlled by the environmental factors. The ownership of agricultural land, agricultural climatology, agricultural region, land classification, land conservation, crop combination and agricultural efficiency are certainly very important and determining issues of agricultural economy.

The concept of agricultural resources is based on the theory of economics :

(1) The use of resources of environment, space, time, energy, property, goods, techniques and information.
(2) The choice of alternative enterprises, farming system, method of agricultural practices, highways network and marketing behaviour of agricultural goods.
(3) The exchange of agricultural goods and landed property right, which play an important role in subsistence agricultural economy.
(4) The scarcity of agricultural commodity or land resource on a particular place and time which provide an

opportunity for an individual farmer to make the best use of what he has.

(5) Improving the quality of agricultural land.

Besides these concepts the idea of supply and demand are very important which are as follows :

(a) Lower demand—lower produce; Higher demand—higher produce;
(b) Higher demand but low produce—higher will be the price; and
(c) Lower demand but high produce—lower will be the price.

In order to understand how demand, supply and prices are related, we should examine the rate of response of demand and supply to change in price. The rate of response is measured by elasticity.

$$\text{Elasticity of demand} = \frac{\%\ \text{change in the quantity demanded}}{\%\ \text{change in price}}$$

$$\text{and Elasticity of supply} = \frac{\%\ \text{change in the quantity offered}}{\%\ \text{change in price}}$$

Elasticity varies from commodity to commodity, from district to district, from village to village and even in the same village. In such a situation the elasticity of rice and wheat are generally low, while it is high for fruit, salad, vegetables and meat.

On the basis of above observations we shall examine the present position, problem and prospects of agricultural resource in relation to growing population of the Munger division.

Sustainable Agriculture Development

- *Sustainable agriculture for sustainable development is one of the plant science industry's key priorities.*

- *The plant science industry contributes to sustainable agriculture by delivering crop protection solutions and technologies that help meet the food, feed and fibre needs of an increasing world population, while at the same time delivering economic, social and environmental benefits.*
- *Sustainable agriculture needs to be economically viable, environmentally sound and socially acceptable.*

'World agriculture cannot be sustainable without science-based technology...'—Per Pinstrup-Anderson, Professor of Food, Nutrition and Public Policy, Cornell University, CropLife International Annual Conference, June 2003. Sustainable development requires a major contribution from agriculture—the basic driver of most of the world's economies and the essential underpinning of even the most advanced societies. CropLife International is a key stakeholder in any debate on agriculture, and has much to offer as a genuine partner of farmers, civil society and governments.

Sustainable agriculture is a key element of sustainable development and is essential to the future well-being of the human race and the planet. Sustainable agriculture needs to be economically viable, environmentally sound and socially acceptable. It is a system of agricultural production that, over the long term, will :

- Satisfy human food, feed and fibre needs.
- Enhance the environmental quality and the natural resource base upon which the agricultural economy depends.
- Make the most efficient use of available technologies, non-renewable resources and on-farm resources, and integrate, where appropriate, natural biological cycles and controls.
- Sustain the economic viability of farm operations.
- Enhance the quality of life for farmers and society as a whole.

Croplife member companies and associations support the aims of sustainable agriculture : to produce sufficient, affordable food and fibre, economically and in an environmentally and socially sensitive manner, maintaining the natural resource base for future generations.

Plant science products and technologies have made a significant contribution to the sustainable development of agriculture and, in turn, to the dramatic economic and social developments that have taken place in most parts of the world during the 20th century.

Our efforts focus on protecting and improving yields, and preserving natural resources such as soil and water, using a wide range of technologies, including chemical crop protection and biotechnology. By making one of humankind's most environmentally disruptive activities more productive and efficient on existing arable land, we are helping prevent the conversion of further virgin land and its precious biodiversity to farmland.

In addition to innovative technology development and transfer, we support integrated farming techniques and capacity building through the transfer of knowledge to less developed countries. Integrated farming, which comprises Integrated Crop Management (ICM) and Integrated Pest Management (IPM) means using the best available technologies and methods to meet the goals of sustainable agriculture.

Technology and proper and effective stewardship are the pillars of sustainable agriculture and sustainable development. We lead programmes to ensure the safe and effective use of our products and technologies throughout their lifespan. Wherever possible, we work in constructive partnerships with relevant stakeholders to find solutions and maximise impact.

Sustainable Development of Indian Agriculture : Green Revolution Revisited

There are many things unique about the story of technological change in Indian Agriculture in sixties, seventies and eighties in India. However, my regret is that when our experience is applied in Africa or other developing countries, all the wrong lessons are learned.

This note has three parts : First deals with the current challenges in Indian agriculture. Part two deals with the historical review of the social, political and economic forces that shaped our policies. Part three deals with the issues to be faced in the nineties.

Part One : Where have we reached?

(a) The growth rate has been 3.1, 2.5 and 2.7% per annum during 1951-52 to 1964-65 (pre-green revolution); 1969-70 to 1987-88 (post-green revolution) and 1951-52-1987-88 (almost whole period after independence in 1947). The growth in pre-sixties was contributed mainly by bringing new area under cultivation either by cutting forests or previously fallow land. In post-sixties, yield increased entirely because of increase in the productivity.

(b) While the jump from 74.2 million tonnes in 1966-67 (a drought year) to 94.0 million tonnes in 1967-68 after introduction of new technology was indeed very dramatic given the fact that post-war economy (wars of 1962 and 1965) was quite sluggish as far as public investments were concerned. However, we had reached a figure of 82.3 million tonnes way back in 1960-61 when monsoon was good. Also, the production level hovered around 100 million tonnes till 1975. It came down to 97 million tonnes in a year of severe drought in 1972-73 and to 99.8 million tonnes in another drought year 1974-75.

(c) The jump of 20 million tonnes in 1975-76, 1982-83 and 1988-89 to take the production levels to 170 million tonnes has been contributed by good monsoon but also better distribution of inputs, improved irrigation infrastructure, technological upgradation, and resurgence of productive potential in Eastern India.

(d) Per capita availability of foodgrains and pulses has remained around 441 grams per day. It was as high as 480 gms in 1965. Population growth rate has kept a pace with the food production.

(e) Per capita income of rural producers (1966-77 to 1978-79) has ranged from Rs. 1,627 and Rs. 1,270 in Punjab and Haryana to Rs. 395 in Bihar. It ranged from Rs. 600 to 750 in 6 states and Rs. 490 to 600 in 12 states.

(f) The growth rate of wheat has been 5.1 per cent per annum between 1969-70 to 1987-88 and has been higher or at same level for several other cash crops and vegetables like potato. The oilseeds have a growth rate around 3.7 per cent, rice 2.4 per cent, pulses around 1 per cent, Jowar 0.3 per cent and Bajra negative. Thus the green revolution has escaped the rainfed crops affecting the poorest people so far.

(g) Input productivity has been declining at an alarming pace. With 1970-71 as base of 100, the index has come down to less than 60 in 1987-88. Fertilizer consumption has increased by about 387 per cent during 1970-71 to 1988-89, gross irrigated area has increased by about 30 per cent, power consumption by 88 per cent, pesticides by about 323 per cent during the same period.
The food production increased by only 57 per cent.
The non-sustainable nature of the green revolution doesn't need a more telling testimony.

(h) Despite increasing production in recent years, government has found it difficult to fulfill its procurement current targets to meet obligations of food stock for national security and public distribution system. The larger irrigated landholders have increased their capacity to stock grains.

(i) The expenditure on drought relief has continuously increased being more than Rs. 2000 crores during 1987. If the bank loans waived or to be waived may be added to it, it may increase by another couple of thousand crores.

(j) The rainfed crops like oilseeds and pulses have been pushed over to more and more marginal lands making technological change in them all the more difficult.

(k) Most of the millet, pulse and rainfed oilseed growers being scattered, vulnerable to seasonal fluctuations and living in the regions with low population density have poor access to formal delivery systems for meeting various minimum needs. The market forces are also quite weak.

(l) Emigration of poor people from drought and flood prone regions and hill areas to cities adds to the problem of urban poverty. This also robs the disadvantaged regions of many able men and women who could have taken some risk to undertake technological transformation. This is in addition to the misery that children and other family members of these migrants have to suffer.

(m) The proportion of women headed or managed households, is quite large in these regions. Given conspicuous neglect of women clients by public agencies, poor infrastructure becomes poorer for these families.

Part Two : Historical Overview of Technological Change

There was a strong research interest among Indian Scientists in forties on finding out best way of conjuctive use of organic and inorganic fertilizers. The agronomic ways of weed control were given importance and overall pest and weed problems in any case were less severe. Dr. N.R. Dhar had pioneered many imaginative experiments on sustainable agriculture in Allahabad.

We begin the story of technological change from 1962, that India had the first major war with China and realized its poor defense preparedness. National consensus on several macro policy issues primarily aimed at import substituting self-reliant economy started breaking down.

The year 1964-65 was a major drought. The war with Pakistan in 1965 increased the pressure on economy. But stoppage of US shipments of cheap food aid provided the right stimuli for searching alternatives for self-reliance in food sector. It may be recalled that America had by then given considerable aid both for setting up so-called community development blocks (of about 80-100 villagers each) all over the country and also for meeting the needs of public distribution and emergency aid for victims of natural disasters. American interest in keeping communists at way was served well.

Distribuion of imported wheat under public employment programmes had created taste and market for wheat even in the non-wheat growing regions.

Further as American publications in Foreign Affairs (David K. Kunkel, 1984) showed, India was a good example of how American "food aid leads to cash sales for US farmers" for edible oil and livestock products. Pakistan was shown to have become largest importer of American vegetable oil under P L 480 and GSM 102, access to irrigation and institutions providing other inputs. Disparities increased and the rural tensions started mounting.

Indian Scientists were aware of the high yield responsiveness of Mexican varieties of wheat and IR varieties of paddy. Decision to bulk import the wheat seeds heralded the massive effort for technological change primarily in the Indo-Gangetic plain with good rich soil and abundant waters.

Even though the extension machinery was quite weak and so were the markets, the alleged resistance of Indian farmer to a visibly 'viable' technology evaporated in the thin air.

The drought of 1965-66 had also witnessed deaths on large scale (the last time it happened) in Bihar. New technology led to high yields and also high incomes but initially to only those who had by 1967, four major events took place :

— The devaluation of rupee and its failure to boost Indian exports—large scale violence in rural areas by left radical group scaled as Naxalite movement.
— Several states had opposition SVD (joint front) governments ruled mainly be the rightist parties—Central government though ruled by Congress was critically dependent upon Communist party or support.
— By this time, several Communist leaders had infiltrated Congress party hoping to capture it from within.
— Around the same time, experiment in social control of banking had started. But with split in the ruling party

> on the eve of election for the President of the country, a new polarized party with relative dominance of socialists ideology emerged on the scene.

A confidential report of the Home Ministry by this time had brought to fore the need for land reforms and pressure for direct attack on poverty if rural violence had to be contained. Bankers by then had begun to see the potential of mobilizing rural savings accruing on account of technological charge. Government wanted to prove its socialistic ideals and thus nationalised the banks among other measures. Rural credit also started being pumped into the agricultural sector to improve incentives for technological transformation.

By early seventies, India had to face another war (1971) which transformed the fortunes of the ruling party. Even though politically it was less dependent upon the leftist allies (because it had huge majority in the parliament), the efforts for containing hopes of small and marginal farmers and landless labourer had to be initiated. Several subsidy linked programmes for provision of credit for irrigation, land development and other assets started. There was a brief spell when several programmes aimed at solving eco-specific problems of the country were launched e.g. Drought Prone Area Programme, Hill or Tribal area development programme etc.

By 1973, the failure of poverty alleviating programmes was obvious. It was also clear that trickle down of growth made possible though new technology may not take place at sufficient pace and scale. The land reforms had got stuck through complicated but easy loopholes left in the acts leading to large scale litigation.

The country side was getting restive. The erstwhile Socialist colleagues of the then prime minister were also getting disillusioned.

The oil price hike in 1973 and resultant inflation made worse by the drought of 1972-74 resulted in a widespread social unrest. The difference this time was that unrest was not restricted merely to rural areas this time. It spread to urban areas too. By 1974, the country was facing a massive industrial unrest, widespread social discontent and glamour for change. But change that followed was a shock therapy type. Emergency was declared in June 1975 and all fundamental rights were suspended.

Among other things, two or three things had crystallised by now :

(i) World Bank had extended massive line of credit for mechanisation and minor irrigation;
(ii) technological change was indeed diffusing rather widely leading to perceptible change in food supply;
(iii) Political tilt of ruling party towards the right was being manifested openly.

While Price Commission had been set up in mid-sixties, the terms of trade were moving against agriculture. Even though there were a few announcements ostensibly to help poor but it was clear that government had lost contact with people.

Emergency excesses led to the change in government. The new political alignment was certainly to the right but also had some committed socialists. The direct attack on the poverty through IRDP (Integrated Rural Development Programme) started in 1978. Agricultural prices were made more favourable and input subsidies continued. Good monsoon coupled with continued supply of new varieties of wheat and paddy helped in achieving new heights in food production.

Emphasis remained on input responsive technologies for increasing the yield.

By 1983-84, thanks to the recognition of seed distribution as a major bottleneck, attention was shifted to Eastern India where growth rate had been low. Good monsoon coupled with good management of input distribution, food grain production increased again.

By now however, two major signals started appearing :

— The input-output ratio was becoming seriously adverse in the agricultural sector.

— Efforts of the government to reduce subsidies to contain budget deficit were not meeting much success.

This government ruled by a coalition of opposite political interests broke up and the old party came back to power.

The political change this time had some very clear implications for technology transfer.

Despite the drought of 1979, the scientists and the public officials were happy that food production had not dipped too low. The mechanization, which was slowed down due to large scale studies on its labour displacing effects was reintroduced in early eighties. Mounting of inventories in tractor industry was the major reason. But pressure from affluent farmer had also made some difference. The credit constraint had been removed. Fertilizer subsidies continued.

Emphasis was shifting towards high tech approach to agricultural development through reliance on Biotechnology and in fact import of seeds made more easily possible than hitherto.

It was realized belatedly that standardized approach for agricultural development relying on the most favoured and well endowed regions may not work any more. The attention started increasing towards regionalization of technological change policies.

But this time next yield barrier was crossed in 1988-89, the non-sustainability of agricultural technology was becoming more and obvious. At the same time need for better soil and water conservation through watershed management in dry regions and restoration of ecological balance in hill areas denuded severely in the past was being felt. But the patience for technologies which could generate surplus in longer period of time but in a sustainable manner was just not there. There was an appreciation for decentralised planning nor was there any attention towards improving the capacity of agricultural scientists to continue to deliver results. The research funds per scientists in the country had gone down in real terms by as much half and in many states more than that. Budget of Indian Council of Agricultural Research was almost equal to or slightly more than the budget of Department of Biotechnology alone.

Sustainability of the institutions in which research for sustainable agriculture could be done was itself under doubt. The private and public corporate sector (particularly agri-input industries) was thriving on the work of agricultural scientists but was hardly making any investment in improving research productivity.

Part Three : Where do we go from here

The following research needs based on the lessons of Green Revolution in India are unlikely to attract attention within India without considerable pressure from relevant but less powerful constituencies.

Some constituencies are historically quite weak such as the ones belonging to graziers or livestock herdsmen, riders of bullocks following extensive systems of resource management; makers of hand tools or bullock drawn implements, small scale manufacturers of herbal non-toxic pesticides; developers of non-monetary input requiring technologies etc.

But then sustainable development requires close involvement of these very constituencies :

1. The time of individual oriented technologies is over. More and more emphasis will need to be placed on the group based technologies. The areas where group action is called for include :

   (a) Synchronization of sowing schedules of certain crop to avoid synchronization with pest reproduction cycle in certain crops where this is a major problem.
   (b) Watershed management of both arable and non-arable land belonging to individuals as well as villages, forest and revenue department.
   (c) Synchronized sprays compelled with pest reducing crop rotations and mixtures to minimize use of chemicals in the short run and total elimination in the long run.

(d) Biological pest control coupled with water management and drainage.

(2) Most of the postgraduate and even other research particularly in the discipline of Agronomy is concentrated on Inorganic Fertilizer. Very little work is being done on green manuring, conjuctive use of inorganic or organic manure and replenishment of micronutrients through ash and other organic supplements.

(3) Genetic uniformity in high growth regions has created another major reason for concern. Very few varieties of wheat occupy large areas. And a large number of rice varieties with IR parents have common genetic source of disease resistance. Diversity in genes cannot be achieved in the short run without making a trade off in yield. Just like Europe and America provide incentives to growers to fallow the land or for cutting the dairy animals, incentives for genetic diversity, and green manuring will have to be provided in India in high growth regions. These will be necessary till alternative genetic sources of resistance and yields are available.

(4) The shift from crop to trees (horticultural and/or timber) is taking place both for reducing the need for outside labour and also for reaping larger commercial gain by larger farmers. As long as the shift is marginal, nothing much may matter. But once it becomes large scale, the imbalances in food supply may emerge. Agro-forestry for high and slow growth regions thus is an urgent priority.

(5) The till towards fashionable technologies rather than relevant technologies has to be reversed. Even though we do not have even ten calves produced under field conditions from Embryo-transfer technology, the scarce resources for dairy research and technology transfer are being misallocated towards such technology. Even the conventional technologies like frozen semen (of

good local breeds) for upgrading cattle haven't yet been provided at large scale. Shifting gears every now and then on technologies popular in West is costing us a lot. Of course the problem is common to most disciplines and not merely to dairy. If elites in third world recognize merit of third world scholar only/mainly when they get approval/notice from the West, then is not it natural that the scholars (undoubtedly highly competent) should work on problems fashionable in West.

(6) Synchronization of maturity was a necessary trait in cereals for improving mechanizations in West. Research on their subject became popular in seventies in India also because many scientists returning from US or UK after training carried on this research. Perhaps alternative ways of combining response to inputs could have been found if searched.

(7) It should be mandatory for every agri-input industry to contribute a part of its profits to a National Fund for Sustainable Research (NFSR) to be managed by a group of inter-disciplinary scientists, NGOs and committed public servants. NFSR should be used to find long range research programme for identifying low/no external input sustainable technologies.

(8) One of the causality of the Green Revolution in India has been the discontinuance or weakening of longitudinal research. Without such research, comprehensive understanding of changes in ecosystem cannot be achieved. There are hardly any experiments in the country running for 30-40 years to provide basis for developing long range forecasts for different type of technologies (particularly for soil and water conservation, but also for input use).

Unfortunately, most foreign aid agencies reinforce such a face long term challenges can't be developed unless basic research on system basis is supported in different agro-climatic regions of the country.

(9) The continued neglect of dry regions is really criminal. It is a pity that when we think of hill areas or drought prone regions, we conjure image of domestic servant or a cheap labour. National Commission on Development of Backward Areas (1981, Planning Commission, New Delhi) went so far as to say that we should not try to create conditions by which supply of cheap labour for large irrigation projects is affected adversely.

(10) It is true that far fewer Members of Parliament are elected from dry regions due to low population density and hence much political pressure is not expected in the short run. But it should be acknowledged that low investment in pasture development in dry regions affects the sustainability in other regions directly. For instance :

— population growth rate of sheep and goat has been several hundred per cent compared to 30-60 per cent increase in cattle in most districts in Western India. Poor have to increase weight of low capital requiring enterprise in their portfolio if they have to survive without any support for other possible technologies or enterprises.

— The cultivation of marginal regions increases siltation rate of seasonal rivulets in these regions further affecting the run off, soil and water conservation. Even snap floods have been experienced in some parts of Rajasthan—a phenomenon unheard of in history.

— Continued emphasis on crop varieties with high harvest index (grain/straw ratio) has intensified pressure on dry matter supply. The pressure on trees for leaves and energy (twigs) is inevitable. Tractorization in dry regions does damage in two ways, (i) it makes escaping spontaneously sprouted tree seedlings while ploughing virtually impossible (this was possible with bullock ploughing) and (ii) it increases the tendency to cultivate river banks, marginal lands which were earlier left as fallow. Once the soil is loosened and grass roots are removed, the rate of soil erosion increases. Strong winds do bring this sand (and some silt) on to the neighbouring regions. This trend will only increase in future.

— Mining of ground water in high growth regions has assumed alarming proportion.

(11) Development of Common Property Institutions for generation of collective restraint is most essential. It is obvious that as the water table goes down lesser and lesser people are able to mine it. Having lost the battle of land reforms the struggle for water reforms must not be lost.

(12) The dilemma of continuing malnutrition and poverty among 30-40 per cent of the people having no purchasing power is manifesting in certain paradoxical ways. For instance, wheat was sought to be exported a few years ago when the food stocks reached a level of 30 million tonnes. It was realised that (a) more and more countries were trying to seek self-reliance reducing the demand for wheat at commercial prices, (b) the quality of wheat was not good such that even Russia—a close ally of India refused to buy a few lakhs tonnes of wheat, and (c) the African countries which had need for this grains did not have the ability to pay in hard currency. Instead of using those stocks for launching massive employment programme for reclaiming denuded forest lands and other eroded lands, thinking of export was obviously a short-sighted answer.

(13) The demand of manufactured goods which are income inelastic is unlikely to increase unless the distribution of purchasing power becomes more even.

Green revolution has certainly proved that Indian farmers were no less enterprising than the farmers anywhere else in the world. However, it is also true that the rate and pattern of growth is not sustainable either ecologically or even politically. Some of the tensions in high growth regions are arising because of the inability of state to continue the subsidy at the rate at which these were available earlier. On the other hand that deterioration in the level of living of the people in drought prone regions and hill areas is leading to demands for separate states and more autonomy. There is no denying the fact that next round of violence could be in the backward regions if neglect continues to be what it has been. The recent trend towards inviting multi-national corporations for initiating agri-business model of contract farming in the high growth regions cannot but bring the experience of Latin America and Philippines into India. It is a pity that planners cannot see that.

Land Utilization

Land use analysis is essential in order to know the distribution growth and utilization of land resources. As such this study aims to investigate the basis of the scientific land resource allocation to various agricultural crops so that maximum productivity may be attained.

Land is classified into seven classes according to use, i.e., net sown area, current and old fallows, barren and uncultivable waste, non-agricultural, permanent pasture and orchards, forest and others. Patterns of land use is determined by two sets of factors (a) the physical factors like topography, climate and soil which broadly determine the capabilities of the land, and (b) the human factors like the length of occupance, density of population, social and economic institutions, which determine the extent to which the resources of the land are utilized.

Taking the net sown area and fallow lands together, cultivation is found to be extended to the farthest limits. The highest amount of forest is found in Dharhara 53%, Jhajha 48%, Khaira 42%, Kharagpur 38%, Chakai 34%, and Sono 32%. The considerable amount of forest exists in Surajgarha, Lakshmipur and Sikandra. Poor forest cover is found in Jamalpur (5%), Sangrampur (8%), and Lakhisarai (2%). The forest area in Khagaria, Jamui and Barbigha is very negligible which cannot be counted in forest region of the district. Similarly, barren and uncultivable wastelands are 6% of the total area. So far as the non-agricultural uses of land are concerned, built-up area, roads and water bodies are major constituents under this head and the picture is practically the same allover the Munger division.

Orchards and pasture occupy just 1% total land of the division, Fallows include two types of land, namely, old fallows (for 2 to 5 years) and current fallows. These are cultivated lands which have been left fallow because of rainfall deficiency or economic reasons in one year or other. Their extent, therefore, varies according to rainfall conditions in different years. Fallows are highest in Alauli 50%, Ariari 33%, Beldaur 30%, Sikandra 25%, Halsi 25% and Sono 24%. Parbatta, Chautham, Khagaria, Jhajha, Jamui, Chakai, Lakhisarai, Surajgarha, Sangrampur, Munger, Jamalpur and Kharagpur have 8% to 16% fallow land. Fallows are found just 2% in Dharhara, 0.84% in Barahiya, 0.76% in Lakshmipur and 3% in Gogri. The district average of the fallow is just 14%.

The net sown area is found 86% in Barbigha, 81% in Sheikhpura, 70% in Sangrampur, 62% in Jamui and 60% in Khagaria and Gogri and 55% in Jamalpur. Here the pressure of population is so intense that nearly every available patch of land is cultivated. Extension of cultivation is often at the expense of mango orchards and pasture lands. The presence of hills and uplands accounts for a lower percentage of net sown area in south hilly tract.

When we consider the resource development as a whole the net sown area constitutes 49% and fallows 13.7% of total land. Thus almost 62.8% of the total land is cultivated and 4% of the total land is under orchards and cultivable waste. Barren and the cultivable land is 5.3% and non-agricultural land covers about 16% of the total area. Thus the intensity of cultivation is highest in the Ganga riparian tract of south and north Munger (Table 6.1).

Table 6.1 : Percentage of Land Utilization in Munger division (1999-2000)

Name of anchals	*Forest*	*Barren and non-cultivable land*	*Non-agri-cul-turable use*	*Culti-vable waste land*	*Pasture and orchard*	*Fallow*	*Net sown area*
Munger	—	14.86	36.21	2.45	0.06	9.42	37.00
Bariarpur	23	7.15	4.05	0.48	0.56	6.98	57.78
Jamalpur	3	18.73	6.95	0.47	0.46	13.58	56.81
Dharhara	50	1.23	3.17	0.70	0.48	0.67	38.75
Kharagpur	32	3.75	6.48	0.60	0.77	7.89	48.51
Asarganj	—	0.64	14.11	0.15	0.31	0.29	84.50
Tarapur	—	0.58	14.25	0.25	0.37	0.41	84.14
Tetiha Bambar	1	5.24	11.25	0.48	0.51	10.87	70.65
Sangrampur	1	5.58	9.73	0.58	0.55	11.01	71.55
Barahiya	—	11.41	12.21	—	—	0.84	75.54
Pipariya	5	6.24	3.87	1.11	2.87	12.68	68.23
Surajgarha	18	8.17	8.97	0.26	0.55	11.37	52.68
Lakhisarai	6	5.59	4.01	1.22	11.28	14.17	57.73
Ramgarh Chowk	—	2.53	18.44	0.25	0.16	24.87	53.75
Halsi	—	2.42	22.38	0.18	0.17	25.56	49.29
Barbigha	—	5.33	7.24	0.08	0.17	0.95	86.23
Shekhopur Sarai	—	5.11	7.03	0.06	0.08	0.87	86.85
Shekhpura	—	4.98	10.11	—	0.16	3.89	80.86
Ghat Kusumba	—	0.46	9.76	0.03	0.21	4.25	85.29
Chewara	—	0.45	13.27	0.53	0.37	33.38	52.10
Ariari	—	0.46	12.98	0.57	0.33	32.98	52.68
Islamnagar Aliganj	13	0.36	11.01	6.98	10.45	25.57	32.63
Sikandra	12	0.44	11.76	6.85	10.23	26.01	32.71
Barhat	18	3.88	7.46	12.87	2.26	0.77	54.76
Lakshmipur	17	3.97	7.24	12.78	2.25	0.76	56.00
Jhajha	39	2.73	14.87	5.41	1.77	16.87	19.35
Gidhaur	18	3.77	7.57	6.97	10.43	26.25	27.01
Khaira	36	8.76	8.93	7.13	—	7.45	31.73
Sono	30	13.48	5.89	2.87	0.51	24.44	22.81
Chakai	33	6.12	4.22	26.13	2.74	8.17	19.62
Alauli	—	1.98	4.32	0.70	0.60	50.78	41.62
Khagaria	0.1	4.98	15.87	0.51	3.33	15.21	60.00
Mansi	—	10.98	11.22	9.14	0.37	14.17	54.12
Chautham	—	11.78	11.19	9.44	0.45	14.46	52.68
Beldaur	—	0.49	12.12	—	0.87	29.94	56.58
Gogri	—	10.57	15.46	5.57	4.28	3.13	60.99
Parbatta	—	4.67	24.48	0.84	4.17	14.09	51.75
District Average	**18.69**	**5.40**	**11.08**	**3.67**	**2.15**	**13.65**	**54.74**

Source : District Statistical Office, Munger, 2001.

From the point of view of social and economic aspects, the agricultural system of the district needs to be given special attention to raise the fertility to its maximum capacity. In this sphere, the role of government as well as public, bears equal importance to mobilize the agricultural machinery with a view to presenting suitable basis for the scientific land resource allocation to various agricultural crops and other uses and planning for dynamic land resource utilization. Traditional methods of cultivations and outdated agricultural implements which result in inefficient use of land resulting in poor yield should be replaced by modern means and farming technology. In this way, if agricultural laud use is intensified and made production-oriented the productivity of sonic crops may be raised to high level.

Table 6.2 shows the intensity of cropping in the Munger division (2004-05). The Munger division has only 25% of double cropped land in comparison with the total cultivable area. Anchals which have higher value of double cropped area in comparison with this average are Munger, Tarapur, Sangrampur, Sheikhpura, Barbigha, Ariari, Khagaria, Alauli, Gogri and Parbatta. The highest double cropped anchal is Tarapur (75.87%) and the lowest double cropped area is found in Chakai anchal (0.70%). The lower percentage of double cropped area is also found in Chautham, Lakshmipur, Khaira, Jamui, Barahiya and Dharhara anchals. This shows that either mountainous forested or flood affected area have lower percentage of double cropped land.

So far as the treble cropped land is concerned the highest percentage of 11.66 is found in Sikandra anchal of South Munger. The lower percentage of treble cropped land is nil in Halsi and Chakai anchals. This shows that in the Munger division most of the cultivable land come under single cropped and double cropped cultivation, whereas treble cropped land has a very poor percentage in almost all anchals.

Out of total cropped area of 342,166 hectares, 103,508 hectares is irrigated. Wholly and partly irrigated holdings area under irrigation in 1999-2000 is given in Table 6.3.

Table 6.2 : Intensity of Cropping in the Munger division (2004-05)

Name of Anchals	*% Doubled Cropped Area*	*% Trebled Cropped Area*
Munger	45.75	0.30
Bariarpur	18.64	0.18
Jamalpur	21.17	0.16
Dharhara	6.48	0.20
Kharagpur	15.79	0.63
Asarganj	66.89	0.98
Tarapur	74.76	0.97
Tetiha Bambor	44.62	0.81
Sangrampur	49.61	0.83
Barahiya	8.23	0.03
Pipariya	14.63	0.185
Surajgarha	13.16	0.12
Lakhisarai	22.58	2.27
Ramgarh Chowk	15.87	—
Halsi	16.90	—
Barbigha	52.87	0.68
Sheikhopur Sarai	49.69	0.71
Sheikhpura	33.89	0.40
Ghat Kusumbha	36.71	0.49
Chewara	36.41	0.15
Ariari	30.41	0.11
Islamnagar Aliganj	16.25	12.05
Sikandra	14.96	11.66
Jamui	6.28	1.96
Barhat	2.95	0.71
Lakshmipur	3.90	0.60
Jhajha	5.06	0.41
Gidhaur	4.11	0.75
Khaira	4.74	0.81
Sono	8.13	0.41
Chakai	0.70	—
Alauli	59.37	0.37
Khagaria	49.93	1.56
Mansi	0.58	0.44
Chautham	0.78	0.45
Beldaur	23.58	0.54
Gogri	49.99	0.90
Parbatta	25.72	0.64

Source : Agriculture Department, Government of Bihar, 2006.

Table 6.3 : Irrigated and Unirrigated Holdings (1999-2000)

	No.	*Area in Hectares*
Wholly irrigated holdings	62,678	52,383
Partly irrigated holdings	62,078	88,006
Wholly unirrigated holdings	2,50,721	2,01,777
Total	3,75,477	3,42,166

Source : Agriculture Department, Government of Bihar, 2001.

As far as irrigation potential is concerned, the facility is available for 8,400 hectares in Summer seasons, while in Kharif and Rabi seasons 131,366 hectares and 52,550 hectares respectively can be irrigated.

In the Munger division there are about 57 seed sales centres, 71 fertilizers sales centres, one agro-service centre, 23 custom service centres, 35 commercial banks branches and 7 cooperative banks. The blockwise information is given in Table 6.4.

Agricultural Holdings

In the Munger division, the agricultural holdings of most of the farmers are fragmented in different parts of the village. With the increase of population the shareholders of the same piece of land are increasing continuously and hence the sub-division of bigger plots of land is continuously in process. It is well-known fact that fragmentation and sub-divisions of holdings are not economical for planned agricultural development. Hence, the lower size class of holdings is economically increasing which resulted into uneconomic agricultural practices.

The economy of the Munger division is traditionally and predominantly agriculture. Out of total 463,162 hectares of cultivated area, the net cropped area during 1999-2000 was 349,166 hectares besides 121,996 hectares was under current fallow land. Hence, the average area per holding in the district is 1.40 hectare. The land holding patterns show that 67.2% holdings are up to 1.00 hectares size and 15% is between 1 to 2 hectares of size class. More than two-thirds of the total holdings of the district are less than 5 hectares size class.

Table 6.4 : Statement Showing Infrastructure Facilities in Munger Division (1999-2000)

Name of Anchals	*Seed sale Center*		*Fertilizer sale Center*		*Agro-service center*	*Plant protection center*		*Nation-alised banks*	*Co-operative banks*
	Govt.	*Private*	*Govt.*	*Private*	*Private*	*Govt.*	*Private*		
Munger	1	15	1	7	4	4	1	10	1
Bariarpur	1	5	-	2	1	1	4	2	-
Jamalpur	1	14	1	1	1	1	-	6	1
Dharhara	1	9	-	2	-	1	4	3	-
Kharagpur	1	11	1	6	3	3	1	6	1
Asarganj	1	8	1	6	2	2	1	4	-
Tarapur	1	7	1	4	2	1	1	4	-
Tetiha Bambor	1	5	-	-	-	-		2	-
Sangrampur	1	7	1	1	1	1	1	4	-
Barahiya	1	11	1	3	1	2	1	3	-
Pipariya	1	4	-	-	-	-	-	-	-
Surajgarha	1	10	1	3	2	1	1	3	-
Lakhisarai	1	16	1	6	2	2	1	4	1
Ramgarh Chowk	1	3	-	-	-	-	-	-	-
Halsi	1	5	1	1	1	1	1	2	-
Barbigha	1	9	1	2	2	2	1	3	-
Sheikhopur Sarai	1	4	-	1	-	1	-	2	-
Sheikhpura	1	13	1	4	3	3	1	7	-
Ghat Kusumbha	1	4	-	-	-	-	-	-	-
Chewara	1	3	-	1	-	-	-	-	-
Ariari	1	5	1	-	1	1	1	2	-
Islamnagar Aliganj	1	6	-	-	-	-	1	2	-
Sikandra	1	8	1	1	1	1	1	3	-
Jamui	1	14	1	2	3	3	1	8	1
Barhat	1	3	-	-	-		-	2	-
Lakshmipur	1	9	1	1	2		1	3	-
Jhajha	1	6	1	1	-	2	1	3	-
Gidhaur	1	3	-	-	-	-	-	2	-
Khaira	1	6	-	-	-	-	1	4	-
Sono	1	9	-	1	-	-	1	3	-
Chakai	1	8	1	1	-	-	1	3	-
Alauli	1	8	-	1	-	-	1	2	-
Khagaria	1	13	1	1	2	5	1	9	1
Mansi	1	8	-	1	-	1	1	3	-
Chautham	1	6	-	1	-	1	1	2	-
Beldaur	1	7	-	1	-	1	-	3	-
Gogri	1	5	1	1	1	2	1	4	1
Parbatta	1	9	-	1	1	1	-	2	-

Source : Agriculture Department, Government of Bihar, 2001.

The distribution of holding according to size class and area is given in Table 6.5.

Table 6.5 : Distribution of Land according to Size Class of Holdings of Munger Division (1999-2000)

Size class in hectares	*Total Holdings*		*N % of total*	*Net cropped area (Hectare)*
	Number	*Hectare area*		
0.7 to 1.0	72,467 (19.2%)	51,107	9.9%	39,014
1.0 to 2.0	57,228 (15.2%)	79,407	15.2%	57,682
2.0 to 3.0	25,476 (6.8%)	61,119	11.7%	42,572
3.0 to 4.0	13,682 (3.6%)	46,454	9.0%	31,684
4.0 to 5.0	8,462 (2.2%)	37,897	7.3%	25,461
5.0 to 10.0	12,729 (3.4%)	87,235	16.7%	55,015
10-20	4,120 (1.1%)	54,251	10.0%	31,463
20-30	884 (0.2%)	20,823	4.0%	11,756
30-40	336 (0.1%)	11,368	2.2%	6,346
40-50	919 (0.1%)	8,527	1.6%	4,302
50-above	224 (0.1%)	20,832	4.0%	9.044
Total	376,277 (100)	520,865	(100)	342,166

Source : Agriculture Department, Government of Bihar, 2001.

The above analysis of size class of holdings shows that for better agricultural production, the smaller size class of holdings should be consolidated either on efforts of the Government or co-operation among villagers. In order to achieve planned economic development of agricultural resources it is necessary to consolidate our land holdings because it was certainly checked subdivision fragmentation and deteriorating conditions of agricultural production.

Agricultural Situation

Paddy, maize, wheat and Khesari are the main crops of the district. Maize is mostly cultivated in Kharif season, production of which depends mainly on monsoon and yield is highly affected by floods. During 1999-2000 production of various crops are given in Table 6.6.

Tabke 6.6 : Trends of Crop Production in the Munger division

Name of the crops	1999-2000		2000-2001	
	Total production M/T	Av. yield kg/ha	Total production M/T	Av. yield kg/ha
1. Autumn Paddy	1,54,637	621	198,742	535
2. Winter Paddy	1,54,637	828	198,742	904
3. Summer Paddy	1,54,637	1,259	198,742	—
4. Wheat	1,14,102	1,045	96,507	11,509
5. Maize	1,17,433	986	21,586	600
6. Khesari	11,235	456	21,708	607
7. Barley	—	—	5,949	—
8. Peas	717	290	1,796	617

Source : Agriculture Department, Government of Bihar, 2002.

Relative Importance of Different Crops

***Rice* :** Rice is the most important crop in the Munger division. It covers largest cropped area and grow in both northern and southern parts of the district. In South Munger it mainly grows in Kharagpur, Lakshmipur, Jhajha, Jamui, Halsi, Lakhisarai, Sikandra, Barbigha and Sheikhpura anchals. In North Munger winter rice grows in almost all blocks. In 2000-01 rice (High Yielding Variety) covers 25,333 hectares and rice (local) covers 12,818 hectares of land. Rainfed variety of rice is also called Sathi because of the period between sowing and reaping is about 60 days and is cultivated mostly in the southern part of the district, ordinarily on high and somewhat poor land of Khaira, Sono and Chakai plateau areas. The total production of rice in 1999-2000 is 154,637 metric tonnes and 198,742 metric tonnes in 2000-01.

***Wheat* :** Another important crop is wheat. It occupies the largest crop area among the cereals of Rabi season. It is grown extensively of the Ganga especially in diara areas. In North Munger wheat is cultivated on extensive farms. In the north, due to inundation of the floods of Ganga alluvial soil is deposited each and every year. In such areas ordinarily heavy cost of cultivation of this crop is to a large extent avoided and the cultivator is then able to bear with comparative equanimity hence the chances of loss through blight, to which crop is particularly liable. The sowing generally starts from the middle of October and continues up to the end of November. The seed rate generally varies from 60 to 80 kilograms per hectare but where mixed cropping is practised the seed rate is reduced to 40 kilograms only.

Maize **:** Maize is an important crop in the district of Munger which grow mostly on flood plains north of the Ganga and in southern diara and tal areas. It is consumed in huge quantity locally. Two crops are raised in the district, one is Garma maize and the other is Kharif maize. The cultivation of Kharif crops is an usual practice, but bumper harvests are taken in years of only favourable climatic conditions. The hybrid varieties have been adopted by the farmers mostly for Kharif cultivation. The maize (Hybrid) Kharif crop covers 12,568 hectares and Maize (local) Kharif crop covers 59,187 hectares in the district in 2000-01 and its output is 117.433 metric tonnes in 1999-2000 and 21,586 in 2000-01 at the rate of 986 kilogram per hectare in 1999-2000 and 600 kilograms in 2000-01.

Gram **:** Gram is the most important winter pulse crop grown here. Its dal is very popular and liked much by the people. The green gram plant locally known as Jhangri is consumed in enormous quantity during the months of January and February. It is mainly cultivated as a diara crop. The total output of gram in the district in the year 2000-01 was 2,866 metric tonnes.

Khesari **:** Khesari is cultivated both as a full Rabi crop and also as a diara crop in the paddy fields in inter months in some pockets of the division. It is also used as fodder crop. The preparation of the land for sowing is similar to that of Mung and Kalai. The total output of Khesari in the division is 21,708 metric tonnes in 2000-01. Area of different crops (2000-01) is given in Table 6.7.

The cultivable area in the division also covers a minor produce of Barley, Masoor, Jwar, Bajra and Urad. The oilseeds grown in the division are linseed, rape and mustard and groundnut. Khagaria district is an important tobacco growing area.

Table 6.7 : Area Under Different Crops in '00 Hectares of the Munger division (1999-2000)

Name of Anchals	*Pa-ddy*	*Maize*	*Wh-eat*	*Bar-ley*	*Gram*	*Ar-har*	*Khe-sari*	*Ma-soor*	*Total*
Munger	1	57	4	6	20	2	1	1	91
Bariarpur	10	38	20	4	14	2	3	5	96
Jamalpur	4	21	8	9	6	-	1	2	51
Dharhara	42	8	25	2	25	2	2	1	107
Kharagpur	141	20	26	10	25	1	13	4	236
Asarganj	18	1	6	1	1	-	12	-	39
Tarapur	21	-	9	-	2	1	16	4	50
Tetiha Bambor	58	7	12	1	4	1	8	-	91
Sangrampur	23	5	9	1	3	1	6	-	48
Barahiya	5	43	33	3	20	1	9	33	147
Pipariya	16	14	13	1	6	1	8	4	63
Surajgarha	44	42	36	3	20	2	26	11	184
Lakhisarai	128	19	37	1	18	3	29	3	238
Ramgarh Chowk	55	1	4	1	5	1	8	-	75
Halsi	35	3	11	1	8	1	14	1	114
Barbigha	72	12	18	1	3	1	24	1	132
Sheikhopur Sarai	32	5	11	1	2	1	16	-	68
Sheikhpura	84	11	20	1	12	1	48	11	188
Ghat Kusumbha	40	8	14	1	5	1	22	5	96
Chewara	46	3	16	1	1	1	9	1	78
Ariari	56	4	18	1	2	1	12	2	96
Islamnagar Aliganj	92	6	18	1	8	1	12	1	139
Sikandra	98	7	20	1	10	1	14	2	153
Jamui	101	2	15	1	5	1	30	1	138
Barhat	45	4	4	2	1	1	3	-	60
Lakshmipur	48	5	5	2	1	1	5	-	67
Jhajha	72	29	16	9	6	1	1	-	137
Gidhaur	30	2	1	1	-	4	2	-	37
Khaira	85	13	19	3	6	-	2	-	130
Sono	72	31	12	4	3	2	1	-	125
Chakai	88	19	52	1	1	2	1	-	163
Alauli	55	86	74	13	14	1	15	-	260
Khagaria	67	81	69	21	14	3	10	-	271
Mansi	14	18	17	1	2	9	1	-	55
Chautham	38	38	40	4	6	1	2	-	131
Beldaur	28	64	34	4	1	3	1	-	133
Gogri	55	68	62	10	1	1	1	-	138
Parbatta	54	68	63	16	18	1	8	-	224

Source : Agriculture Department, Government of Bihar, 2001.

The horticulture comprises the cultivation of fruits, vegetables and flowers which grow in 592 hectares. Other important fruits are mango, litchi, guava, citrus, banana and other minor fruits. The mango commands the maximum hectares which is more common in the north of the Ganga and along its southern bank. About 20% of the existing hectares is under grafted varieties of early mid-season and late ripening types, viz., Bombai, Malda and Fazli, etc. and remaining 80% is being commanded by seedling and inferior types. The guava belt is situated on the north Ganga where the market is over-flooded with fruits during the peak season and in the southern side the hectares is not much except in certain pockets of Munger anchals.

The vegetable command nearly 7,300 hectares in the district including the root crops. The important vegetable tracts are Munger, Lakhisarai, Sheikhpura, Jamalpur and Khagaria. Potato occupies nearly 1,200 hectares. The other winter season vegetables extensively grown in district are onion, cauliflower, cabbage, brinjal, tomato, radish, carrot, turnip and spinach. The hot weather and rainy season vegetables are grown in abundance in every nook and corner of the district specially in pockets where surface percolation wells have been sunk by the Agriculture Department.

Rotation of Crops

The cultivators from times immemorial are conscious of the beneficial effects of rotation of crops. Crops are generally sown in rotation but there are certain tracts especially in the tal and diara areas where rotation is not strictly followed. In the tal area only Rabi crops are grown. Crop rotation maintains the fertility of soil by supplying organic matters and nitrogen, increase in the yield of crop, improve the quality of crops and the physical condition of soil help in the conservation of soil, control the incidence of diseases and pests, keep the land free from weeds and ensures constant employment of labour and livestock throughout the year. The crop rotation practice is generally followed by the cultivators in the division to ensure the desired yield of crops, though they do not understand the scientific principles underlying the rotation of crops. Some of the rotation of crops followed in the district are given below :

	Kharif	*Rabi*
(i)	Maize	Wheat, Barley, Mustard, Gram, Peas, etc.
(ii)	Late Paddy	Diara, Gram or Khesari.
(iii)	Early Paddy	Gram, Khesari, Peas, Wheat, Barley and Onion.
(iv)	Jowar for fodder	Wheat or Barley or Wheat and Mustard.
(iv)	Fallow	Chilli, Tobacco.
(vi)	Maize	Arhar.
(vii)	Maize	Sugarcane.

In most parts of the district cultivators generally sow mixed crops through open cast method especially in small holdings. The crops usually grown together are :

(i) Maize, Arhar and Turmeric;
(ii) Wheat and Gram;
(iii) Maize and Moong;
(iv) Barley and Gram; and
(v) Wheat and Mustard.

Hence, it is necessary to conserve the fertility of fields. But as the holdings are small, the farmer does not want to leave the land fallow for successive seasons. However, most fields left fallow in between the gap of Rabi and Kharif harvests.

The National Extension Service and Community Development Blocks have been spreading agricultural education to follow rotation of crops by using legumes and chemical fertilizers to improve the fertility of soil.

Crop Combination

On account of the over-dependence on rainfall, Indian agriculture is said to be a 'gamble in monsoon'. But agriculture is the predominant occupation of the people of the region.

Nearly 80% of the total population derives sustenance from agriculture as compared to 69% in India as a whole. Munger division forms a poor fertile agricultural tracts in India and grows a variety of crops.

The alluvial and upland plains of the Munger division which intervened with rivers and ridges are rich in all sorts of crops. In some tracts nothing but an enormous stretch of rice fields meet the eye, but in others the level plain is dotted with numerous clusters of bamboo and groves of mango and sisam trees.

Rice is the principal Aghani crop and its greatest concentration lies in a triangular belt between the Ganga and the Kosi in the north and mountainous parts of South Munger. Wheat and maize are the main Bhadai crops. They are predominant in the entire part of north Munger and Diara and Tal areas in the south along with Sangrampur, Lakhisarai, Ariari, Chakai and Khaira anchals.

Gram, khesari and other Rabi crops become more important in Munger, Dharhara and Jamalpur anchals.

A new method has been devised by R.B. Mandal for crop combination on regional level. The formula is as fallows :

$$\frac{HA}{1} - 1 = \text{1st ranking crop}$$

$$\frac{HA}{2} - 2 = \text{2nd ranking crop}$$

$$\frac{HA}{3} - 3 = \text{3rd ranking crop}$$

$$\text{C.C. } \frac{HA}{4} = - 4 = \text{4th ranking crop}$$

$$\text{Lower limit of diversification for good combination} = \frac{TA}{N}$$

where C.C. is crop combination.

HA is highest hectares of crop in the study area. 1, 2, 3 and 4 are denominators for different crops and -1, -2, -3 and -4 are diversification index in order to relegate lower area of crops.

TA is the total hectares of all crops and *N* is the number of crops. The computation of crop combination on the basis of above formula is as follows :

Munger Anchal

Paddy-1, Maize-57, Wheat-4, Barley-6, Gram-20, Arhar-2, Khesari-1 and Masoor-1. (in '00 hectares)

Thus according to *TA* in case of Munger anchal, is 91 hectares for 8 crops = 11.3 (average).

Hence the lowest limit of a crop to enter into the combination is 11.3 hundred hectares of land under particular crop. In such a situation only Maize and Gram will enter into the combination.

$\frac{57}{1}$ - 1 = 56; 1st ranking crop (Maize).

$\frac{57}{2}$ - 2 = 26.5; 2nd ranking crop (nil).

$\frac{57}{3}$ - 3 = 16; 3rd ranking crop (Gram).

Lakhisarai

Paddy-128, Maize-19, Wheat-37, Barley-1, Gram-18, Arhar-3, Khesari-29, Masoor-3. ('00 hectares).

Thus according to *TA* in case of Lakhisarai is 238 hectares for 8 crops = 29.7 (average).

Hence the lowest limit of a crop to enter into the combination is 29.7 hectares of land under particular crop. In such a situation only Maize and Gram will enter into the combination.

$\frac{57}{1}$ - 1 = 127; 1st ranking crop (Paddy.)

$\frac{57}{2}$ - 2 = 63; 2nd ranking crop (nil).

$\frac{57}{3}$ - 3 = 39; 3rd ranking crop (nil).

$\frac{57}{4}$ - 4 = 28: 4th ranking crop (Wheat)

Barahiya

Paddy-5, Maize-43, Wheat-33, Barley-3, Gram-20, Arhar-1, Masoor-33 ('00 hectares),

Sum of the area of all crops is 147 hectares.

$\frac{147}{8}$ = 18.3 (average for crops).

Therefore, crop ranking is

$\frac{43}{1}$ - 1 = 42; 1st ranking crop (Maize).

$\frac{43}{2}$ -2 = 19.5; 2nd ranking crop (Wheat, gram and masoor).

Hence Barahiya is a case of 4 crops association.

The result for other anchals are given in Table 6.8. The crop combination map is prepared according to Table 6.9 (Fig. 6.1).

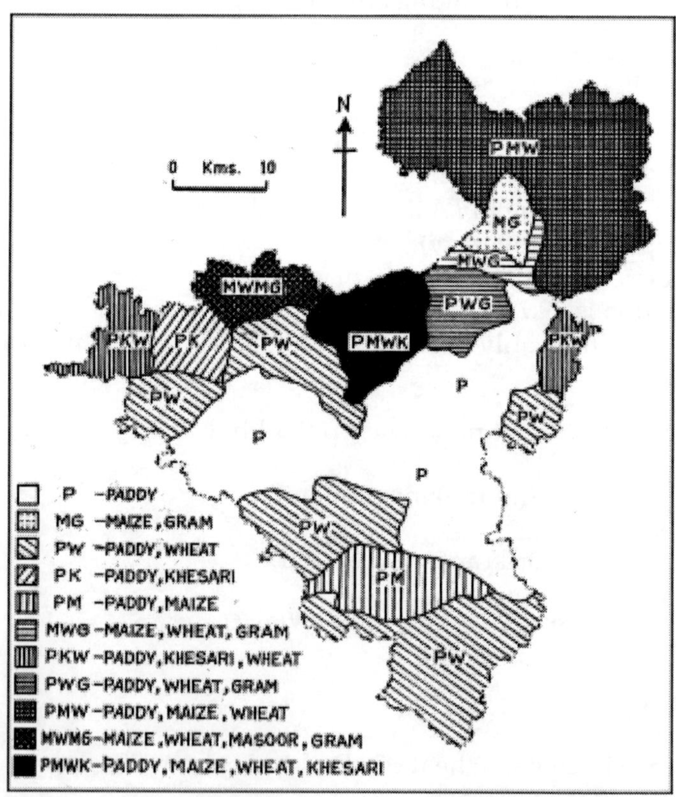

Fig. 6.1 : Munger Division: Crop Association Region.

Table 6.8 : Crop Association and Ranking in the Munger division (1999-2000)

Name of Anchals	*1st Ranking crop*	*2nd Ranking crop*	*3rd Ranking crop*	*4th Ranking crop*
Munger	Maize (57)	—	20g	—
Bariarpur	Maize (30)	Wheat (20)	14g	—
Jamalpur	Maize (21)	Barley (10)	Wheat (8) Gram (5)	Paddy (3)
Dharhara	Paddy (42)	Wheat (25), Gram (25)	—	Maize (8)
Kharagpur	Paddy (141)	—	—	—
Asarganj	Paddy (18)	Kesari (12)	Wheat (6)	Barley (1) Gram (1)
Tarapur	Paddy (21)	Kesari (16), Wheat (9)	—	Gram (2)
Tetiha Bambor	Paddy (58)	—	—	Wheat (12)
Sang-rampur	Paddy (23)	—	Wheat (9) Khesari (6) Maize (5)	Barley (1) Arhar (1)
Barahiya	Maize (43)	Wheat (33), Masoor (33), Gram (20)	—	Khesari (9)
Pipariya	Paddy (16)	Maize (14), Wheat (13), Keasri (8)	Gram (6) Masoor (4)	Barley (1) Arhar (1)
Surajgarha	Paddy (44)	Maize (42), Wheat (36), Kesari (26)	Gram (20)	Masoor (11)
Lakhisarai	Paddy (128)	—	—	Wheat (37) Khesari (29)
Ramgarh Chowk	Paddy (55)	—	—	—
Halsi	Paddy (85)	—	—	—
Barbigha	Paddy (72)	—	Khesari (24)	Wheat (18)
Sheikhopur Sarai	Paddy (32)	Kesari (16)	Wheat (11)	Maize (5)
Sheikhpura	Paddy (84)	Kesari (48)	—	Wheat (20)
Ghat Kusumbha	Paddy (40)	Kesari (22)	Wheat (14)	Maize (5)
Chewara	Paddy (46)	—	Wheat (16)	Wheat (20)
Ariari	Paddy (56)	—	Wheat (18)	Maize (8)
Islamnagar Aliganj	Paddy (92)	—	—	Khesari (9)
Sikandra	Paddy (98)	—	—	Khesari (12)
Jamui	Paddy (101)	—	Khesari (38)	—
Barhat	Paddy (45)	—	—	—
Lakshmipur	Paddy (48)	—	—	—
Jhajha	Paddy (72)	—	Maize (29)	Wheat (16)
Gidhaur	Paddy (30)	—	—	—

(Contd.)

Name of Anchals	1st Ranking crop	2nd Ranking crop	3rd Ranking crop	4th Ranking crop
Khaira	Paddy (85)	—	—	Wheat (19)
Sono	Paddy (72)	—	Maize (31)	—
Chakai	Paddy (88)	Wheat (52)	—	Maize (19)
Alauli	Paddy (86)	Wheat (74), Paddy (55)	—	—
Khagaria	Maize (81)	Wheat (69), Paddy (67)	—	Barley (21)
Mansi	Maize (18)	Wheat (16), Paddy (14)	—	Barley (21), Gram (2) Khesari (2), Arhar (2)
Chautham	Wheat (40)	Paddy (38), Maize (38)	—	Gram (6)
Beldaur	Maize (64)	Wheat (34)	Paddy (28)	—
Gogri	Maize (68)	Wheat (62), Paddy (55)	—	—
Parbatta	Maize (68)	Wheat (63), Paddy (68)	—	Gram (18), Barley (16)

Figure 6.2 and Table 6.10 show the flood and drought affected areas in the Munger division. There is a clear cut division of flood and drought affected areas in which northern part is a flood prone area whereas southern part is drought affected one.

The flood affected anchals are Munger, Jamalpur, Dharhara, Kharagpur, Surajgarha, Barahiya, Khagaria, Alauli, Chautham, Beldaur, Gogri and Parbatta. All these anchals are chronically flood affected where occasional flood affects a larger portion of land in all these anchals. In 1976-77 in the Munger division chronically flood affected area was 46,511 hectares whereas occasionally affected area was 37,546 hectares.

The drought affected anchals are Halsi, Sikandra, Jamui, Khaira, Lakshmipur, Jhajha, Sono, and Chakai. There is also a division of annually and occasional drought affected areas. Annually affected area occupies 41,500 hectares and occasional affected area 27,500 hectares.

In northern part of Munger due to the meeting point of several streams like the Ganga, the Harohar, the Kiul, the Burhi-Gandak, the Kosi, the Kareh, the Ghugri, etc. flood water accumulates during the rainy season. This excess of water which accumulates over the ground come from the vast tracts of North Bihar and Himalaya and Hazaribagh plateau in the

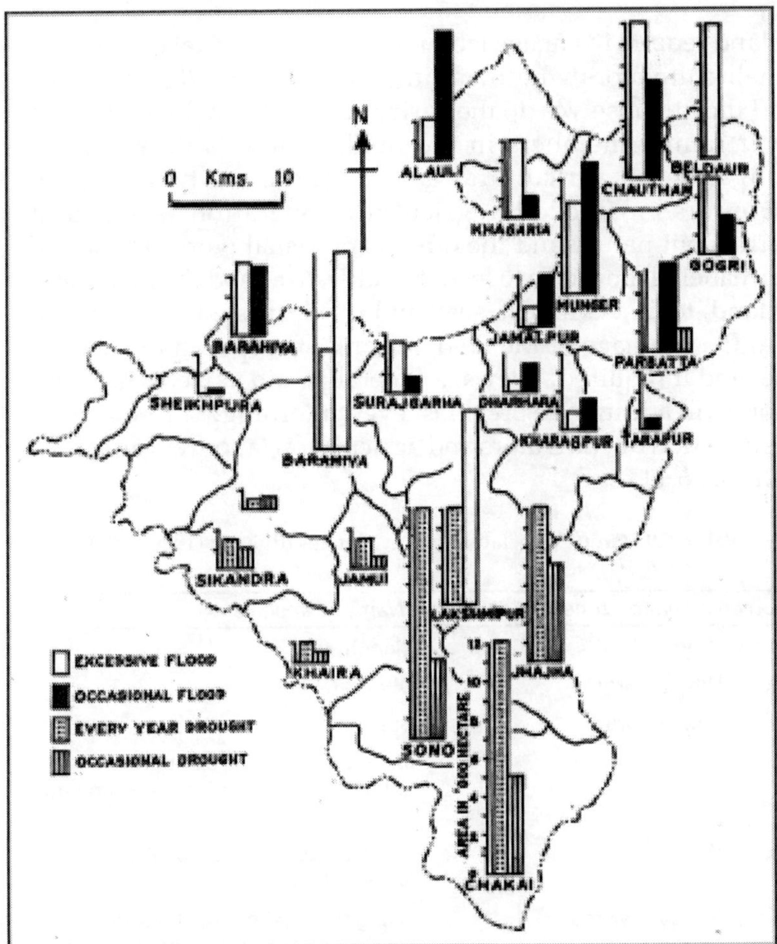

Fig. 6.2 : Munger Division : Distribution of Flood and Drought Affacted Area.

South. On the other hand, the southern portion of Munger is an upland plain where only rainfed streams are found. Hence, southern part is a drought affected area where the Government is concentrating on the construction of several irrigational schemes as a measure of drought planning area. Generally, landless people who reside in villages offer themselves for agricultural labour. The cultivators who possess a large area of land require to engage labour for the agricultural operations. Small cultivators who own little land do not engage labour and they themselves do the agricultural work with the help of their family members in the fields. There is no statutory obligation on agricultural labour. There are two types of labourers engaged for agricultural operations—one is of permanent nature and the other is of casual work. The job of permanent labourer is to feed the cattle, to attend the ploughing of land, to cast compost, seed and the harvested crop, to look into the drainage of the field, sowing of crops, and irrigating the land if required. They are generally paid a fixed salary per month, in addition to breakfast every morning. The blockwise distribution of cultivators and agricultural labourers are given in Table 6.11.

Table 6.9 : Crop Association in the Munger division (1999-2000)

Crop combination zone	*Name of anchals*
Maize, Gram	Monghyr
Paddy, Wheat, Gram	Dharhara
Maize, Wheat, Gram	Bariyarpur
Paddy	Jamui, Kharagpur, Sikandra, Lakshmipur, Halsi, Jhajha, Ramgarh Sarai
Paddy, Wheat	Sangrampur, Lakhisarai, Ariari Chakai, Khaira
Paddy, Khesari, Wheat	Asarganj, Shekhpura, Tarapur, Barbigha, Ghat Kusumba, Lakhisarai
Maize, Wheat, Masoor, Gram	Barahiya
Paddy, Maize, Wheat, Khesari	Surajgarha
Paddy, Khesari	Jamui
Paddy, Maize	Sono
Paddy, Maize, Wheat, Kesari	Surajgarha, Pipariya, Sangrampur
Paddy, Maize, Wheat	Khagaria, Gogri, Parbatta, Alauli, Chautham, Beldaur, Mansi

Table 6.10 : Statement Showing Flood and Drought Affected areas of Munger division (1999-2000)

Name of the Blocks	*Flood affected*		*Drought affected*	
	Chronically affected	*Occasionally affected*	*Every year*	*Some year*
Pipariya	1,463	500	—	—
Monghyr	74,269	6,800	—	—
Bariyarpur	1,000	2,734	—	—
Dharhara	500	1,435	—	—
Kharagpur	809	1,605	—	—
Tarapur	—	500	—	—
Surajgarha	2,610	800	—	—
Barahiya	3,712	3,500	—	—
Sheikhpura	—	200	—	—
Khagaria	4,000	1,000	—	—
Alauli	2,200	6,860	—	—
Beldaur	8,000	—	—	—
Parbatta	4,100	4,400	—	—
Halsi	—	—	500	500
Sikandra	—	—	1,500	1,000
Jamui	—	—	1,500	500
Khaira	—	—	1,000	10,000
Lakshmipur	—	—	5,000	10,000
Jhajha	—	—	8,000	5,000
Sono	—	—	12,000	4,000
Chakai	—	—	12,000	5,000
Total	46,511	37,546	41,500	27,500

Agriculture Market

In any planned economic development programme, exchange of goods assumes a very important role in maintaining an equilibrium between production and consumption. The importance of marketing agricultural produce is subject to innumerable natural and economic limitations and is, therefore, of paramount importance, especially in the district of Munger. Agricultural marketing is one of the manifold problems which have a direct bearing upon the prosperity of the cultivator in the district. Agricultural marketing in its widest sense comprises of all the operations involved in the movement of food and raw materials from the farm to the final consumer. It

Table 6.11 : Distribution of Cultivators and Agricultural Labourers (2000)

Name of Anchals	*Total workers*	*Cultivators*	*Agricultural labourers*
Munger	71561	6091	14390
Bariarpur	29120	4818	15342
Jamalpur	43384	3807	7638
Dharhara	31894	6042	16068
Kharagpur	60843	11256	31835
Asarganj	21154	4528	11277
Tarapur	26894	5716	14097
Tetiha Bambor	19712	5648	9922
Sangrampur	26889	7639	15413
Barahiya	43444	11232	22637
Pipariya	10786	4873	4389
Surajgarha	78856	21778	33233
Lakhisarai	97732	29049	10697
Ramgarh Chowk	25671	38320	12406
Halsi	36145	16904	16475
Barbigha	38470	10201	17962
Sheikhopur Sarai	21437	8237	9515
Sheikhpura	57799	19079	20779
Ghat Kusumbha	15660	5240	8817
Chewara	23543	10488	10870
Ariari	37487	19736	14073
Islamnagar Aliganj	44218	18552	12440
Sikandra	49283	17001	24512
Jamui	62628	15913	26728
Barhat	30375	9534	15528
Lakshmipur	47200	17802	18747
Jhajha	96619	17071	6536
Gidhaur	27632	15442	9429
Khaira	83978	27102	25229
Sono	80426	19617	17214
Chakai	74258	31496	26072
Alauli	86328	23884	53454
Khagaria	95767	18723	47236
Mansi	24242	5458	11184
Chautham	48088	14410	27616
Beldaur	63280	18802	38785
Gogri	82791	19342	41801
Parbatta	66744	20106	33821

Source : District Statistical Office, 2000.

includes the handling of the product at the farm, initial processing, grading and packing in order to maintain quality and avoid wastage. Storage is another important feature of agricultural marketing, together with the method of packing and presenting the product to suit the requirements of final consumers. Hence, an efficient marketing system is of vital importance to a country under all conditions, at each stage, in its development.

At present South Munger and North Munger both are not agriculturally developed due to its peculiar situation and agro-climatic condition. Since it has been prepared to take up some important items of agricultural development in the region as for facility for irrigation, soil conservation and allied agricultural industries, it has become necessary to develop marketing side of the agricultural production in order to ensure farmers due share of the price paid by the traders and consumers and also to get rid of the unscrupulous traders who at every stage victimise the innocent farmers.

With the aim in view, some of the important markets of the region, should be taken up under the Bihar Agricultural Produce Market Act, 1960.

For the present, following primary markets are being selected under the above Act :

Name of Anchals	*Name of the Market*
Jamui	Jamui
Jhajha	Jhajha
Sikandra	Sikandra
Aliganj	
Kharagpur	Kharagpur
Tarapur	Asarganj Tarapur
Sangrampur	Sangrampur
Jamalpur	Bariarpur
Surajgarha	Surajgarha
Barahiya	Barahiya
Barbigha	Barbigha
Sheikhpura	Sheikhpura
Parbatta	Parbatta
Alauli	Alauli
Chautham	Chautham
Beldaur	Beldaur

All these are primary markets and farmers of the neighbouring villages bring their agricultural produce into these markets for disposal.

It is proposed to acquire 3 hectares of land for the construction of a market yard which will contain shops, godowns of traders, where the cultivators would sell their produce. Other shops would be meant for the sale of agricultural inputs such as seeds, fertilizers, insecticides, implements including consumer needs. Some other building structure will be sub-platform, office building, office of the market committee, cattle shed, nightshed for the farmers with feeding for the cattle, etc.

Livestock

Munger division is chiefly an agricultural tract where livestock forms an important part of agricultural economy. Total breedable cows and she buffaloes of the area are 60,950. The cows and she-buffaloes of the area are potentially good as the average milk yield of the cow is 3 kg. and that of she-buffalo is 4 kg. as is evident from the milk recording done under key village scheme of the Animal Husbandry Department. There are several Cattle Development Schemes operating in the area for a long time. During the pre-plan period some Haryana Bulls were located in the Munger Veterinary Hospital for natural breeding. In the Fourth Five-Year Plan key village scheme, and scheme for All India Centres like Bariarpur, Jamalpur and Munger are under implementation. Munger proper is a good market for milk produce.

There is one Veterinary Hospital at Munger and 4 Veterinary Dispensaries at Nowagarhi, Surajgarha, Jamalpur and Sultanganj. There are 16 Field Veterinary Centres in the area. In each command development Blocks there is provision for one block Animal Husbandry Officer whose main function is to implement the Animal Husbandry and marginal farmers scheme in the area who supplement their income by keeping cows and she-buffaloes. Even well-to-do farmers of the locality are practising mixed farming and possess high yielding cows and she-buffaloes. There is a good scope for marketing of milk in this area as Munger town itself is growing very fast a milching centre and the present human population of Munger and Jamalpur is about 197,000. Two fairly big industries (Tobacco Factory, Munger and Jamalpur Loco Works) are located in the area. The Government of Bihar has established a dairy plant at Barauni and its handling capacity is 1 lakh litres of milk per day. There would be no difficulty in the disposal of surplus milk to the said plant after meeting the local demand.

It is, therefore, proposed to establish one medium size Intensive Cattle Development Project with a milk plant in this area for producing more and more milk by adopting cross-breeding and by giving incentives for cattle development to farmers which in turn will improve socio-economic condition of the farmers. Under the I.C.D. Project, 50,000 breedable cows and she-buffaloes will be covered through artificial insemination technique. The provision for the following has been made under the project :

1. Training of farmers.
2. Breeding of cows.
3. Veterinary aid and diseases control.
4. Feeds and fodder development.
5. Cooperative societies.
6. Registration and milk recording.
7. Introduction of high yielding milch cattle.
8. Dairy extension work.
9. Assessment and evaluation.
10. To prepare an organizational set up.

1. Training of Farmers

It is a well known fact that there are certain places like Khaira, Munar, Bangalore, etc. in India which have made considerable progress in the field of Animal Husbandry by adopting improved scientific practices. These places can be good training ground. It is, therefore, proposed to send 250 enterprising farmers @ 50 farmers per year to receive practical training for

10 days. These farmers will get the benefit of actual travelling expenses and daily allowance of Rs. 15 per day. For better supervision one experienced Officer of the Animal Husbandry Department will accompany them. These farmers will be acquainted with the improved method of dairy development under field condition which will enable them to acquire proper scientific background for breeding cattle wealth on scientific lines.

2. *Breeding*

As stated earlier, 50 thousand of cows and she-buffaloes are to be brought under controlled breeding programme under this project for which a central scheme of collection centre will be established with 16 Bulls of Jersey breed and 8 Bulls of Murrali breed. To put more milk in the local cattle cross-breeding with bulls of exotic breed will be practised. The local she-buffaloes will be up-graded with buffaloes, bulls of Murrah breed. There will be 50 breeding centres which will be called stocking centre each catering to the breeding needs of one thousand cows and she-buffaloes.

3. *Veterinary Aid and Disease Control*

There is one Veterinary Hospital at Munger and Veterinary Dispensaries at Nowagarhi, Surajgarha and Jamalpur. There are also 16 dispensaries in the area. These institutions will cater to the animal health needs of the area. Provision has also been made for medicines, sera and vaccines also. In addition, stocking centres to be established under project will also attend to animal health work.

4. *Feeds and Fodder Development*

Cross-breeding programme needs massive support of fodder production programme. It is necessary to provide assistance for free supply of planting materials. An arrangement will be made to supply balanced cattle feed to farmers on no loss no profit basis. Initially readymade feed concentrates will be purchased from Amul or other agencies. Subsequently feeds will be made available from the feed mixing plant given to small and marginal farmers for cross-breeding calves.

5. Cooperative Societies

Hundred milk cooperative societies will be organised in the area. Dairy plant will arrange to collect milk from the milk cooperative societies.

6. Registration and Milk Recording

To identify the breedable cows and she-buffaloes in the area, an initial survey will be conducted, and milk recording will be regularly done. Detailed record of each breedable cow/she-buffalo will be maintained.

7. Introduction of High Yielding Milch Cattle

Since cattle development programme is of long gestation, it is proposed to induct high yielding cows and she-buffaloes in the area possibly from outside the state, so that there may not be reshuffling of the milch animals within the state itself. Institutional finance will be utilised for advancing loans to the farmers for the purchase of high yielding cows and she-buffaloes.

8. Dairy Extension Work

To identify the problems of individual farmer, provision has been made for dairy extension work so much so, that, the farmers may get all possible help in time in respect of veterinary aid, disease control, disposal of milk, cattle feed and loan for purchase of animals.

9. Assessment and Evaluation

To assess the initial level of production of milk and other, animal husbandry inputs are available in the area for which a benchmark survey has been conducted in the beginning. Subsequently repeat surveys will be carried out at regular intervals to assess the progress made under the programme in various directions. For this purpose, provision has been made in the scheme for a statistical unit with necessary staff.

10. Organisational Set up

The project will be headed by a project officer who will be in the rank of special officer in special class I of Bihar Animal Husbandry Service. He will be assisted by different specialists and office staff for which provision has been made in the project.

Dairy Development

For procurement, transport, processing and supply of milk to the citizens of Munger and Jamalpur, a milk supply scheme will be established at Munger. A milk processing plant with a handling capacity of 25,000 litres per day will be set up at Munger.

Surplus milk if any, will be diverted to the composite milk plant, Barauni, where it will be converted into milk powder and other milk products.

As the production points are situated within the radius of 30 kilometres, therefore, provision has not been made for chilling centre. It is proposed to set up 50 assembling centres with an average collection of 500 litres per day, per centre, to feed the central dairy.

The proposed milk collection area will be as follows :

1. Munger to Surajgarha.
2. Munger to Sitakund.
3. Munger to Sultanganj, Kharagpur and Tarapur.
4. Munger to Jamalpur. All these routes have all weather roads.

Table 6.12 : Blockwise Break up of Diara Land

(Area in Hectare)

Name of the Block	*Total Diara Land*	*Total Cultivable Diara Land*	*Total Irrigation in Diara*
Munger	15,213	9,059	840
Bariyarpur	11,258	4,057	316
Pipariya	1,720	1,480	540
Surajgarha	3,528	2,710	15
Barahiya	9,999	5,800	340
Khagaria	2,726	1,684	182
Gogri	12,165	568	99
Parbatta	2,080	1,000	234
Total	**59,142**	**26,509**	**2,567**

It will provide direct employment to about 128 persons and indirect employment to 1,500 persons.

The total outlay of the dairy scheme is 103 lakhs. This scheme will be operated under the cooperative sector. The government share will be Rs. 10 lakhs, another 10 lakhs will be the contribution of the cooperative sector and the remaining 83 lakhs will be arranged through different financial agencies.

Problems

Problems of Diara areas regarding agriculture are as follows :

(a) A new cropping pattern is to be developed for this newly created area having irrigational facilities. Till now no research has been done towards this aspect.

(b) There is no scientific recommendation available for enhancing the production in cucumber growing zones of diara area. Parwal is the main crop besides cucumber. These crops occupy an area of about 3,000 hectares.

(c) Sweet-potato is also grown in 1,500 hectares.

(d) During floods, there is no fodder available for cattle wealth.

(e) When flood recedes, soil remains cultivable only for a brief period.

Suggestions

To combat the existing problems, the following steps are suggested :

(a) A 50 to 60 hectares investigation farm needs to be established in the heart of diara-area in order to have practical view and experience of the diara problems and to develop ways to solve them, for both irrigated and non-irrigated conditions given in an agricultural farm (Figure 6.3);
(b) Besides food crops, research works on green fodders should also be taken up, which can be made available to farmers especially during floods;
(c) Works on vegetable, parwal, misrikand and sugarbeet should also be undertaken which are the cash crops of diara area;
(d) Researches on growing of chillies need to be initiated under diara conditions;
(c) Farm practices should be developed taking into consideration the prevalent agro-ecological situation of the diara land;
(f) Storage problems during floods needs to be solved; and
(g) Cattle rearing, fodder management, animal health and other allied aspects should he given due attention.

Prior to the coming of the Operational Research Project in Taufir Diara, Munger following traditional crops were being grown by the cultivators of that area, with practically very little or no irrigation or fertilizers in different seasons of the year.

Summer

During summer, cultivation of small millets like Cheena, Khesari, Sewan was done in tiny plots near homestead land where irrigation was available. The row of the lands, after Rabi harvest, were left fallow where stray cattle used to roam.

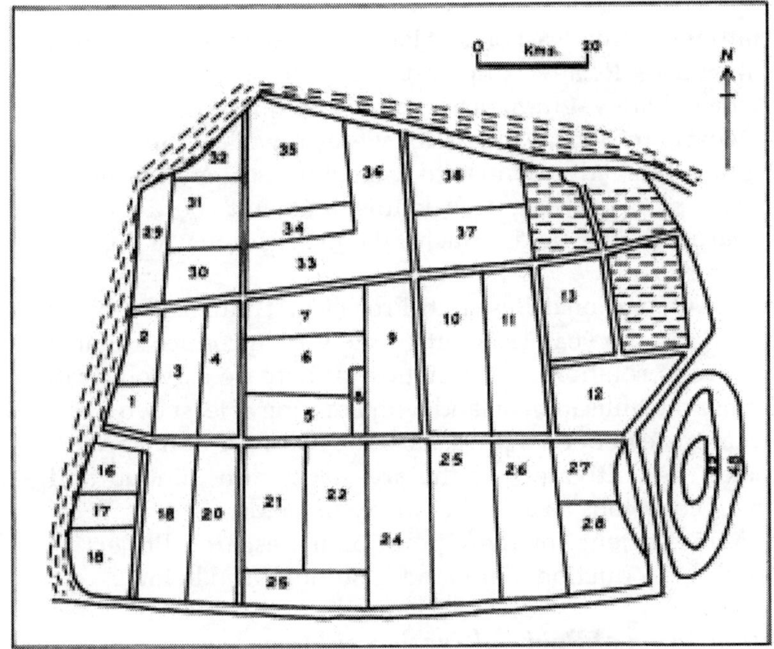

Fig. 6.3 : Munger Division : Agricultural Farm.

The sand soil near river beds, cultivation of cucurbits like melons, parwal and sweet-potato, all sown during the month of November-December, was being practised.

Kharif

With the onset the first shower which generally comes in the month of May and June, a local short duration, low yielding variety of maize known as "Tulbulia" was sown. However, the cultivators were able to harvest a full crop even of this short growing variety only once in three or four years as the crop at the various stages of maturity was damaged by floods.

Rabi

The main cropping season for the diara area was Rabi, with recession of floods in the last week of September, kalai and mustard were broadcast on wet lands near abandoned streams or in their beds. Relatively uplands were left for sowing of crops like wheat, barley, gram, peas, lentils and linseeds, etc.

Most of the wheat and barley were grown as a mixed crop with gram. Linseed and mustard were either scattered as mixed crop with plots growing peas, lentils or sown as border crops of wheat and barley fields. Practically no irrigation was applied to these crops.

The Operational Research Project in Taufir Diara came into being in the year 1975 with a view to introduce suitable crops, crops rotation and varieties to increase as well as to maximise the utilisation of land with growing at least two crops in a year instead of one. It was also the objective of the project to make the cultivators of the area irrigation minded and fertilizer conscious as an aid to increase yields.

Achievements of the Operational Research Project in improving production diara land is summarised below :

1. *Introduction of Rabi Cultivation of Maize*

Winter maize cultivation was unknown in the diara area. Being influenced and impressed by the Project trials, the cultivators have taken up the cultivation of winter maize like Hi-starch, Laxmi, Ms, Swan, Pool 17 and Diara composite giving yields of 40 to 65 quintals per hectare. This year, about 400 hectares of land have been brought under winter maize by the cultivators themselves as against 20 hectares of last year and practically nil year before last. This area under maize is expected to further increase in the coming years.

2. *Replacement of the local 'Tulbulia' by higher yielding Diara Composite in summer*

Diara composite maize, having a yield of 30 to 40 quintals per hectare, has replaced 'Tulbulia' the local maize variety yielding 12-15 quintals per hectare. An area of about 1,000 hectare was covered by Diara composite during last year under expansion programme. The cultivators were also induced to sow maize in April just after harvest of Wheat to ensure harvest before the onslaught of flood. This idea is gaining ground and more and more bamboo boring is being sunk by the cultivators for this purpose.

The cultivators are also now using recommended dose of fertilizer both for winter and summer maize as they have realised the practical benefit of fertilizer application.

3. *Introduction of Improved Varieties of Wheat*

The better performance of High Yielding Varieties in the project trials have led to the cultivators to sow varieties like HP 1102, HD 1533, Janak and RR 21 with recommended dose of fertilizers and irrigation. In 1975 an area of about 450 hectares has come under the high yielding varieties. Further, the cultivators have abandoned the area where irrigation is not available in order to grow mixed crop of wheat and gram.

4. *Introduction of Better Yielding Varieties of Oilseeds*

Cultivation of only low yielding local varieties of mustared was done in the diara area previously. Only last year, the superiority of "Varuna" variety of rai was in demonstration farm and this year more than 20 hectares of this variety have been sown by the cultivators and this area is expected to increase in the coming years.

5. *Increasing Irrigation Facilities*

There were only about 50 bamboo borings prior to the coming of the Project. The number of bamboo borings have considerably increased in those areas where cultivators have taken tip growing winter maize, improved and high yielding varieties of wheat Diara composite maize. Such bamboo boring is being sunk in increasing numbers by the cultivators with the aid of different agencies of which Commercial Banks are leading ones.

6. Bringing out Importance of Plant Protection Measures

Previously, cultivators did not attach any importance to plant protection measures to save their crops against pests and diseases. But now, being impressed with plant protection trials conducted on different crops, they have realised the beneficial effect of pesticides in increasing crop yields and have started using Aldrin 5% dust, BHC 5%, Aldrin 30 E.C., Sevin 4G, Rogor, Dithane M45, etc. against pests and diseases of crop. They are also using zinc phosphide and aluminium phosphide for rat control. The cultivators have been enlightened to some extent that they are asking the district authorities for opening of Plant Protection Centre in Diara areas. Besides, they have been trained for using Malathion 5% dust and aluminium phosphide tablets for safe storage of their seeds and grains.

Present Activities of the Project to Increase Crop Production

Activities of Operational Research Project in this direction are as follows :

1. Introduction of Suitable High Yielding Rainfed Wheat and Barley

Out of 4,000 hectares of cultivable land, only twenty per cent land has yet come under irrigation. For rest of the area which is unirrigated at present, field trials on improved high yielding varieties of rainfed wheat and barley has great possibility. Trials with high yielding rainfed varieties of wheat and barley are in progress.

2. Summer Paddy for Local Consumption

No paddy in any season was grown in Diara area but people are desirous to grow some paddy for their home consumption. Hence, during summer 1979, three Pilot trials on paddy variety Pusa 2-21 were conducted and yields of 30-39 quintals per hectare were obtained. This has caught up with the imagination of cultivators and they are asking for more and more varieties suitable for their lands. This year, trials with different short duration rice is being contemplated during summer, specially in the low lying areas where Rabi cannot be grown due to prolonged wetness.

3. *Summer Groundnut as a Cash Crop*

During summer 1999, one pilot trial on groundnut (1C-12-24) was conducted which gave an yield of 17.5 quintals per hectare. So, there is greater crop as well as demand for groundnut cultivation in Diara land. Trials with different varieties of groundnut both as pure crop and as inter crop in winter maize are being tried.

4. *Testing for High Yielding and Increasing Yields of Gram and Peas with Rhizobium Inoculation and Fertilizers*

Last year, trials on gram and peas were practically lost due to heavy winter rains. However, some of the varieties were found promising even under this condition. There was also a marked effect of inoculation. Trials are being conducted this year also with more varieties of grain and peas with improved aeronomical practices to ward off damage due to rains as well as to increase yields.

Extension of Area Under Summer Pulses Trials on summer moong and kalai, done for the first time, revealed good results, yielding 8 quintals and 9 quintals per hectare, respectively. This new introduction has found favour with cultivators and on their demand, more trials with moong and kalai during summer have been proposed.

5. *Introducing New and High Yielding Varieties of Parwal*

Parwal is an important cash crop for this area. To find out the most suitable varieties, 5 varieties of Parwal were tried last year. Nimia variety among the five, was found to be the best, giving an yield of 56.7 quintals per hectare. The variety is getting popularity among Diara cultivators. Some other new varieties have been included in trials this year.

6. Testing for Better and High Yielding Varieties of Cucurbits

Field trials with different varieties of kaddu (bottle gourd) and karela (bitter gourd), have been tried last year and are being tried this year also to find the best varieties under Diara condition.

7. Trials on Water Melon and Musk Melon

Water melon and musk melon are important cash crop of Diara area. Trials with improved varieties like Durgapur Mitha and Sugarbaby for water melons and two varieties of musk melon have been laid this year to test their suitability under Taufir conditions.

8. Introduction of Improved Varieties of Sweet Potato

Uptil now, only local low yielding varieties of sweet potato is grown. Trials with seven improved varieties of sweet potato against the local have been introduced to test the suitability of new varieties in the Diara areas.

9. Pilot Trial on Potato Cultivation

Pilot trials to test the suitability of different varieties of potato has been laid out and regular trials will follow if varieties will be found unsuitable.

10. Pilot Trials with Summer Moong, Dalai and Groundnut as Cash Crops with Winter Maize

Winter maize has been sown and kalai, moong and groundnut will enter as cash crops, after providing last irrigation at tusselling stage. These cash crops will be safely harvested on full maturity well ahead the coming of flood.

11. Trials on Plant Protection

On the basis of last year's finding, trials have been simplified. Each trial has been divided into two sets. In one set, all possible plant protection measures are taken and in another set, no protection is given to the crops against pests and diseases.

(a) Plant Protection Trial on Maize

In protected set, soil application with Aldrin 5% dust at the time of sowing, spraying with Aldrin 30 E.C. + Dithane M45 and granular application of seven 14 G have been given against various pests and diseases of maize. No plant protection measures have been taken in unprotected set so that the cultivators may see the beneficial effect of plant protection measures.

(b) Plant Protection Trial on Wheat

Soil application with Aldrin 5% dust at sowing time and spraying with Aldrin 30 E.C. have been given and spraying with Endosulfan is to be given in protected set as against no plant protection in unprotected set.

(c) Plant Protection Trials on Pulses

(i) ***Pea*** : Soil application with Aldrin 5% dust and spraying with Sumithion against pests and spraying with Karathane and Bavist in against powdery mildew have been given in protected set as against no plant protection in unprotected set.

(ii) ***Gram*** : Soil application with Aldrin 5% dust and drenching and Dithane M45 should be given and two spraying with Malathion 50 EC, one at flowering stage and second at pod stage will be given against pod borer in protected set as against no plant protection measures in unprotected set.

If we succeed in protecting pea against powdery mildew and gram against root rot and stem rot, production of both the pulse crops will be considerably increased.

Rat Control Drive

There has been heavy infestation of rats due to negligible flood in Diara this year.

Compact demonstration in rat control on newly sown maize and other Rabi crops has been done in about 100 hectares by the application of aluminium phosphide in living rat holes and baiting with zinc phosphide and Warfarin.

The Munger division grows a varieties of crops in different seasons and regions according to the extent of cultivation and improvement in techniques of irrigating the land. In South, the cultivation of rice and other inferior varieties of crops are important, whereas in North Munger the cultivation of rice and maize is important in supporting the high population density of the area. In Diara land under Operational Research Project, the cultivation of new varieties of crops is under current but due to lack of irrigation the same is not giving a good dividend.

REFERENCES

Buck, J.L., Land Utilization in China, London, Vol. 3, 1937.

Ghosh, A.B., "Some Aspects of Input Technology in Agricultural Production", Presidential Address at the 66th Session of Indian Science Congress, 1979, p. 1.

Gupta, A.K., "Sustainable Development of Indian Agriculture : Green Revolution Revisited", Indian Institute of Management Ahmedabad-380 015 (India), 1990, W.P. No. 896.

Mamoria, C.B., 'Agricultural Problems of India', Kitab Mahal, Allahabad, 1973.

Quintal, Oswald "Agriculture and Sustainable Development".

7

Water Resources and Sustainable Development in New Millennium

In socio-economic development of the country, the conservation and effective utilization of water resources, have played a vital role. Whether it is surface water or underground water, its utilization for irrigational purposes, power generation, industrial consumption and for domestic purposes is imperative for planned development of our economy. It is also essential to identify problem areas keeping both long term and short term planning perspectives in view. Water resource planning and management would be desirable with special reference to National Water Plan in collaboration with hydrologists, civil engineers and planners.

The historical accounts of Vedas, Smritis and other Classics, including Ramayana and Mahabharata and the subsequent works also indicate a scientific approach to water resource management in the past.

The Sustainable Development of Water Resources

The sustainability of water is critical to local, regional, national, and global security. It is impossible to think of a resource more essential to the health of human communities or their economy than water. Water runs like a river through our lives, touching everything from our vigour and the fitness of natural ecosystems around us to farmer's fields and the production of goods we consume (Adler, 2002). Inequitable access to resources, like water, cause poverty and environmental degradation that results in human conflict. And with conflict comes regional and national disputes that can best be alleviated by the sustainable use of these resources.

Unfortunately, societies worldwide have not always appreciated the need to protect adequate sources of freshwater. Water consumption has nearly doubled since 1950 (UNESCO, 2003). Thus, much of the world suffers greatly from inadequate access to potable water. According to United Nations statistics, more than 200 million people every year suffer from water-related diseases, and about 2.2 million of them—mostly the poor—die. About 20% of the Earth's population of 6.2 billion lacks access to safe drinking water (Hall, 2003).

The human demand for water has been particularly devastating to natural ecosystems, such as wetlands, lakes, and rivers. For example, globally the world has lost half of its wetlands, mostly in the last 50 years. One-fifth of the world's freshwater fish—2,000 of the 10,000 species identified so far—are endangered, vulnerable, or extinct (GreenBiz.com, 2003).

Besides being an integral part of the ecosystem, water is a social and economic good. Demand for water resources of sufficient quantity and quality for human consumption, sanitation, agricultural irrigation, and industry will continue to intensify as populations increase and as global urbanization, industrialization, and commercial development accelerates (Flint and Houser, 2001). The conservation of biodiversity, aquatic habitats, and complete ecosystems will likewise, from an ecological perspective, demand requirements for water resource planning and management, such as the need to maintain minimum in-stream flows and to anticipate the impact of hydrologic modifications on downstream environments (Flint, et al., 1996). Therefore, it is critical that management efforts intended to be sustainable fully consider the health and operation of aquatic ecosystems on which they are based, and that the environmental value of watersheds be recognized when making decisions on water allocation and use.

This presentation provides an overview of how to approach the sustainable use of water resources from a holistic perspective by considering a set of principles that can promote a multidimensional way of protecting water resources and achieving their recovery to improve quality of life for everyone. Discussion will focus upon how achieving sustainable development requires groups and organizations to pursue an evolving and ever-changing programme of activities, including a continuous process of evaluating current and emerging trends, an ongoing means of encouraging citizen participation and negotiating conflicts, and an updating of plans. And because plans for Water Security in the 21st Century are intended to reflect substantive policy outcomes, principles of sustainability will be defined to guide future water resources management actions. With these guiding principles a framework will exist for incorporating the concept of sustainability into decision-making at all levels of the public and private sector that finds itself focused on watershed-wide development issues.

New requirements for natural resource assessments, stipulated for example in the 1992 Rio Conference, Agenda 21, include : (a) integrated and timely access to data and information from many different sources and disciplines; (b) analysis of environment-development interactions and policy/ management options; (c) identity of cause and effect relationship as well as emerging issues of potential international importance; and (d) assessment of potential impacts and long-term sustainability of alternative development, policy, or management scenarios. To address these needs, we must develop and implement new frameworks for organizing and integrating water resources environmental information, and new tools to assess the information and communicate results to decision-makers. The framework discussed here for achieving sustainability of water resources will include the following elements :

(1) articulating a general concept for what sustainable water resources means;
(2) identifying the overall goals for developing sustainable water resource strategies;
(3) defining the characteristics that identify criteria for sustainability; and

(4) delineating the indicators that will tell us how we are doing in achieving sustainability criteria.

The sustainable development of water resources is a multi-dimensional way of thinking about the connections among natural, social, and economic systems equally and simultaneously in the use of water (the three overlapping circles model). For example, consideration of the value of ecosystem services is a method for blending social sciences and environmental science to better understand the economic benefits of watershed protection (Flint, 2003). Integration across sectors will be essential to meet growing challenges of the 21st century. In fact, as population and economic growth place increasingly greater demands on finite aquatic resources, it is doubtful that many activities can afford not to integrate. Likewise, sustainability advocates consideration of space and time scales. It recognizes interactions among different geographical ranges—globally, nationally, regionally, and locally. Acting in a sustainable way also challenges us to look to the future, and to fully assess and understand the implications of the decisions made today on the lives and livelihoods of future generations and the natural ecosystems they will rely upon. This view of sustainable water resources development implies that the conventional economic imperative to maximize production is accountable to an ecological imperative to protect the ecosphere, and a social equity imperative to minimize human suffering.

Goals can be established to begin the in-depth, integrated assessment of watershed resources that lead to sustainability. These goals should be formulated to address a number of fundamental principles that underlay the conservation, protection, remediation, and longevity of water resources. Such goals might include :

- Provide safe, adequate water supplies at times and of the quantity needed for domestic, municipal, industrial, agricultural, and hydropower uses.
- Allocate effectively and fairly freshwater among diverse uses and users.

- Protect freshwater supplies in jeopardy from depletion of groundwater aquifers and/or salt water intrusion.
- Reduce discharges of pollutants into surface waters and eliminate contamination of groundwater equivalent, to integrate a watershed approach.
- Decisions on water quantity and water quality are considered concurrently, because when evaluated separately these issues pose challenges for water resource management and protection (i.e. in determining whether aggregate pollution from multiple sources exceed standards, both the total amount of pollution reaching the water body and the amount of water present to "dilute" or "assimilate" the wastes affect the determination).
- Water conservation strategies are regularly relied upon for reducing waste of water, using water more efficiently, and meeting new demands using existing water supplies.
- Holistic, integrated assessment strategies with regards to water allocation and transfers are adapted to limit impact on the economic stability of rural communities.

After a consensus is developed with regards to criteria that describe the future longevity of healthy water resources, indicators to measure sustainability can be defined. Indicators will tell decision-makers and society in general, how we are doing toward the achievement of sustainable use with regards to water resources. Indicators represent standards for measuring characteristic criteria (conditions) of sustainability.

Sustainable development is the centerpiece and key to water resource quantity and quality, as well as to national security, economic health, environmental stability, and societal well-being.

Following the guidance of the overlapping circles model (three-legged stool analogy), and incorporating an adaptive management approach to decision-making, allows us to transform the rhetoric of sustainable development into actions that better align economy, environment, and society, by increasing the area of circle overlap. In this way we can test our making of choices and the consequences they represent, to guarantee each act, project, or programme implemented will not degrade the environment upon which economic prosperity and social stability rest, by asking the following :

- Does this choice provide an economic benefit? What is it?
- Does this choice provide an environmental benefit? What is it?
- Does this choice offer equal benefits to all sectors of society? What are some?
- Was this choice agreed to through the participation of all stakeholders?

If the answer to any one of these questions is NO, then the project, programme, or act should be rethought to better address the core principles of sustainability.

It is equally important to recognize the fact that sustainability is not strictly a problem of science, engineering, or economics, but is also founded on values, ethics, and the equal contributions of different cultures. Additionally, all members of a community have a shared future; they are dependent on each other in ways that are both complex and profound. Thus, ideals of preservation and protection on the one hand, and of economic vitality and opportunity on the other are not in conflict, but rather in a sustainable future they are linked together. And, we recognize our ability to see the needs of the future are limited; therefore, any attempt to define sustainability should remain as open and flexible as possible, through the guidance of adaptive management.

Water resources management is one of the most important challenges the world faces. It is difficult to think of a resource more essential to the health of human communities or their economies than water. Humans cannot live for more than several days without water, shorter than for any source of sustenance other than fresh air. In meeting their demand for water, societies extract vast quantities from rivers, lakes, wetlands, and underground aquifers to supply the requirements of cities, farms, and industries.

The summer of 2002 in the United States will be remembered for Americans from coast to coast going through one of the worst droughts in decades. While experts discussed the links between water shortages, erratic weather conditions, and population growth, there is also evidence that the way we grow—development patterns—can exacerbate problems with both water quality and quantity. And ironically, water supply is no longer just a western issue in the United States. We are drinking, irrigating, and using water faster than precipitation can replenish groundwater from the Great Plains to the Chicago suburbs to the Florida Everglades.

There is also growing recognition that functionally intact and biologically complex freshwater ecosystems provide many economically valuable commodities and services to society (ecosystem services) beyond simply direct water supply. These services include flood control, transportation, recreation, purification of human and industrial wastes, habitat for plants and animals, and production of fish and other foods and marketable goods. These ecosystem benefits are costly and often impossible to replace when aquatic systems are degraded. Deliberations about water allocation should therefore, always include provisions for maintaining the integrity of freshwater ecosystems, including the need to maintain minimum in-stream flows and to anticipate the impact of hydrologic modifications on downstream environments (Flint, et al., 1996). Otherwise, we have few safeguards that will protect the systems that sustain us.

Besides being an integral part of the ecosystem, water is a social and economic good. Demand for water resources of sufficient quantity and quality for human consumption, sanitation, agricultural irrigation, and manufacturing will continue to intensify as populations increase and as global urbanization, industrialization, and commercial development accelerates (Flint and Houser, 2001). Water runs like a river through our lives, touching everything from our vigour and the fitness of natural ecosystems around us to farmers' fields and the production of goods we consume. It is critical that efforts intended to be sustainable fully consider the health and operation of aquatic ecosystems and that the environmental value of watersheds be recognized when making economic and social decisions on water allocation and use.

Why We Concerned About Water Quantity and Quality?

Table 7.1 : Why We Concerned About Our Water Resources?

Water Resource Availability	*How Safe is Our Water?*	*Water and World Security*
Average U.S. household uses about 50 gal/ person/day, nearly triple Europe's level and more than 7 times the rest of the world (ENS, 1999b).	In 1996, 263 million tons of Nitrogen and 18 million pounds of Phosphorus ran into the Chesapeake Bay (ENN, 1998).	Israelis and Palestinians have argued for years over how to share the Mountain Aquifer beneath the West Bank (Edie Summaries, 2000).
The World Health Organization says good health and cleanliness require a total daily supply of about 8 gal/ person/day (Collier, 1999).	Fish advisories for risks to human health have become a standard practice of the 1980s and 1990s (ENN, 1999).	Pakistan and India have been in conflict for centuries over water in the Indus and Ganges Rivers, which both originate in Kashmir (Mustikhan, 1999).
Two-thirds of residential interior water is used for toilet flushing (4 gal/ flush) and bathing (15-50 gal/shower or bath) while a dish-washer uses 8-12 gal and a toploading clothes washer 40-55 gal (ENS, 1999b).	In 1986 a study statistically linked children with leukemia in Woburn, Massachusetts to contaminated drinking water affected by a nearby waste site (Montague, 1998).	While the Syrians press for an Israeli withdrawal from the Golan Heights, water, not land is the crucial issue between the two countries. The Golan Heights provides more than 12% of Israel's water requirements (Edie Summaries, 2000).
Land irrigation pumping extracts underground water much faster than it is replaced and spray irrigation loses 1/3 of water to evaporation before reaching plant roots (Lazaroff, 2000).	Clusters of child leukemia are occurring in regions where drinking water has been contaminated by carcinogenic volatile organic compounds from industry (Sutherland, 1999).	The Nile River in Africa runs through ten countries (ENS, 1999b).
The Ogalalla aquifer (1/5 of U.S. irrigated land) is overdrawn by	In 1995, 29 cities and towns in U.S. corn-belt had herbicides in	Malaysia sells water to neighbouring Singapore and is now demanding an

(Contd.)

Water Resource Availability	*How Safe is Our Water?*	*Water and World Security*
12 billion cubic m/yr causing more than two million acres of farm-land to be taken out of irrigation (Center for New American Dream, 2000).	drinking water that exceeded federal safely levels (Grossman, 1998).	increase in the price of this water (ENN, 2003).
China is draining some of its rivers dry and now mining ancient aquifers that take thousands of years to recover (Brown, 1999).	Between 1976-1996, annual rates of harmful algae blooms—leading indicator of health risks for marine animals and people—increased from 74 to 329 (Barker, 1997).	The Mercosul countries of Brazil, Argentina, Uruguay, and Paraguay launched a project for preservation of the Guarani Aquifer that serves all four countries (Muggiati, 2003).
Africa's Lake Chad has shrunk from a surface area of 25,000 sq km in 1960 to only 2,000 sq km today (GreenBiz.com, 2003).	Aging infrastructure, source water pollution and outdated treatment technology increase human health risks in 19 US cities (ENS, 2003).	Canada and the U.S. signed a treaty approxi-mately 10 years ago that states no water can be removed from the Great Lakes basin (ENS, 1999a).
Mexico City is sinking as residents pump water beneath them—elevated train tracks, built flat in the 1960s, look like roller coasters now (Center for New American Dream, 2000).	Mass fish kills and disease outbreaks went from nearly unheard of before 1973 to almost 140 events in 1996 (Borenstein, 1998).	Mexico and the U.S. have a longstanding treaty for maintaining water flow in the Colorado River (Stevenson, 2003).
One-fifth of the world's freshwater fish—2,000 of the 10,000 species identified so far—are endangered, vulnerable, or extinct (GreenBiz.com, 2003).	Stranding of whales, dolphins, and porpoises, linked to poor oceanic environmental condi-tions jumped from nearly zero in 1972 to almost 1,300 in 1994 (Borenstein, 1998).	Along the Missouri River, there is conflict among navigation, power generation, and environmental concerns (Quaid, 2003).
Globally the world has lost half of its wetlands, mostly in the last 50 years (Wilson and Yost, 2001).		Maryland is in control of Virginia's water destiny (IATP, 2003).
Two of every 3 persons could live in water-stressed conditions by the year 2025 (GreenBiz.com, 2003).		

In total, less than three-tenths of 1% of Earth's freshwater is in the lakes and rivers that have served as the major sources of water through most of human history (Adler, 2002), but societies worldwide have not always appreciated this easily accessible freshwater and the need to protect it. Thus, water consumption has nearly doubled since 1950 (UNESCO, 2003), and much of the world suffers greatly from inadequate access to potable water. About 20% of the Earth's population of 6.2 billion lacks access to safe drinking water (Hall, 2003). According to the United Nations, more than 200 million people every year suffer from water-related diseases, and about 2.2 million of them—mostly the poor—die.

The demand for water resources is continuing to increase. This increase is being driven not only by a growing world population but also by the aspirations of that population for an ever-increasing standard of living (Bartlett, 1999). At the same time, the capacity of the planet to meet this demand is in decline because of over-harvesting, inappropriate agricultural practices, and pollution, to name just a few. These impacts on Earth are occurring because humans are not in line with the way the natural world functions.

Currently a large proportion of the world's population is experiencing water stress (Table 7.1). Rising population demands for water from irrigation (70% of all water uses), industrial (20%), and residential (10%) uses greatly outweigh greenhouse warming affects on world water supplies (Vorosmarty, et al., 2000). Likewise, humans use more than 50% of the available freshwater in our world, 60% of which is wasted, leaving less than half for all other life forms on Earth. The average quarter-pound hamburger requires 616 gallons of water to create its meat; the cheese requires 56 gallons; and the making of the bun 25 gallons of water (Ryan, 1997). The average U.S. household uses about 50 gallons per person per day, a rate more than seven times the per capita average in the rest of the world and nearly triple Europe's level. Yet the World Health Organization says good health and cleanliness require a total daily supply of about 8 gallons per person per day (Table 7.1).

Other signs are just as frightening (Table 7.1). In 1986, a study statistically linked children with leukemia in Woburn, Massachusetts to contaminated drinking water affected by a waste site nearby. A study released in 1995 has shown that herbicides in drinking water exceed federal safety levels in 29 cities and towns in the U.S. corn-belt. Israelis and Palestinians have argued for years over how to share the Mountain Aquifer, which lies beneath the West Bank. Pakistan and India have engaged in conflict for centuries over water rights to the Indus and Ganges Rivers.

Society is "hitting the limits" or hitting the wall of a funnel, as in *The Natural Step* (Robert, 1991; Gips, 1998), in its never-ending use of natural resources and production of waste. The situation of the people on Earth can be viewed as a funnel with ever diminishing room to maneuver. Life-support systems for our continued existence on the planet are in decline. At the same time, the global population and global demand for these resources are increasing, leading us to "hit the wall" of the funnel. Increasing water shortages or inequitable access to safe water can cause poverty and environmental degradation that can lead to global hunger, resulting in civil unrest and human conflict. With conflict comes regional and national disputes, even war, that can best be alleviated by the sustainable use of these resources.

How Do We Act Sustainably?

Society consistently faces issues related to economy, environment, and fairness among people. Each of these human concerns is in some way impacted by the forces that drive the natural world. However, development models intended to tackle societal problems have traditionally taken a piecemeal, singular approach, addressing issues of economics, environment, or social health, sometimes in isolation from one another (Flint and Danner, 2001). For example, socio-economic systems often become caught up in the adversarial "economy versus environment" debate and begin operating in a linear direction : taking resources from the Earth, making them into products, and throwing them away to produce large amounts of waste (take-makewaste). This process leads to communities being unsustainable.

Sustainable development is the centerpiece and key to water resource quantity and quality, as well as national security, economic health, and societal well-being. The word *sustainability* implies the ability to support life, to comfort, and to nourish. For all of human history, the Earth has sustained human beings by providing food, water, air, and shelter. *Sustainable* also means continuing without lessening (Flint, et al., 2002). *Development* means improving or bringing to a more advanced state, such as in our economy.

Thus, *sustainable development* can mean working to improve human's productive power without damaging or undermining society or the environment—that is, *progressive socio-economic betterment without growing beyond ecological carrying capacity : achieving human well-being without exceeding the Earth's twin capacities for natural resource regeneration and waste absorption* (Flint, 2003). By acting under the principles of sustainable development, our economic desires/demands become accountable both to an ecological imperative to protect the ecosphere and to a social equity imperative to create equal access to resources and minimize human suffering. These requirements are the foundation of sustainable development as represented by the three circle model (principle elements) of sustainability in Figure 7.1. These three elements interact with each other so continuously that we cannot make decisions, make policy, manufacture, consume, essentially do anything without considering the effects and costs upon all three simultaneously. Each circle (sustainability principle) is defined as follows (Flint, 2003):

Economic Vitality (Compatible with Nature) is development that protects and/or enhances natural resource quantities through improvements in management practices/policies, technology, efficiency, and changes in life-style.

Ecologic Integrity (Natural Ecosystem Capacity) is understanding natural system processes of landscapes and watersheds to guide design of sound economic development strategies that preserve these natural systems.

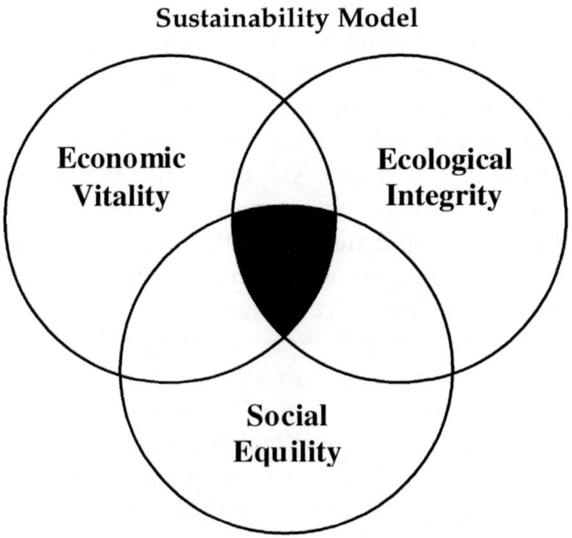

Sustainability Model

Economic Vitality

Ecological Integrity

Social Equility

Fig. 7.1 : This conceptual model of sustainable development illustrates the relationship among economic, ecologic, and social issues of concern in decision-making. The black overlap of the three circles represents the nexus of connection among issues.

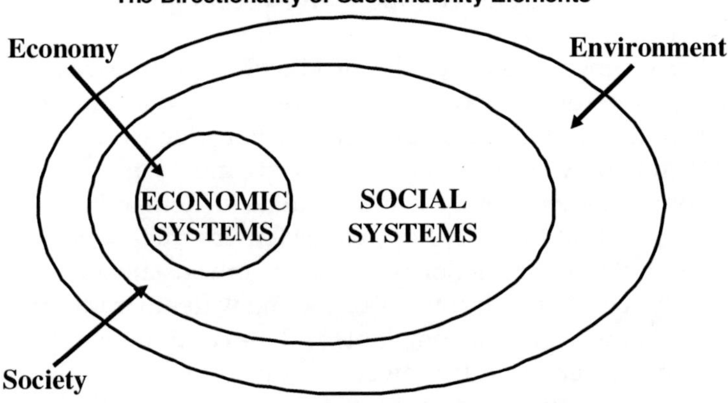

Fig. 7.2 : The directionality of each sector element's dependence upon the other elements of the sustainability model, where financially viable and cultural activities are integrated into natural processes in a cyclic fashion so as not to degade the environment upon which economic prosperity and social stability rest.

Social Equity (Balancing the Playing Field) is guaranteeing equal access to jobs (income), education, natural resources, and services for all people : total societal welfare.

Carrying out activities that are sustainable requires simultaneous, multi-dimensional thinking about the *consequences* of present actions in a cause and effect pattern on future public and environmental health through examination of the *connections* among environmental, economic, and social concerns when we make *choices* for action.

In understanding the three overlapping circles, it is also critical to recognize there is "directionality" to each circle's dependence on the others (Figure 7.2). It is true that all life depends on natural resources (Wackernagel and Rees, 1996). However, economy and society are no less important to humanity than ecology. Rather, there is a "directionality" of dependence. Sustainable development does not try merely to attain a "balance" between economics and environment as if they were two distinct entities. Rather, it considers directionality, where economic and cultural activities are integrated into natural processes in a cyclic fashion so as not to degrade the environment upon which economic prosperity and social stability rest.

For example, consider the production of electricity. To have a prosperous economy, society demands the continued and added production of electricity. For electricity to be produced to power our economies, society must both develop the appropriate technologies and regulate its demand for this electricity so that the supply of water (environmental issues) used for electrical production is consumed in a sustainable way. Electricity requires both sources of cooling water in traditional fossil-fuel power-production plants and the continuous supply of flowing water in hydropower production facilities. Thus, the directionality of this scenario (Figure 7.2) is that our economic ventures cannot be driven by electricity if society does not provide the human capital resources and there are not adequate supplies of freshwater. In a feedback process, the use of water as a natural resource for making electricity must not impair other users of the water by the activities of power production releasing polluted or in other ways degraded water as an output. In sum, the existence of economies is based solely on the existence of societies and their capacity to add value to natural resources. Furthermore, society cannot exist without an acceptable environment and the resources that the environment provides for basic human needs. This is the directionality of water resources sustainability (Figure 7.2). In terms of a three-stage rocket : natural capital (environment) is built or enhanced to power human capital (society) propelling financial capital (economy) through the engines of society and the resources to which society adds value.

Evaluation Strategy to Determine Water Resource Sustainability

Current piecemeal and consumption-oriented approaches to water policy cannot solve the problems confronting our increasingly complex world.

Traditionally, we apply a sectorial approach to the evaluation of water (e.g., the present conflict over the water resources of the Missouri River among navigation, power generation, and environmental concerns) (Quaid, 2003). The only equitable solution to these problems, however, is a systemic approach that considers ecological integrity and the ecosystem services that natural resources can provide. By considering the ecosystem services that water resources offer, our deliberations become able to more fully integrate the social and economic issues.

Thus, *sustainable development* can mean working to improve human's productive power without damaging or undermining society or the environment—that is, *progressive socio-economic betterment without growing beyond ecological carrying capacity: achieving human well-being without exceeding the Earth's twin capacities for natural resource regeneration and waste absorption* (Flint, 2003). By acting under the principles of sustainable development, our economic desires/demands become accountable both to an ecological imperative to protect the ecosphere and to a social equity imperative to create equal access to resources and minimize human suffering. These requirements are the foundation of sustainable development as represented by the three circle model (principle elements) of sustainability in Figure 7.1. These three elements interact with each other so continuously that we cannot make decisions, make policy, manufacture, consume, essentially do anything without considering the effects and costs upon all three simultaneously. Each circle (sustainability principle) is defined as follows (Flint, 2003) : *Economic Vitality* (Compatible with Nature) is development that protects and/or enhances natural resource quantities through improvements in management practices/policies, technology, efficiency, and changes in life-style.

Ecologic Integrity (Natural Ecosystem Capacity) is understanding natural system processes of landscapes and watersheds to guide design of sound economic development strategies that preserve these natural systems that would elude us if our only concern was the environmental aspects of water.

A conceptual framework of a system's perspective to evaluating water resources through development of criteria and indicators could be represented by the theoretical model in Figure 7.3. A systems perspective should be used to understand the interactions among the various forms of environmental, social, and economic capital and the processes that most directly affect them in order to guide better decision-making for the sustainable development of water resources (Figure 7.3A).

Using the principles of sustainable development outlined above, the following illustrates the steps one might take in pursuing the guidance of this conceptual framework.

1. *Develop a conceptual view of sustainable water resources (Figure 7.3B)*

Because there is fear that too many different perspectives will interfere with the overall goal of developing a strategy, groups of people often avoid the discussion of sustainability in their dialogue about how best to manage resources. Others contend

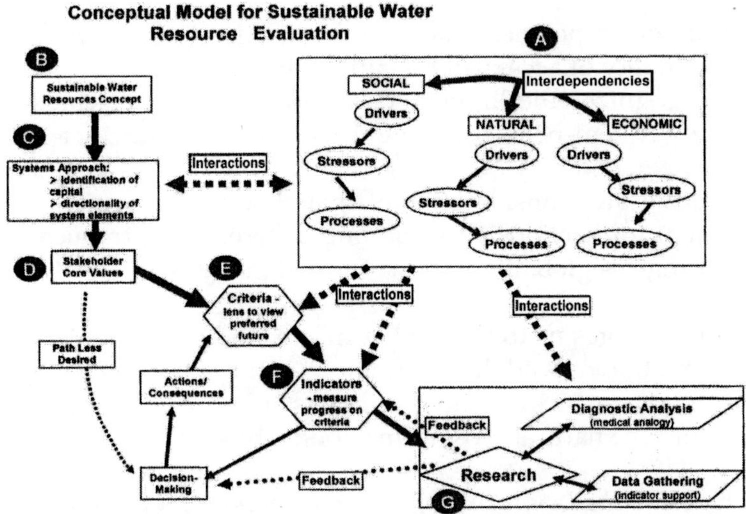

Fig. 7.3 : Systemic framework for evaluating water resources through the development of criteria and indicators : (A) illustrates the systems approach. (B) shows the development of a conceptual view of sustainable water resources. (C) defines water resources "capital" and assesses directionality of issues. (D) elaborates the sustainability goals of stakeholders. (E) develops criteria to judge water resource sustainability. (F) identifies indicators to measure sustainability. (G) demonstrates the research support required by this evaluation strategy.

that this discussion is necessary to provide everyone with the perspective of others and to honour the ideas of all who are participating in the discussion. Having this discussion can produce a point of reference that is absolutely necessary for there not to be a constant "moving target" to the focus of discussions. For example, an agreeable statement might include the following : "The sustainable development of water resources is a multi-dimensional way of thinking about the connections or interdependencies among natural, social, and economic systems in the use of water." This view suggests that attempts to achieve economic vitality are done in the context of the enhancement and preservation of ecological integrity, social well-being, and security for all (Conceptual Model group of the Sustainable Water Resources Roundtable, 2003). The sustainable development of water resources :

- involves policies, plans, and activities that improve equality of access and quality of life for all;
- recognizes the limits and boundaries beyond which ecosystem behaviour might change in unanticipated ways;
- advocates consideration of spatial scales, recognizing that interactions occur among different geographical ranges—globally, nationally, regionally, and locally; and
- challenges us to look to the future, and to fully assess and understand the implications of the decisions made today on the lives and livelihoods of future generations and the natural ecosystems upon which they will rely.

2. *Categorize the key forms of natural, social, and economic capital that need to be sustained to identify stakeholder core values with regard to water (Figure 7.3 C)*

Capital refers to the condition and capacity of any stock, inventory, or accumulation of materials or resources found in economic, environmental, or social systems yielding a flow of goods and services that possess a value directly, or may be devoted to the production of other goods (Daly and Cobb, 1994; Wackernagel and Rees, 1996b). For example, natural capital refers to any stock or inventory of natural resources found in our environment that yields a flow of valuable goods and services into the future (e.g., an underground water aquifer or fish stock that can provide a harvest or flow).

Natural capital might include the following : surface/groundwater quantity and quality; precipitation/climate trends; biodiversity; fisheries production; wildlife habitat; energy production; watershed/ecosystem services; watershed functional integrity; ecological infrastructure; land-use conversion; or environmental aesthetics (tourism quality). Social capital could include : subsistence rights; drinking water supply; community capacity; fiscal spending ability; regulatory framework (governance capability/resource policies); resource policies; institutional infrastructure; access to knowledge;

quality of life; equal resource access; beauty and play; or security. Likewise, economic capital might include : commercial/recreational fisheries; forestry production; energy supply; agricultural production (irrigation); industrial use; resource ownership; true-cost pricing; waste as resource; value-added production; transportation support; waste treatment; flood control; or tourism/recreation.

3. *Develop achievable goals for water resource sustainability that reflect the various stakeholder core values (Figure 7.3 D)*

Goals can be used to begin the in-depth, integrated assessment of watershed resources that lead to sustainability. These goals, which will represent stakeholder core values, should be formulated to address a number of fundamental principles that underlay the conservation, protection, remediation, and longevity of water resources. Such goals might include :

- Provide safe and equal access to water supplies needed for domestic, municipal, industrial, agricultural, and hydropower uses.
- Provide sufficient water quality/quantity to support ecological function.
- Measure and protect against biological and ecological degradation in aquatic ecosystems and restore integrity of degraded ecosystems.
- Reduce discharges of pollutants into surface waters and eliminate contamination and overconsumption of groundwater.
- Prevent human health risks due to the spread of waterborne diseases, water contamination, and hostile actions.
- Prevent physical modifications from land use/cover changes or hydrologic disturbances within watersheds that cause risks to humans, natural systems, and property.
- Encourage a holistic, watershed-based approach to evaluating all water resource issues that is

participatory, democratic, equitable, and socio-economically sensitive.

- Develop.

4. *Define criteria that establish the conditions to protect and maintain all the perceived beneficial uses of water assets (Figure 7.3 E)*

Following the development of goals all stakeholders can agree to concerning their efforts at sustaining water resources, criteria should be identified that establish the conditions deemed necessary to protect and maintain all the perceived beneficial uses of water assets. In essence, criteria provide a "lens" through which to evaluate the preferred future status of water (i.e., characteristics that best define water sustainability) (Flint, et al., 2002).

The definition of criteria is extremely important in this conceptual framework because, by choosing to develop criteria, stakeholders and managers are deciding to pursue a specific path. This path differs from making decisions based upon expressed stakeholder values (Figure 7.3). The choice of appropriate criteria can guide communities toward their anticipated outcomes, as defined by their goals, and introduce a process for establishing expected outcomes as well as a means of measuring progress toward those outcomes. Such criteria for water resources might include :

(a) The quantity of groundwater is monitored, and these reserves are protected from pollution and depletion.
(b) The water resource in question meets the quality and quantity for designated uses.
(c) Fish taken from recreational and commercial fisheries are not contaminated.
(d) Fish populations and other wildlife that rely on aquatic habitats and on the assemblages of species that inhabit aquatic ecosystems are healthy according to standards established by science.
(e) In-stream flows are enhanced and protected for environmental benefits.

(f) No actions are taken that will harm or threaten endangered species.
(g) Industrial and municipal point sources of pollution are less than the Total Maximum Daily Load (TMDL) standards for the water resource in question.
(h) Non-point sources of polluted run-off and erosion from intensive land uses are monitored and sources eliminated through the use of best management practices.
(i) Watershed-wide assessment programmes exist to identify the full range of pollution sources within the watershed, and they use the TMDL approach or its equivalent to integrate a watershed approach.
(j) Decisions on water quantity and quality are considered concurrently because they pose potentially contradictory challenges for water resource management and protection when evaluated separately.
(k) Water conservation strategies are regularly relied upon for reducing waste of water, using water more efficiently, and meeting new demands upon existing water supplies
(l) Holistic, integrated assessment strategies with regards to water allocation and transfers are adapted to limit impact on the economic stability of rural communities.

5. *Define indicators to measure sustainability progress (Figure 7.3F)*

Communities need a believable means of setting sustainability goals and then determining the degree to which these are reached. Policy-makers also need "early warning signals" of poor performance that can enable appropriate adjustments. After a consensus is developed with regard to criteria that describe the future longevity of healthy water resources, indicators to measure sustainability can be defined. The role of an *indicator* is to make complex systems understandable and perceptible (tangible). It clarifies a problem or condition by showing how well a system is working. Indicators point the way and mark progress toward a community vision of sustainable development. An indicator creates a *snapshot* of a resource's economic, social, and environmental system conditions and provides the opportunity to better understand past trends so that the decision-makers can influence future directions of development. A good indicator alerts one to a problem before it gets too bad, and it helps you recognize what needs to be done to fix the problem.

An effective indicator or set of indicators helps a community determine where it is, where it is going, and how far it is from chosen sustainability criteria that reflect established water resource criteria.

"Where Do We Want To Be" will reveal the goals or issues that are important. Indicators of sustainability examine a resource's long-term viability based on the degree to which its economic, environmental, and social systems are efficient and integrated in striving to reach community goals. Before time is spent gathering and reporting data for an indicator, it should be compared to the community's perception of sustainability to make sure that the chosen indicator is measuring the right thing. If data do not exist for some chosen indicators, try to define the best indicators and only settle for less as an interim step while developing data sources for better indicators.

Indicators of sustainability are not the traditional indicators of economic success or environmental quality. Because the achievement of sustainability requires a more integrated view of the world, indicators of sustainability should link economy, environment and society as well as point to where these links are weak. For example, an economic indicator that does not include environmental and social effects will not help move water resource protection in a sustainable direction (*e.g.*, the Missouri River conflict). Likewise, an environmental indicator that does not take into account economic and social impacts will not provide adequate insight into the best way to improve water resource health and vitality. A perfect example of what is being said here is the following : when the Exxon Valdez ran aground, the spilled oil killed millions of animals and cost millions of dollars to clean-up. The jobs created from clean-up activities made the U.S. Gross National Product (GNP), a much-relied upon national indicator, go up (Flint and Houser, 2001). In this case, using the GNP as an indicator suggests that we should get more oil tankers to run into rocks more often.

Indicators will tell decision-makers and society in general, how we are doing toward the achievement of sustainable use with regards to water resources. Indicators represent standards for measuring characteristic criteria (conditions) of sustainability, and they are as varied as the types of systems they monitor. However, there are certain characteristics that effective indicators have in common :

- Relevant to sustainability and link economy, society, and environment
- Developed and accepted by the people in the community
- Understandable to the community at large and reflect stakeholder's concerns : important to the lives of the audience
- Attractive to the media and can be used to monitor, analyze, and communicate local trends.
- Accurately measure the issue or goal in a scientifically defensible way
- Focus on long-range view : reliable up to two decades or more
- Flexible enough to incorporate new scientific information and changing public perceptions
- Can be compared to existing and past measures to define trends and identify stresses
- Advance local sustainability, but not at the expense of other regions
- Measure an appropriate geographic area and/or an appropriate time interval
- Provide early warning of changes
- Can measure movement towards or away from a specified target/goal

- Based on reliable and timely information that is easy to gather at modest cost
- Outcome (results) oriented : focus on measuring achievements instead of efforts or expenditures.

After identifying key indicators and corresponding databases, we must conduct the exercise of setting benchmarks or targets for each indicator. A target is the threshold used to define sustainability from unsustainable practices. These targets will help identify water resource criteria that are sustainable. Unsustainable criteria are long-term problems for the region of concern.

From this conceptual framework evolves the need for added research activities (Figure 7.3G). Such activities are an important form of feedback for social learning and adaptive management. In following the criteria/indicator model, there will probably be a need for system diagnosis to explain undesirable trends that may be shown by indicator measures. Such diagnosis, as characterized by the medical analogy example (Heintz, 2003), is a key element in adaptive management processes that should be designed to direct the use of water resources within a sustainable framework and to better help us understand what the system conditions are alerting us to when indicators tell us something is wrong (*e.g.* high body temperature in humans). With time and continued application of this strategy, a dialogue will also evolve on research needs to address recognized data gaps for identified indicators and to build our understanding of ecosystem processes.

More than one-sixth of the world's population does not have access to safe water supplies. The potential conflicts from this disparity are frightening.

The escalation of a water crisis in the world is due essentially to the unsustainable use and management of water resources and to the destruction of ecosystems such as forests, wetlands, and soil that capture, filter, store, and release water. Through our evaluation of water resource sustainability, we must not only increase public awareness about the challenges the world is facing in relation to water, but we must also change the way the water issue is perceived : from being a driver of conflict to being a catalyst for collaboration. In doing so, we must not only view sustainability as a problem of science, engineering, or economics; it is also founded on values, ethics, and the equal contributions of different cultures. Additionally, all members of a community have a shared future; they are dependent on each other in ways that are both complex and profound. Thus, ideals of preservation and protection, on the one hand, and of economic vitality and opportunity, on the other, are not in conflict. Rather, in a sustainable future, they are linked together. Moreover, we recognize our limited ability to see needs of the future; therefore, any attempt to define sustainability should remain as open and flexible as possible through the use of adaptive management.

In summary, it might be helpful to consider the following. Although we need only about 1.5 to 2.0 quarts of water per person per day to stay alive, the total human population needs the balance of the water resources in the atmosphere, oceans, ice, wetlands, and other aquatic systems to buffer emergencies (Adler, 2002). These masses of water provide crucial functions by absorbing and redistributing energy and waste products from life forms. They shield us from the atmosphere's fluctuations in gaseous content, and they offer transportation and provision of conversion sites for nutrients in food chains. If such resources are spoiled, conditions for human life will inevitably deteriorate. A person's 1.5 to 2.0 quarts of water alone will not save them because we do not know specific quantities that constitute a sufficient buffer. Policies must reflect and be built on the conservative natural distribution and use of the world's total water resources.

Hydrological Cycle

The science which deals with the occurrence, distribution and movement of water resource on the surface of the earth is known as hydrology. Water accumulates in the atmosphere in the form of vapour, on the earth surface as sub-surface or stagnant water and in the earth as underground water. The total water supply of the earth is constantly in circulation from ground to the atmosphere and back to the earth. This circulatory system is known as hydrologic cycle. Hence hydrologic cycle means the process of transfer of moisture from the atmosphere to the ground in the form of precipitation and the flow of water through rivers to ocean and lakes (Figure 7.4).

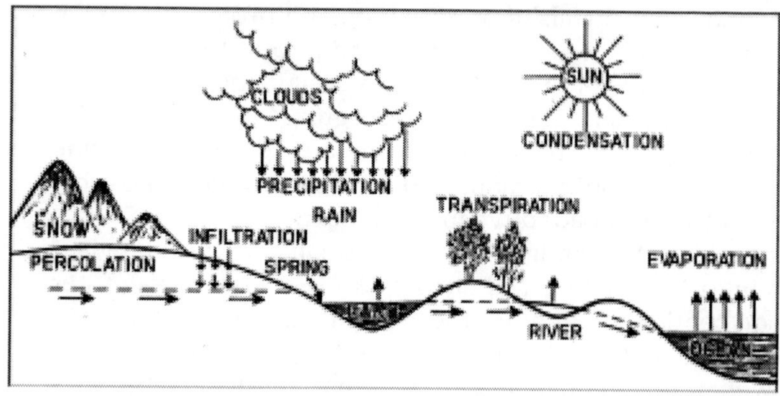

Fig. 7.4 : Hydrological Cycle.

Following are the processes of hydrologic cycle :

1. Evaporation and transportation;
2. Precipitation; and
3. Run-off.

The hydrologic cycle may be expressed as P = E+R, where P is precipitation, E is evaporation and R is run-off. So far as the underground reserves of water is concerned North Munger has enough storage of water in comparison with the rocky sub-stratum of South Munger. But the southern part is rich in hot springs due to uranium and sulphur beds inside the sub-stratum and availability of volcanic ejections especially during Cretaceus period.

Underground Water

In various parts of the Munger division a variable quantum of underground water is found. Northern part of Munger has high potential of underground water whereas the rocky sub-stratum of South Munger has poor water potential. A micro-level assessment of the ground water resources was done by the Geological Survey of India during 2000-2005 (Table 7.2). The result is that north of the Ganga has a shallow zone of acquifer comprising fine to medium sand usually occur between 6.0 to 12.0 metres depth whereas potential acquirers with medium to coarse sand occur at a depth of 40 metres. A 170 metres deep tubewell in the area can yield about 4000 litres water per minute without any trouble. South of the Ganga has also variable groundwater potential such as west of Sikandra-Lakhisarai axis which merges with the deeper alluvial zone of Patna district. In this tract the extraction of ground water through tubewell is possible at a depth of 147 metres at Lakhisarai but decreases rapidly to the east near Munger. Although the area between Khaira and Jamui, shallow acquifer zone has been encountered within a depth of 30 metres, yet this area is not suitable for groundwater development through tubewells. Same is the situation in Jamui-Jhajha belt. In all these areas open wells have been suggested for small scale irrigation. East of Munger up to Sultanganj along the narrow coastal tract south of the Ganga, tubewell irrigation is feasible. Medium duty tubewells may be successful locally between Sultanganj and Tarapur and between Bariarpur and Lohchi.

Hot Springs

In the Munger division there are seven hot springs which need closer attention for their development as health resorts and for proper utilization of water as it has medicinal value. The seven hot springs of this area are given in Table 7.3.

Table 7.2 : Water-table in the Munger Division

Name of Place	Date of measurement	Depth of water table in metres	Depth of water bearing bed in metres
Gurmaha	January, 2005	12.60	4.95
Chormara	January, 2005	13.50	3.60
Gurmaha	February, 2005	14.70	3.15
Lakhisarai	February, 2005	8.40	7.20
Gurmaha	March, 2005	16.35	1.50
Chormara	March, 2005	15.45	1.80
Kharagpur	April, 2005	9.20	1.70
Kharagpur	May, 2005	9.50	1.20
Kharagpur	June, 2005	5.20	5.50
Kharagpur	July, 2005	1.17	9.00
Kharagpur	October, 2005	1.40	9.40
Jhajha	February, 1965	4.04	1.00
Balia	February, 2005	12.08	3.30
Balia	March, 1965	12.10	3.28
Jhajha	March, 2005	4.06	0.98
Jhajha	July, 2005	4.20	0.84
Balia	July, 2005	12.63	2.75
Jhajha	May, 2005	14.14	0.90
Balia	May, 2005	12.38	3.00
Balia	June, 2005	12.38	3.00
Balia	October, 2005	13.17	2.21
Chakai	February, 2005	6.60	1.22
Dullanpur	February, 2005	12.74	4.44
Dullanpur	May, 2005	14.74	2.44
Chakai	May, 2005	7.19	0.63
Chakai	July, 2005	6.90	0.95
Chakai	October, 2005	7.84	1.52
Dharhara	March, 1965	3.50	3.00
Dharhara	April, 2005	3.70	2.80
Kajra	June, 2005	5.40	3.10
Kajra	July, 2005	4.20	4.00
Dharhara	October, 2005	2.50	4.00
Dharhara	December, 2005	3.00	3.50

Source : Geological Survey of India, Calcutta.

Table 7.3 : Some Relevant Aspects of Hot Springs of Munger Division

Name of Hot Springs	*Temperature in °C*	*Resistance in ohms*	*Discharge in litres/Sec.*
1. Sitakund	62	05	21
2. Rishikund	65	—	—
3. Bhoura Kund	47	31	1
4. Rameshwar Kund	45	42	20
5. Barhi Garam Pani	63	32	400
6. Bhima Bandh	63	24	1500
7. Bhata Garam Pani	72	24	500

Source : Personal Library of Sri D.P. Yadav, Former Deputy and Education Minister, Government of India, 1980.

It will be seen that the resistivity of water decreases with the increase in temperature. This suggest that the water of hot springs tend to be pure with the rise in temperature. Sitakund hot spring is located in the plains near the Ganga river. Thus there is a certain amount of contamination. It is to be noted that these thermal springs constitute one of the purest form of water associated with the hot springs in the world.

Structure : The hot springs of Kharagpur hills are located at the point of structural dislocation (Figure 7.5). They appear to occur along a fault plain in the quartzite formation. Occurrence of the hot springs may be due to the rise of water along the fault plain from greater depth. Since the quartzite formations are stable there is less mineralisation as depicted in the chemical analysis of the water of these hot springs.

Another explanation may be that fault zone is associated with the tectonic movement of the earth. The convectional heat inside the earth is able to migrate through the fault zone in the form of thermal springs. The absence of gaseous emanation suggest a remote possibility of the above phenomenon. The cause of these hot springs should be determined through a detailed investigation of geophysical probing and hydrological survey to estimate the geothermal potentialities.

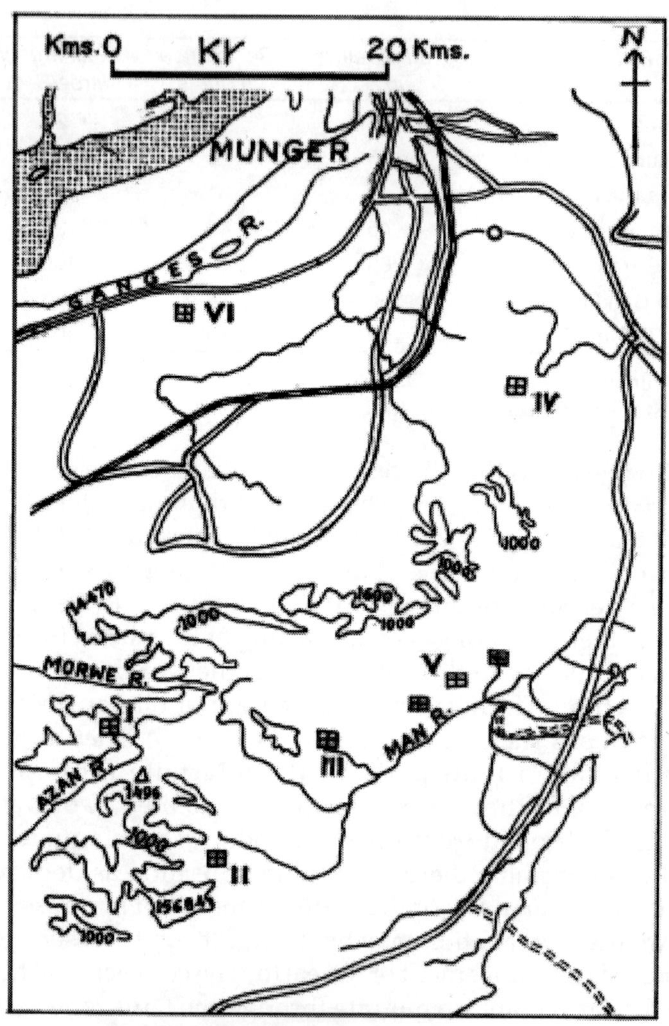

⊞ NAME OF SPRING GROUPS
I BHARARI (CHORMARA)
II BHIM BANDH III HINGUANIA
IV RISHIKUND VI SITAKUND
V RAMESHWAR LACHMISHWAR - BHAWRAH KUND

Fig. 7.5 : Location of Thermal Spring, Munger.

Irrigational Use of Water

Irrigation has proved economically gainful for a country. In fact "Irrigation form the datum line for sustained successful agriculture". It alleviates suffering, preserves life and averts famine. Water is more valuable than land, because when water is applied to land it increases the productive capacity of the land sixfold which otherwise would produce next to nothing. Knowle writes, "the irrigation works have made security of life, they have increased the yield and the value of the land and revenue derived from it. They have lessened the coat of famine, relief and have helped to civilize the whole region. In addition, they yield handsome profit to the government".

Gadgil Survey of the economic effects of the Godavari and Pravada Canals in Deccan, has shown that the total direct and indirect effects of irrigation projects were very favourable. Due to irrigation, farmers could make additional investment in cattle, farm implements and on more valuable crops like sugarcane.

As a result of the studies undertaken in 1958 and 1961, it was observed that canal irrigation has helped in promoting the greater utilization of land, enlarging the average size of the farm, increasing demand for additional productive investment in farm business, favourable input-output ratio, widening the scope for increase in land revenue and other local receipts. In addition to direct benefits there are also secondary and tertiary benefits, e.g., expansion of secondary and tertiary activities, increasing of greater work opportunities, and more employment to both family and hired labour, higher value of output per industrial unit, and higher turnover of business establishments in the project area.

The purpose of irrigation is to help increase agricultural production from the land served. Irrigation helps in fulfilling moisture deficiency in soils during the crop season so as to ensure proper and sustained growth of crops grown. In additional land use second or third crop being raised on the land having irrigation facilities which could otherwise may not be cultivated. The productive aspect of irrrigation helps in establishing agriculture production against drought.

The third aspect of irrigation is changing soil sterility caused by drought, i.e. overcoming low productivity due to dryness or excessive water supply. Thus irrigation may be defined as the process of artificially applying water to the soil for raising crops. It is a science of planning and designing efficient low cost economic irrigation system tailored to fit natural conditions.

Irrigational History in the Munger Division

There are no specified records from which the history of irrigation is found in the Munger division. The division comprises a number of big or small rivers such as the Ganga, the Kiul, the Burhi-Gandak, the Kosi, the Harohar, the Barnar, the Ajai, the Nagi, etc. All the rivers go dry in the summer season except the perennial rivers Ganga, Kosi and Gandak. In this way, it can be said that the primitive people utilised the water of different rivers in their own interest to a little piece of land. It may be possible that irrigation had been facilitated with the help of hot springs which are abundantly found in Kharagpur hills. The only large irrigation works found in the Munger division is Kharagpur lake which was constructed about 100 years ago by the Maharaja of Darbhanga. Height of the dam is 23.16 metres, width 6.09 metres, length 220.98 metres and the storage capacity of the reservoir is 412.8 hectare metres. The construction of the lake had taken place in 1870 and was completed in 1877 with an expenditure of Rs. 6,84,916. Most of the rivers of the division are seasonal and carry sufficient water only in the rainy season. It is obvious that forests and hills are sufficient in southern hilly area of Munger which are major constraints in the utilisation of river water systematically because the presence of undulation, valley, gorge, etc. were the main barriers which existed in different historic periods.

Types of Irrigation

Munger is basically an agricultural division and all its resources depend on agricultural output. There are various types of irrigations as per requirement and availability of water resource. Irrigation has the following main types :

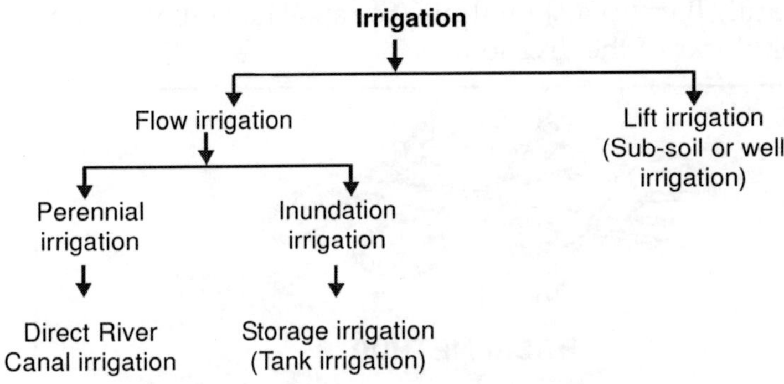

Flow Irrigation

This is the type of irrigation in which the supply of water is available to the land by the gravity flow (Figure 7.6).

(i) Perennial Irrigation System : In this very system water required for irrigation is supplied in accordance with the crop requirement throughout the crop period.

In such a system the storage had works such as dam, barrages are required to store the excess water during floods and release it to the crop as and when it is required. This system exists in Dharhara, Kharagpur, Bariarpur, Lakshmipur, Barhat, Gidhaur and Sono block in Munger division.

(ii) Inundation Irrigation : It is carried out by deep flooding and through situation of the land to be cultivated which is then drained off prior to planting of the crops. This system is prevalent in North Munger, Tal area and Diara belt of Munger.

Depending upon the source from which the water is drawn, flow irrigation can be further sub-divided into two types.

(A) Direct Irrigation or River Canal Irrigation : In this very system, water is directly delivered to the canal without attempting to store the same. For diversion purpose barrage is constructed across the river. Thus head-work raises the level of water in river and thus directs water into the canal taking up stream of the weir. Direct irrigation system exists in Lakhisarai, Barahiya, Jamalpur, Surajgarha, Pipariya and Munger blocks of the division.

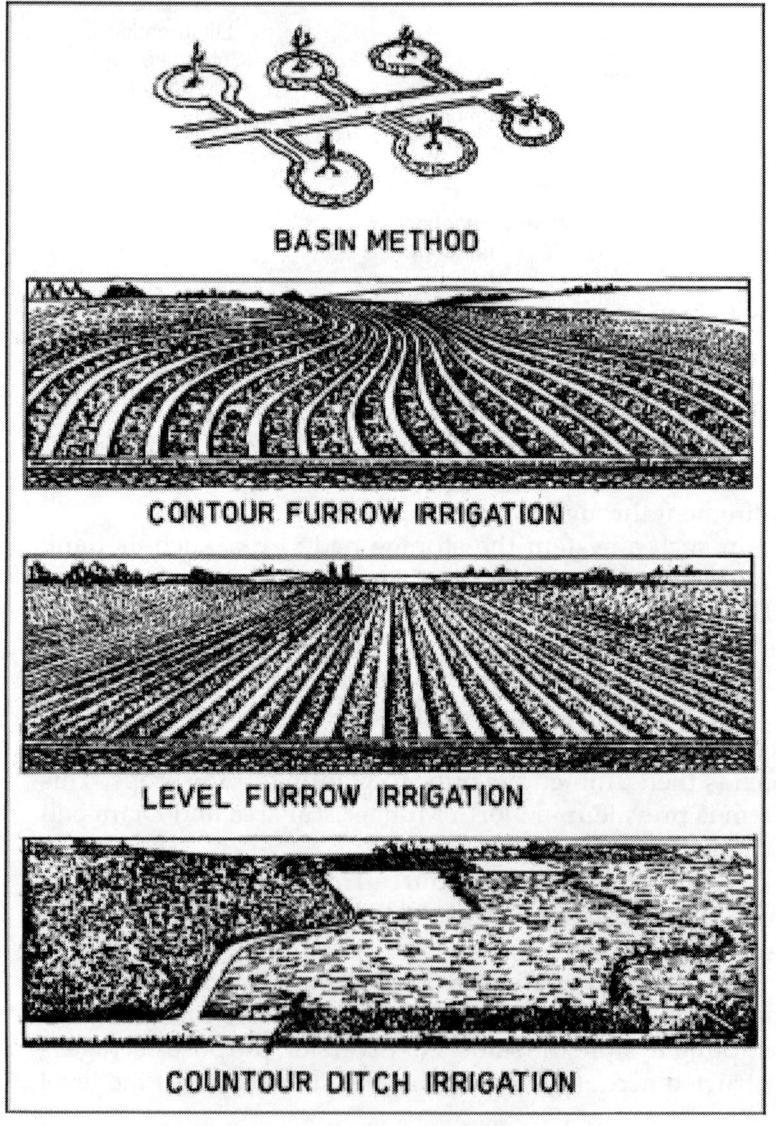

Fig. 7.6 : Modes of Irrigation.

(B) Storage Irrigation or Tank Irrigation : In this very system a solid barrier, such as a dam or as a storage weir is constructed across the river and water is stored in the reservoir and lake so formed. The storage irrigation scheme is comparatively of a bigger magnitude and involved much more expenditure than a direct irrigation scheme. This irrigational system also lies in the hilly areas of Munger. But in the northern part of Munger which consists Khagaria, Alauli, Beldaur, Chautham, Mansi, Gogri and Parbatta anchals irrigation facility have been made available through State and private wells or tubewells.

Methods of Applying Irrigation Water

There are various methods of irrigation which may be classified as follows :

(i) Surface Irrigation Method;
(ii) Sprinkler Irrigation Method; and
(iii) Sub-Surface Irrigation Method.

(i) Surface Irrigation Method

Surface irrigation is accomplished by allowing water to flow over the surface. There are two criteria for supplementing the good surface irrigation that are thorough and sufficient distribution of water. Summary of the conditions through which the various surface irrigation methods almost suited is given in Table 7.4.

(ii) Sprinkler Irrigation Method

The area where surface irrigation methods are not supplied due to uneven topography, sprinkler method is widely used. In this method water is applied in the form of rain, as it is done in the gardening. There are various systems of sprinkler method :

Furrow system, filt system, fully portable system, fully permanent system, semi-permanent system, perforated pipe system, self-propelled system and travelling sprinkler system. If *Ea* is water irrigation efficiency of the land, *Ws* is irrigation water stored in a route zone and *Wp* equal to water :

Pump into the system : Then, $Ea = 100 \times \frac{Ws}{Wp}$,

about 80% irrigation is possible through these systems. But the main drawback is wind pressure. This very system is not applied in the Munger division due to costly affairs and the abundant availability of sub-surface water.

Table 7.4 (a) : Surface Irrigation Methods and Conditions of Use

Irrigation method	*Crops*	*Topo-graphy*	*Water Supply*	*Soils*	*Remarks*
Border Strip method	Close Crops : Wheat, barley, bajra and berseem	Uniform and gentle slopes less than 1%	Moderately large flows	Soils with moderately low to moderately high intake rates	Accurate cross-levelling required between the bunds. It takes care of surface drainage.
Level Border method (Check Basin method or Check method)	It can be used for grain and fodder crops as well as for vege-tables	Best suited to gentle and uniform land slopes	Can be adopted for uniform sized streams. The stream of water should be at least of such a size that it will apply the gross appli-cation in half of the time requi-red for intake	Suitable for soils of moderate to slow intake rate and mode-rate to high available water holding capacity	Since the method requires accurate levelling and shaping, it involves high labour cost. However, high efficiencies can be obtained in water use. The method enables uniform water distribution and reduces wastage and soil erosion. It is suitable where leaching of soil is required.

(Contd.)

Irrigation method	*Crops*	*Topo-graphy*	*Water Supply*	*Soils*	*Remarks*
Small Basin method	Orchards as well as field crops	Relatively flat land suitable for levelling	Can be adopted to various sized streams	Suited for sized having moderate to slow intake rate and moderate to high available water holding capacity	It is a method specially suited for orchard irrigation on sloping lands deep cutting is required for land preparation. It requires special provision for drainage.
Straight or graded or large	Vegetables, row crops and orchards	Uniform slopes not exceeding two per cent for cultivated crops	Requires medium and large sized streams	All soils excepting sands with very high intake rate	This method is not suitable for applying very light irrigation and for soils with very high intake rate. It requires proper land grading to achieve uniform water distribution. Well adapted to mechanised farming. It is best suited to crops which cannot be flooded.

Table 7.4 (b) : Surface Irrigation Methods and Conditions of Use

Irrigation method	*Crops*	*Topo-graphy*	*Water Supply*	*Soils*	*Remarks*
Contour Furrow method	Vegetables, field crops and orchards	Undulating lands with slopes up to 5%	Small and medium sized streams	All soils excepting very light sandy soils which crack	Should have graded terrace installed for erosion control. This method can be used to irrigate safely land too steep for straight furrows. There is hazard from heavy rains or water breaking out of furrow. Rodent control is essential.
Corrugation	Close growing crops such as	Uniform slopes up to 8%	Samll and medium to heavy textured	Best on medium to heavy textured	The method is suitable where erosion form high rainfall occurs. It is not suitable

(Contd.)

Irrigation method	*Crops*	*Topo-graphy*	*Water Supply*	*Soils*	*Remarks*
	grains, pasture and lucern except rice		soils. Good for soils	soils. Good for soils.	to gentle slopes, less than one per cent. Care must be taken in limiting the size of the stream in corrugation to reduce soil erosion.
Con-tour Ditch	Pasture, grain and legumes	Can be used for irregular topo-graphy (suited to slopes ranging from 0.50 to 4%)	Needs abundant water supply at cheap rate	Soils with moderate to high intake rate	The method is suited to and is recommen-ded only if other methods cannot be adopted due to low efficiency in irrigation. They are seldom used on slopes less than one per cent. Small streams are not usually used.

(iii) Sub-surface Irrigation Method

In sub-surface irrigation method water is supplied directly to the route zone of the crop. The favourable conditions for the sub-surface irrigation practices comprise in previous sub-soil at reasonable depth, moderate slopes, good quality irrigation water and uniform topographic conditions. If all the above favourable conditions are not fulfilled then sprinkler irrigation system is suitable.

In the Munger division most of the irrigation is being carried out by means of surface and sub-surface flood.

Suitability of Irrigation

The suitability of irrigation depends upon the following factors :

1. Kinds of soil available with its chemical composition;
2. Topography of the land;
3. Sources of water;
4. Wind pressure;
5. Economic conditions of inhabitants of the area;

6. Total cultivable command area;
7. Agricultural produces;
8. Network of communication or transportation; and
9. Power (Electrical power available in the area).

The aforesaid factors play a vital role for the suitability of irrigation.

North Munger comprising seven blocks, viz., Khagaria, Alauli, Beldaur, Gogri, Chautham, Mansi and Parbatta have abundant number of State and private tubewells. In this area there is no possibility of canal network from the Kosi and the Ganga rivers of this zone which has sufficient humous contents for raising crops. Hence flow irrigation is prevalent here in the form of state and private tubewells.

In comparison with North Munger, the middle zone of the Munger division enjoys the facility of the water resource of Kiul and Harohar rivers. This area is electrified to a great extent and hence there is no problem of tubewell irrigation as well.

Southern hilly area is the third classified zone of the Munger division. This area is full of forests and hill and due to this undulation, the prevalent irrigation is carried out through canals and ahars.

The Geological Survey of India reported that the greater diameter to boring (10 cms-30 cms.) are impossible in Kharagpur, Bariarpur, Tarapur, Tetiha Bambor, Sangrampur, Gidhaur, Barhat, Lakshmipur, Jamui, Khaira, Sono, Jhajha and Chakai belt. Hence small and big reservoirs are only solution for irrigation purposes. The sandy and clayee soils are the main barriers in production of varieties of crops. The main source of water of this area is precipitation and flooding. Underground layer is so hard that geological experiment had showed that there is no possibility of the efficient operation of a tubewell, cavity boring, open boring, etc. for irrigational purposes. There is also dearth of electrification. Hence irrigation is not possible through State and private tubewells. Thus in this zone irrigation can be made possible only through major medium and minor reservoirs.

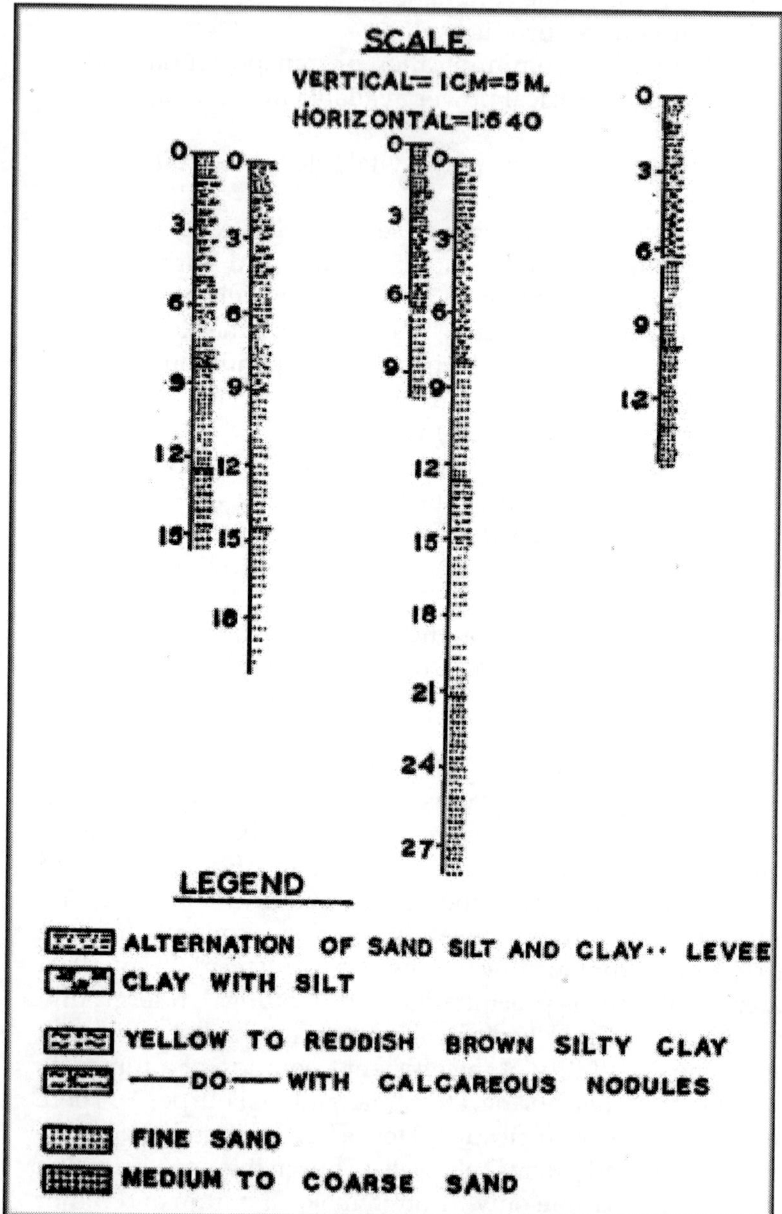

Fig. 7.7 : Munger Division : Sub-surface Hydrology of the Area Sunderpur and Parham.

The underground strata found by the geologists could be seen in Figure 7.7 which shows the strata chart of five state tubewells in South Munger where in almost all cases at the depth of 8 metres and below water is available in fine sand or in medium to coarse sandy layer : In upper part or 3 metres depth the layers of sand, silt and clay are found under which there is a layer of clay with silt. This is followed by yellow to red calcareous nodular beds, found as the sub-surface geological conditions.

Sources of Irrigation

Irrigation is the prime force in a developing agricultural economy like that of the Munger division. Under the vagaries of Monsoonal rain, droughts and flood conditions, the peasant of this region like to get an assured crop harvest with the help of irrigation after adopting the prevalent sources in the region under study (Table 7.5).

In the Munger division the chief sources of irrigation are canal, tank, tubewell, masonry well and kutcha well.

Among other sources ditch, pyne, ahars, rahat pump, latha and koor, karin, dhois, etc. are important. Altogether 25% land is irrigated through canal, but it is almost absent in North Munger. It constitutes 33.6% in South Munger plain and it has a share of only 25.7% in southern hilly margins. Tank is the poorest source of irrigation in the district which constitutes only 1.10% of irrigated land and it has the highest percentage of 1.90% in southern hilly margins. Tubewell is the chief source of irrigation in the Munger division where the average value is 43.20% and the highest percentage of tubewell service is 83.0% in North Munger whereas it is just 29.7% in southern plain and 17.3% in southern hilly region. The irrigation from other wells is just 85.3% in the Munger division and the highest 14.3% comes from southern hilly region and the lowest 2.3% from North Munger. Other sources of irrigation are most dominant in southern hilly region where an average value is 40.8% and just 14.8% in North Munger.

Table 7.5 : Sources of Irrigation

Name of Blocks	*Canals*	*Tanks*	*Tube-wells*	*Other wells*	*Other sources*
Munger	2	-	2.0	11.0	85
Bariarpur	30	0.5	1.5	1.7	45.5
Jamalpur	-	-	1.4.	43.4	55
Dharhara	-	-	-	0.4	99.6
Kharagpur	88	4.2	4.0	3.6	6.8
Asarganj	60	3.0	26	5	2.0
Tarapur	62	3.0	24	6	2.0
Tetiha Bambor	60	-	2.5	2.5	33
Sangrampur	63	-	2.8	2.2	13
Barahiya	5.9	0.3	80	8.8	5
Pipariya	7	0.3	12	0.9	18
Surajgarha	7	0.4	11	0.5	17
Lakhisarai	18	-	15.5	0.5	66
Ramgarh Chowk	71	1.1	18	4.7	5.1
Halsi	69	1.4	22	4.4	3.2
Barbigha	42	-	25	24	9.0
Sheikhopur Sarai	41	-	26	23	10
Sheikhpura	-	8.2	55	19	16
Ghat Kusumbha	-	8.3	56	18.2	15.5
Chewara	3.9	-	45.5	15.5	35.1
Ariari	3.6	-	47	17	32.4
Islamnagar Aliganj	40	-	12	-	-
Sikandra	89	3	10	1	-
Jamui	14	8.7	41	13.5	29
Barhat	-	10.7	29	12.4	51
Lakshmipur	-	3.0	27	11.2	52
Jhajha	31.1	9.4	7	49.9	8
Gidhaur	-	-	28	11.2	52
Khaira	10	-	14	8	68
Sono	-	-	15.9	16	69
Chakai	-	2.0	75.8	0.9	23
Alauli	-	-	90	-	8
Khagaria	-	-	87	8.2	4.8
Mansi	-	-	45	-	54.5
Chautham	-	-	43.5	0.5	56
Beldaur	-	-	99.5	0.5	-
Gogri	-	-	78.5	4.8	16.7
Parbatta	-	-	99.6	-	0.4

Scurce : District Statistical Office, Munger, 2001.

Table 7.6 : Net Irrigated and Gross Irrigated Areas in the Munger Division

Name of Blocks	*Area irrigated (in Hectares)*	
	Net Area	*Gross area irrigated more than once*
Munger	3300	5500
Bariarpur	2200	4100
Jamalpur	460	1050
Dharhara	470	975
Kharagpur	3400	5800
Asarganj	7500	15900
Tarapur	7900	16300
Tetiha Bambor	2600	8200
Sangrampur	2500	7500
Barahiya	1100	3000
Pipariya	2200	5700
Surajgarha	2000	5000
Lakhisarai	10000	30000
Ramgarh Chowk	9500	19000
Halsi	9000	18000
Barbigha	10000	3100
Sheikhopur Sarai	1050	2500
Sheikhpura	4300	13500
Ghat Kusumbha	4200	12000
Chewara	10000	19000
Ariari	11000	20000
Islamnagar Aliganj	5000	9000
Sikandra	4800	8800
Jamui	6900	9800
Barhat	3200	5800
Lakshmipur	3300	6000
Jhajha	400	1000
Gidhaur	3000	5500
Khaira	9000	18000
Sono	400	700
Chakai	300	500
Alauli	3100	4700
Khagaria	4100	12300
Mansi	2200	4800
Chautham	2600	5300
Beldaur	2800	4800
Gogri	2700	5500
Parbatta	4700	9500
Division total	**163180**	**328125**

Source : District Statistical Office, Munger, 2001.

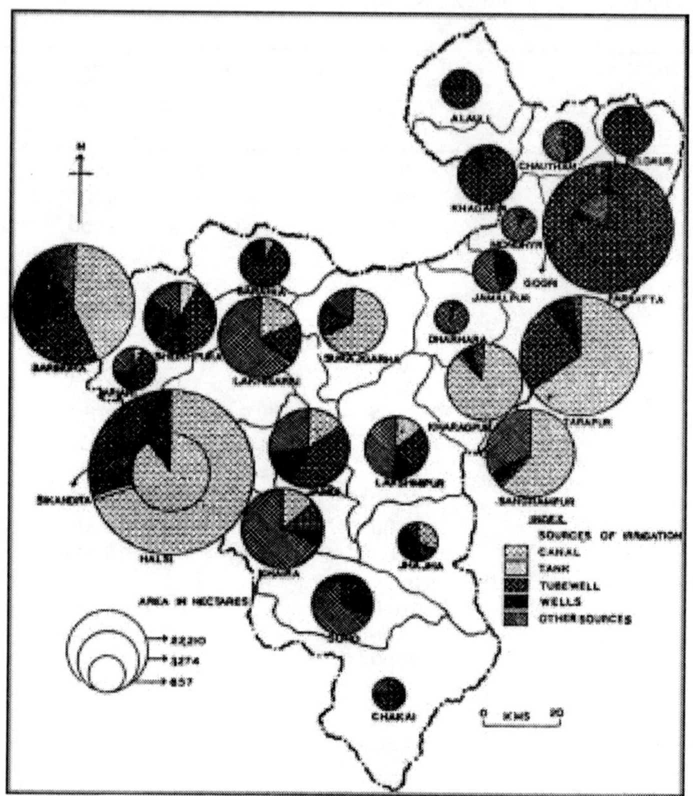

Fig. 7.8 : Munger Division : Source of Irrigation and Irrigated Area.

So far as the canal irrigation is concerned, the anchalwise picture is quite different, from the regional average. The highest percentage of canal irrigated land is available in Sikandra, Kharagpur, Halsi, Tarapur and Sangrampur anchals of South Munger where it is 89%, 88%, 69%, 62% and 62% respectively. In North Munger and Lakshmipur, Sono, Chakai and of South Munger, Khaira, the canal irrigation is almost absent but a very low percentage is reported from Ariari, Sheikhpura, Barahiya and Munger anchals.

Tubewell irrigation is a dominating source of irrigation throughout the Munger division. But it is more dominant in North Munger where in Parbatta and Beldaur anchals 99.5% land is irrigated from tubewell, and the lowest percentage of

North Munger is 43.5% in Chautham anchal. Alauli and Khagaria have also 96% and 87% tubewell irrigated land in different crops.

In southern-hilly anchals 75.8% land is irrigated from tubewell in Chakai anchal whereas it is 14% in Jamui, 27% in Lakshmipur and 25% in Tarapur. But in Jhajha and Kharagpur it is just 7% and 4% respectively.

In this region due to the presence of rocky sub-stratum it is not possible to dig deep tubewell having rocky sufficient amount of water. In plain anchals of South Munger 30% irrigated land is in Barahiya and the lowest percentage of tubewell irrigation is found in Munger anchal followed by 10% in Sikandra, 11% in Surajgarha, 15% in Lakhisarai and in other more than 20% land is irrigated through tubewell.

Tank irrigation is almost negligible in almost all anchals in Munger division in which the highest percentage of 10.9% is found in Lakshmipur, 8.2% in Sheikhpura and 3% in Jamui anchals. In other it is below 1%. So far as the other wells are concerned 49.9% land is irrigated in Jhajha whereas it is 16% in Sono, 13% in Jamui, 19% Sheikhpura, 24% in Barbigha and 17% in Ariari. In other anchals it is less than 5%. The other sources of irrigation is more prevalent in hilly areas where due to sloppy ground the peasant is more skilful in irrigating their fields through different sources, whatever may be possible in different seasons, whereas the highest 85% of irrigated land is found in Dharhara anchal, 69% in Munger anchal, 68% in Sono and 66% in Khaira and Lakhisarai anchals.

In other anchals the irrigated land from other sources is very low (Figure 7.8).

Figure 7.9 (a, b) and Table 7.6 show the graphical representation of net and gross irrigated area in the Munger division during the period 1979-80. This shows the intensity of irrigation in different anchals of Munger division where the highest intensity area of irrigation is found in Sheikhpura, Khagaria, Sangrampur, Barbigha and Lakhisarai anchals. Here the gross irrigated area is more than 3 times in comparison whit net irrigated area. The lowest value of gross sown area is found in Chakai, Sono, Munger, Kharagpur, Jamui, Sikandra, and Ariari anchals. Just double gross irrigated area in comparison with net irrigated area is found in Chautham,

Gogri, Parbatta, Khaira, Halsi, Surajgarha and Jamalpur anchals.

This map clearly represents that in southern hilly area due to intensive canal irrigation and in the plain area of North Munger due to intensive tubewell irrigation, the gross irrigated area is comparatively higher in these areas in comparison with the southern middle plain of the Division.

The distribution of total irrigated crop area and unirrigated crop area according to size class of holding is given in Table 7.7.

Figure 7.10 show the irrigated and unirrigated area according to size class of holding in the Munger division in 1979-80. It has been found that 30 hectares and over size class of holdings in irrigated area. So far as the unirrigated area is concerned the least figure is available for 20 and over hectares size class of holdings where the share of unirrigated land is higher in comparison with the irrigated land. Similar position is available in almost all size class of holding except 4-5 size class where irrigated area dominates over unirrigated area. The highest unirrigated area is found in 10-20 hectares size class in which it is above 51,000 and similar is the position in 5-10 hectares size class of holdings for the unirrigated crop.

Table 7.7 : Distribution of Irrigated Cropped Area and Unirrigated Cropped Area According to Size, Class of Holding (1999-2000)

Size class in hectare	*Total irrigated area*	*Total unirrigated cropped area*
0.0-0.5	10936	30057
0.5-1.0	13152	36222
1.0-2.0	20144	51185
2.0-3.0	14313	43287
3.0-4.0	10736	29605
4.0-5.0	8427	23928
5.0-10.0	17142	49045
10.0-20.0	8393	29741
20.0-30.0	2436	11818
30.0-40.0	808	6642
40.0-50.0	685	4371
50.0 and above	744	10143
Total	**107886**	**326125**

Source : District Statistical Office, Munger, 2001.

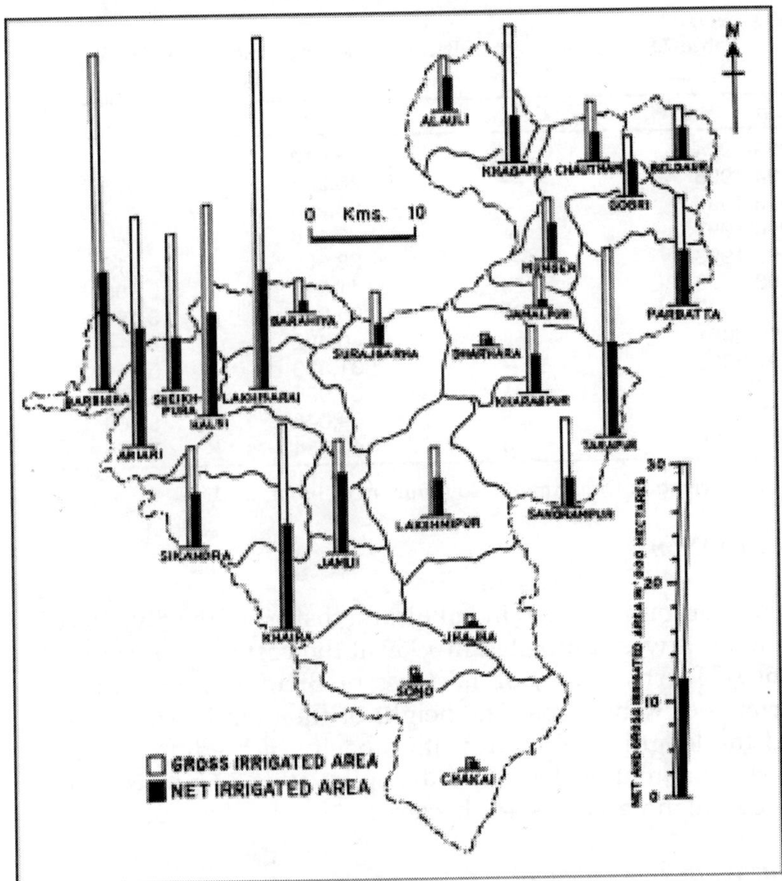

Fig. 7.9 : Munger Division : Variation of Net and Gross Irrigated Area.

Description of Some Important Reservoirs

Badua Reservoir

Badua reservoir is located in Sangrampur, Tatiha Bambor and Tarapur anchals. It was completed in 1968. It has been constructed at the cost of Rs. 608 lakhs. The total command area of this reservoir is 18,980 hectares. The achievement from this period project in 1968-69 is as given in Table 7.8.

Table 7.8 : Achievement of Badua Reservoir Project (1993-2005)

Year	*Garma*	*Kharif*	*Rabi*
1993-1994	—	28,010	—
1994-1995	—	26,238	216
1995-1996	—	26,284	—
1996-1997	—	30,178	—
1997-1998	—	38,300	513
1998-1999	229	41,514	3,995
1999-2000	325	39,941	3,167
2000-2001	368	40,492	—
2001-2002	—	31,833	5,053
2002-2003	421	39,025	5,100
2003-2004	836	39,048	8,000
2004-2005	916	40,984	7,500

Source : Irrigation Department, Government of Bihar, 2001.

Morwe Dam

This reservoir is located in Lakhisarai district of the division of Munger. It was completed in 1969 at the cost of 1.26 crores of rupees. It irrigates 4,532 hectares of Bhadai crops and 447 hectares of Rabi crops. The height of the dam is 523 metres, and the length is 56 metres. Its capacity of holding water is 12.34 cubic metres, length of the spillway is 33.5 metres and the command area is 4,480 hectares.

Kharagpur Lake Scheme

Kharagpur Lake planning was started 100 years back by Darbhanga Raj in Kharagpur block of Munger district. Some improvements were made by raising the height of the dam in order to improve the water storage capacity by the irrigation department. As a result of these measures 4,906 hectares of Kharif and 1,012 hectares of wheat fields and about 407 hectares of Garma crops have been brought under cultivation in Kharagpur block. This scheme irrigated 445 hectares of land and some of the irrigational development from this scheme is shown in Table 7.9.

Besides these, various working reservoirs exist in the division of Munger. For example Amriti Dam Reservoir is found in Sikandra block whose command area is 4,484 hectares.

Gidheshwar Pyne Scheme is situated in Khaira and Jamui blocks whose irrigational command area is 4,047 hectares. Nagi Dam Reservoir Scheme is situated in Jhajha and Lakshmipur blocks. The total command area of this irrigation scheme is 2,428 hectares. Tati River Irrigational Scheme is located in Barbigha and Sheikhpura blocks. The command area of this irrigational scheme is 2,114 hectares. Mohne Irrigation Scheme has already been executed in Sangrampur and Kharagpur blocks whose command area is 1,617 hectares. Balio Dam-cum-Reservoir Scheme is located in Lakshmipur block which covers 1,457 hectares of land. In addition to these there are Bajan Weir Scheme, Firangi Bigha Irrigation Scheme, Jai Kund Dam, Sikhandi Irrigational Scheme, etc. which exist in Barbigha, Sheikhpura, Kharagpur and Halsi anchals respectively (Figure 7.10).

Table 7.9 : Land Irrigated from Kharagpur Lake Scheme (1993-2005)

Year	*Garma*	*Kharif*	*Rabi*
1993-1994	—	1000	1350
1994-1995	875	12000	400
1995-1996	—	12000	11000
1996-1997	600	9000	1100
1997-1998	—	12000	2500
1998-1999	800	12000	2000
1999-2000	164	11000	2100
2000-2001	234	1095	1831
2001-2002	235	11000	2000
2002-2003	450	11000	660
2003-2004	450	11000	2500
2004-2005	814	125	1950

Source : Irrigation Department, Government of Bihar, 2001.

The following irrigational schemes in the district of Munger are under construction :

Dakra Nala Scheme

This is a major irrigation scheme and located just west of the city of Munger where water will be taken from the Ganga through electrically driven high power pumps installed on floating barges. The Project envisages lifting of 304 cusecs of water from the river Ganga to irrigate areas in Munger District. This pump canal scheme is proposed to be done in two phases. In Phase I, irrigation is proposed on the western side of Munger city up to Kajra-Surajgarha road. In Phase II, separate branch canal will have to be taken from the main canal. The gross command area of the Scheme under Phase I will be 17,466 hectares and in Phase II 7,631 hectares on eastern portion of Munger including Bariarpur. Hence the annual irrigation proposal is for 27,303 hectares of irrigation (Figure 7.11).

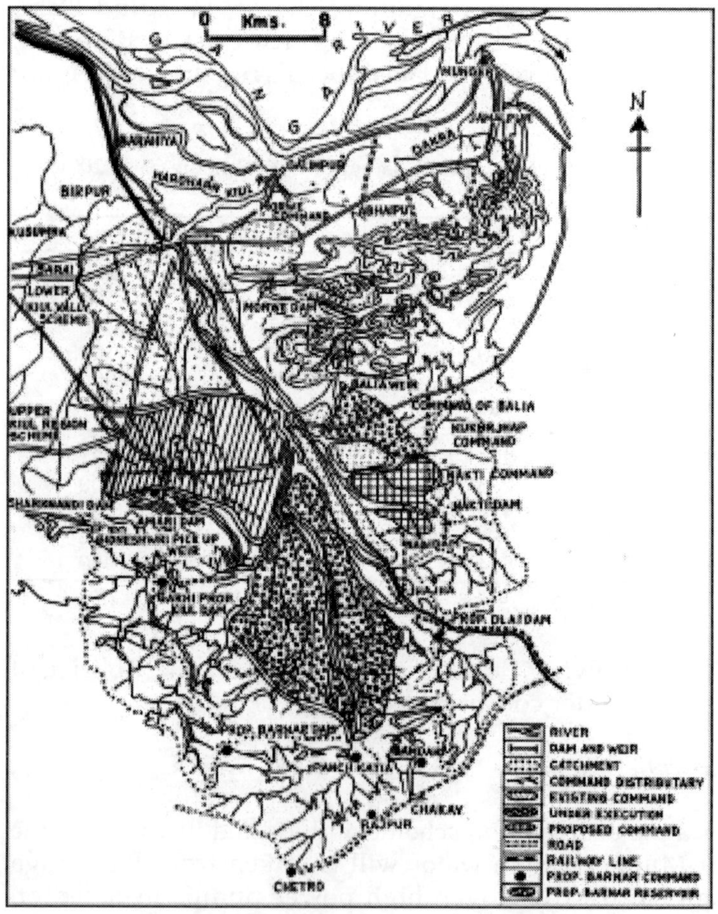

Fig. 7.10 : Munger Division : Different Irrigational Projects.

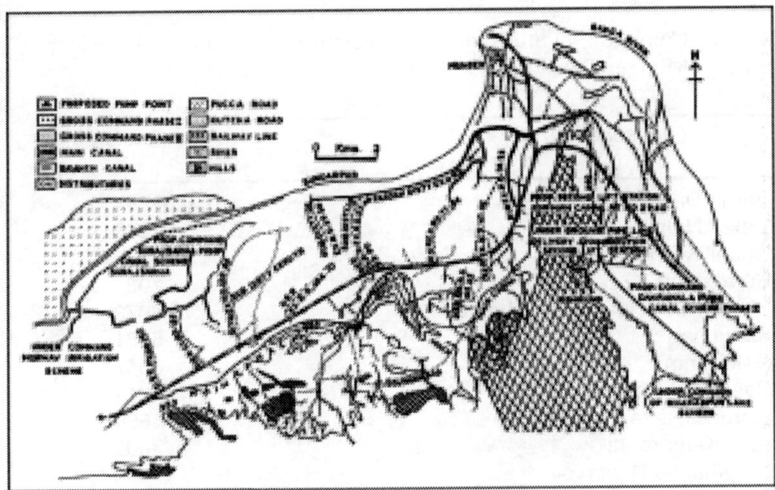

Fig. 7.11: Darka Nala Project: Phasewise Irrigation Scheme.

Table 7.10 : Comparative Study of all the Completed Irrigation Schemes of Munger Division (1999-2000)

Sl. No.	*Name of the schemes*	*Command area (hectares)*
1.	Badua Reservoir	18980
2.	Kaurihari River Scheme	11650
3.	Sakri Canal Scheme	5140
4.	Kharagpur Lake Scheme	4654
5.	Amriti Dam Reservoir	4484
6.	Morwe Dam	4480
7.	Gidheshwar Pyne Scheme	4047
8.	Nagi Dam Reservoir	2428
9.	Tati River Irrigation Scheme	2114
10.	Mohne Irrigation Scheme	1617
11.	Balio Dam Reservoir	1457
12.	Bajan Weir Scheme	1457
13.	Firangi Bigha Irrigation Scheme	1376
14	Jalkund Dam	1214
15.	Sirkhandi Reservoir Scheme	777
16.	Sogra Irrigation Scheme	445
17	Kailash Ghati Yojna	348
	Total	**66668**

Source : Irrigation Department, Government of Bihar, 2001.

Table 7.11 : Comparative Study of all the Proposed Irrigation Schemes of Munger Division (1999-2000)

Sl. No.	*Name of the schemes*	*Area (hectares)*
1.	Barna Dam Project	33994
2.	Dakra Nala	27303
3.	Upper Kiul Reservoir	23876
4.	Surajgarha	7034
5.	Sindhwasni Reservoir Scheme	6073
6.	Kukurjhap	5261
7.	Nakli Dam	3047
8.	Bas Kund Reservoir Sheme	1214
9.	Goraiya Reservoir Scheme	664
10.	E.R. Kamarganj	607
11.	Satgtharwa Reservoir	486
12.	Bagra Reservoir	239
	Total	**109798**

Source : Irrigation Department, Government of Bihar, 2001.

The command area of this scheme covers nearly 210 revenue villages of Surajgarha Pipriya, Dharhara, and Jamalpur anchals.

The intensity of cropping will increase with the availability of assured irrigation. Taking this into consideration the proposed cropping pattern has been finalised in consultation with the District Agricultural Officer, which is given below :

(A) For Flood Prone Area
  (i) Kharif Nil
  (ii) Rabi (high yielding varieties) Wheat
    Wheat (ordinary)
    Other Rabi
  (iii) Hot weather
(B) For Flood Free Area
  (i) Kharif
    Paddy (high yielding varieties) Paddy (ordinary)
    Kharif (others)
  (ii) Rabi
    Wheat (high yielding varieties) Wheat (ordinary)

Rabi (others)

(iii) Hot weather

Early maize.

Thus it is expected that 70% of the total crop command area will have, double cropping. For hot weather early maize has been considered as there is no practice in the locality for sugarcane cultivation.

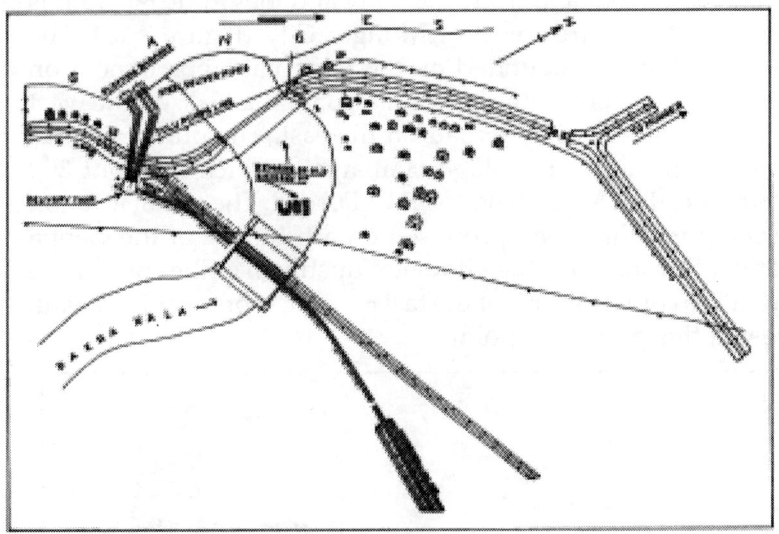

Fig. 7.12 : Environs of Darka Nala Irrigation Scheme.

Figure 7.12 shows the proposed irrigation scheme of Dakra Nala Project which flows from south to north-west of Munger city. Just on the southern bank of Ganga a road runs from Lakhisarai to Munger and Bhagalpur. In the middle south of this road, Gosai tola of village Heru Diara is situated where the dispersion of houses is shown for each and every family. In the north-western part of this map floating barge is shown and the delivery tank of water is situated south of the road from this tank. Eastern section of this main canal runs towards south-east and an approach road is going to be constructed which will be named as Munger-Bhagalpur Bye-Pass Road. Flood gate will be constructed with the help of R.C.C.

Surajgarha Pump Canal Scheme

This is a medium irrigation scheme in Lakhisarai District where irrigation is proposed by lifting water from the river Ganga by electrically driven high power pumps installed on floating barge.

The gross command area of the scheme is 4,586 hectares which is located at latitude 25° 15′N and longitude 86° 15′E is about 29 kilometres west of Munger City (Figure 7.13). The command area is bounded by the marginal embankment on river. Harohar on northern side and Garakha Nala on the west and south and a village road on the east. The pump site has been proposed near village Haibatganj which is about 300 metres north of Munger-Patna P.W.D. road. The actual location of the pump has been proposed on a tributary of the Ganga named Harohar just one kilometre upstream of the confluence point to avoid direct hit of the fast current, moreover firm bank exist at this particular point.

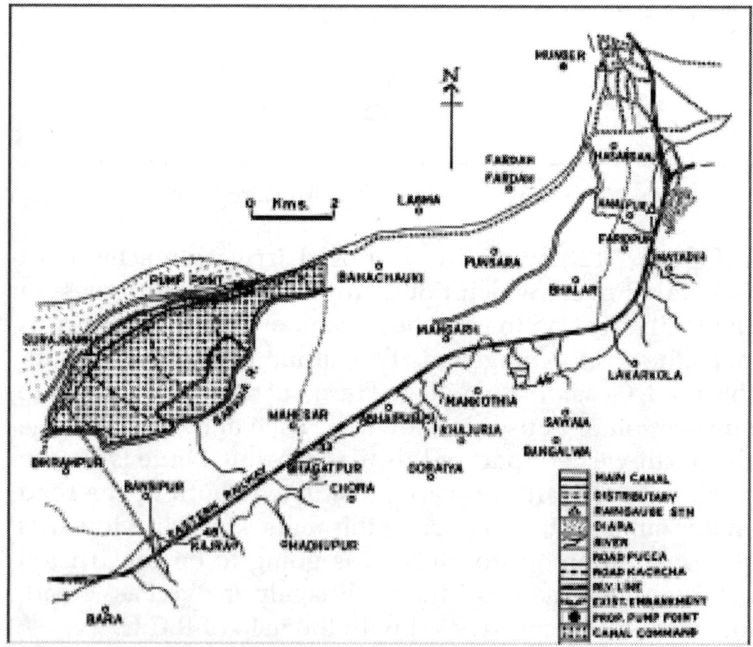

Fig. 7.13 : Munger Division : Showing Surajgarha Pump Canal Scheme.

The river Harohar is perennial and is linked to the Ganga about one kilometre downstream of the proposed pump site. The river Harohar at the point of making confluence has a slope less than 15 centimetres per kilometre and in fact the Ganga water itself is always available at this point. In case of any deficiency the Ganga water will get sucked in through the channel. The minimum water level of the channel has been calculated on the basis of gauge observation for 10 years Kastaharni ghat in Munger.

The command area of the scheme covers nearly 40 revenue villages. From records of District Statistical Office, total cropped area of different crops has been obtained for each village and by taking average the percentage of total cultivable area has been calculated to be 90%. This occasional inundation will also be eliminated by means of a flood protection scheme

The existing crop pattern of the command area on the basis of district statistical records of the year 2000-01 are as follows :

(A) Kharif	*Percentage of cropped command area*	*Area in hectares*
(i) Paddy	0.5%	20
(ii) Dalhan	0.1%	4
(iii) Sarai	8.0%	329
(iv) Maize	42.0%	1724
(v) Jowar	1.5%	62
(vi) Other Bhadai	1.5%	62
Total	**53.6%**	**2201**

(B) Rabi	*Percentage of cropped command area*	*Area in hectares*
(i) Dalhan	6%	246
(ii) Potato	0.5%	21
(iii) Telhan	0.25%	11
(iv) Chilli	0.25%	11
(v) Barley	1.00%	41
(vi) Wheat	12%	494
(vii) Wheat (G)	46%	1886
Total	**66%**	**2710**

From the above Table it will be seen that in Kharif 53.6% of cropped command area is cropped and intensity of paddy cultivation is particularly poor. This is mainly due to absence of assured irrigation. In Rabi, the intensity of cropping is 66%. The yield is poor on account of absence of irrigation.

After completion of this Scheme the flood of the Ganga will be checked and irrigation is assured. The intensity of Kharif will definitely rise from 53.6% to 70% than that of Rabi from 66% to 70%. The following crop pattern is expected to develop :

(A) Kharif	*Percentage of cropped command area*	*Area in hectares*
(i) Rice high yielding	25%	1032
(ii) Rice local	25%	1032
(iii) Other Kharif	20%	826
Total	**70%**	**2890**

(B) Rabi	*Percentage of cropped command area*	*Area in hectares*
(i) Wheat high yielding	25%	1032
(ii) Wheat local	25%	1032
(iii) Other Rabi	20%	826
Total	**70%**	**2890**

(C) Hot weathers	*Percentage of cropped command area*	*Area in hectares*
Early Maize	30%	1239

It is expected that 70% of the total cropped command area will, have double cropping. For hot weather, only early Maize has been considered as there is no practice in the locality for sugarcane cultivation.

Upper Kiul Reservoir Project

The Upper Kiul Reservoir Project report was received from Central Water and Power Commission in July, 1971. The project was estimated to cost Rs. 4.94 crores envisaging the construction of Dam to extend irrigation to 20,235 hectares of Rabi including stabilising the existing command of 6,475 hectares of Gidheshwar Canal.

The Scheme was examined in the Directorate of Central Water Commission and Ministry of Finance (Technical Section and Water Management Division of Agriculture and the comments were sent to different States in December, 1971 and March, 1972).

Barnar Reservoir Project

The estimated cost of Barnar Reservoir Project is Rs. 8.04 crores and envisages construction of a masonry dam of 73 metres high and two canals taking off from either side of the dam will provide irrigation to an area of 22,672 hectares annually in Chakai, Sono, Jhajha and Khaira anchals. The project was submitted to the Technical Advisory Committee for consideration. The T.A.C. in its 18th meeting held on 7.1.1974 had made the following observations in respect of this project :

> "The Advisory Committee noted that whereas points regarding water availability and drainage had been satisfactorily examined, the information regarding soil surveys and cropping pattern are still inadequate. Greater attention has been required to the Agricultural aspect of the project planning if productivity from irrigation project; has to improve."

Besides these the command area of Sindhwasni Reservoir Project is 6,073 hectares, Kukurjha is 5,261 hectares, Nakti dam 3,047 hectares, Baskund Reservoir Scheme is 1,214 hectares, Goraiya Reservoir Scheme 664 hectares, Satgharwar 486 and Bagra reservoir 239 hectares.

Minor Irrigation Schemes

Minor Ahars and Pynes : The system of irrigating land from water stored in Ahars is practised from times immemorial. People uptil now are dependent for irrigation of their fields mainly on water stored in these irrigation sources. Nowadays the condition of these Ahars has deteriorated a lot. A drive for their improvement and repair has been taken up under World Food Programme. It is proposed that almost all the useful and big Ahars and reservoirs which will be about 1,000 in numbers be repaired and improved within a few years.

State Tubewells : Percentage of poor inhabitants is high in this region. They are not in a position to invest money in making their own arrangement of irrigation for their fields.

Hence it is essential that sufficient number of state tubewells be installed so that farmers having small holdings of land may be able to grow double and treble crops. The scheme of sinking state tubewells has proved very successful as it provides an assured source of irrigation.

Private tubewells of 10 cm and 5 cm diameters : This is also an assured source of irrigation increase of irrigation potential by these automatically increase the area of double and treble cropping pattern. People are also very much enthusiastic to get their own tubewells sink for irrigation of their land by taking loans from land mortgage bank and other commercial banks.

Private pumping sets : There are two perennial rivers viz. the Ganga and the Harohar. The installation of high power pumping set on the bank of these two rivers are being explored. This will benefit a vast area in which double and treble cropping pattern will be successful. There is also a proposition to install high power pumping sets nearabout the storage of water of the major irrigation schemes on other rivers in which water is available throughout the year.

State and private wells : There are about 8,000 state and private wells out of which boring without strainer has been completed in a little more than 1,000. As it is an assured means of irrigation for the farmers having small tract, it is therefore, essential that maximum number of state and private wells are bored and farmers are given assistance to purchase pumping set and get them energised for irrigation purposes. It is also proposed that sufficient number of new State and private wells be sunk on the capital received from the government.

Problems in Reservoir Areas

The constructions of reservoirs inevitably lead to other associated problems including failure of rims, leakage or excessive siltation. There is no reason why these aspects cannot be predicted with reasonable accuracy provided adequate investigations have been undertaken. The present generation of geographers and geologists are in a position to suitably assess the extent of silting in the design life of a dam not only from stream sediment but also by using aerial photographs and ground traverses to determine the existence of landslide or rock-falls may take place and contribute abnormal quantities of sediment into stream systems, and ultimately into reservoirs. A critical eye of the investigation can prevent future disaster.

Major Irrigation Problems

Some of the more important problems associated with the existing water utilization pattern for irrigation include :

(1) Development of additional storage at reasonable cost (since many lands now irrigated require additional water for an adequate supply).

(2) Limitation of areas to be irrigated to the available water supplies. Many irrigation enterprises have been developed in which irrigated area is overexpanded so that during dry years, serious losses, troublesome litigation and urgent requests for public assistance occur.

(3) Existence of too many small irrigation organizations on a given stream, pooling a complexity of water rights would simplify administration, reduce operating costs, and eliminate much of the litigation which would otherwise result from attempts to preserve priority of individual water rights.

(4) Failure to utilize for its highest and best use in many areas. In many regions of the West, irrigation is the highest and best use for much of the water, but in

some instances water use projects constructed for navigation and flood control have hindered irrigation development.

(5) Failure to apportion irrigation costs properly. In the past there has been use of the principle of apportioning costs of irrigation development on the basis of benefits received from such development. Arguments advanced for government irrigation enterprises are based largely on benefits to communities where projects are located, rather than merely on benefits to settlers who have to irrigate the land. Businessmen of the communities who receive benefits not only from the construction of the irrigation works, but also from bringing the land into cultivation.

In project involving irrigation and other purposes such as generation of power, flood control, navigation or furnishing of municipal water supplies, cost of development should be apportioned as accurately as possible to various uses in terms of estimated benefits.

The proposition about the supply of need based water for successful cropping is known as irrigation planning. In some areas, too much water causes serious damages through floods, erosion or waterlogging over the soil which make successful crop production impossible. Too much water is harmful for plant growth, and means must be devised to remove this excess.

In many parts of the Munger division lands are so poorly drained that they are unsuitable for crop production. In many cases, these swampy areas are comparatively fertile, because of the accumulated decay of aquatic vegetation, but this is not always true.

The central government of India has followed a policy of increasing the reclamation of swampy land since long and this policy is best suited for the reclamation of Diara and Tal lands in Munger.

It may be concluded that water resource of the Munger division has multiple source of occurrences such as underground reservoir, water stored in river, chaurs, etc including precipitation, running streams and artificial reservoirs constructed by the government or private agencies for the purpose of irrigating land and increasing the agricultural production of the district. This study delineated about the hydrological conditions of the Munger division, besides the location of hot springs and measuring the irrigational use of water with special reference to completed and proposed irrigational schemes and a light description about the irrigational schemes and potentialities of the study area.

In a study of planned development of water resource the district of differential water potentialities in different physiographic regions of the district such as (a) flood prone area of North Munger has abundant resource of undeveloped running water, (b) the Gangetic riparian tracts and upland plains of South Munger where stored water in Tals and Chaurs is abundant, and (c) the forested hilly and undulating lanscape of south-south-west Munger has very poor sub-surface water resource. Ultimately, it can be said that from historic times up to present, the irrigational history of Munger shows almost a planned development of water resource through irrigational use, way of transport, and for the supply of drinking water especially to the city of Munger, which may be called the 'nerve-centre' of the district.

REFERENCES

Dewett, K.K. Singh, Gurucharan and J.D., Verma, 'Irrigation and Power', Indian Economics, S. Chand and Company Limited, New Delhi, 1971, p. 157.

Dutta, Sujit Kumar, 'Geohydrological Report on Thermal Springs : Munger', Rural Electrification Corporation of India, Undertaking, 1973, pp. 2-3.

Flint, R.W, "The Sustainable Development of Water Resources : Water Security in the 21st Century", Five E's Unlimited, 28 Randolph Place, NW, Washington, DC. 20001 (202)232-4591, rwflint@eeeee.nethttp://www.sustainabledevelopment solutions.com

Flint R.W., "The Sustainable Development of Water Resources", Universities Council on Water Resources Update, Issue 127, pages 41-51, February 2004 Five E's Unlimited, Washington, DC.

Gadgil, D.R., 'Economic Effects of Irrigation', 1945, p. 173.

Knowles, "Economic Development of British Empire Overseas", Vol. I, pp. 367-68,

Mamoria, C.B., 'Irrigation and Floods', Agricultural Problems of India, Kitab Mahal, Allahabad, 1973, p. 161.

Mamoria, C.B., op. cit, p. 162.

Mithal, R.S., "Geotectonic Evaluation for Development of Water Resources in India", Presidential Address at the 66th Session; Indian Science Congress Association, Calcutta, 1979, pp. 1-2

Murthy, Y.K., 'Irrigation Research'. Irrigation and Power in the Fourth Plan, Publication Division Ministry, Patiala House, New Delhi, 1970, p. 13

Punmia, B.A. and Bansilal Brij Pande, 'Irrigation and Water Power Engineering', Standard Publishers and Distributors, Delhi-6,

The United States Army Corps of Engineers in Oregon, 'Water Resources Development', North Pacific Division, Custom House, Portland, 1975, p. 37.

Vohra, B.B., 'Hand Book on Irrigation Water Management', Water Management Division, Ministry of Agriculture, New Delhi. 1971, pp. 5-8.

8

Urbanization and Industrialization for Sustainable Development of the Region

Either on international or regional perspective, the growth of urbanization or urban economy and its impact on regional development is highly relevant today for developing countries which are now passing through various phases of economic development. Urbanization being a dynamic force involves urban elements that suffer considerable temporal and spatial variations. In case the process is viewed with a definite regional background, urbanization influence the surrounding rural areas, which can readily be seen in rural urban linkages. Hence, urbanization is equipped with the potentialities of development of non-agricultural production to solve in a better way the requirements of urban manufactured goods in the rural areas. The term 'urban' has developed much wider connotation now in India than ever before. In the context of regional development at the district level, reference has been made to the various economic and social problems which are closely connected with the process of urbanization and growth. Urbanization naturally creats a strange behaviour and one would find that there is more impersonality, less mechanical consensus, more control on behaviour, less informality and high specialisation.

Some notable contributions have been made during the last three decades or so on the process of urban growth and regional development by geographers, location economists and urban economists. In economic literature, the problems of urbanization and their regional development are seen to have developed from the theory of location of industries. Though the present interest in Von Thunen's isolated state traces the growth of the interest in spatial distribution of economic activity back to a purely agricultural economy, yet it is usual to trace the growth of such theory from Weber's classic work on the location of industries. This approach developed originally by Christaller and Losch. Christaller's model of central places hierarchy can be applied with special reference to the economies where urban centres are well-developed and cities regulate the economic life of a country. With poor urban development over large areas, this model may prove to be applicable in practice. Collin Clark studied the density of urban population and its growth with great zeal and received interests of others also to this aspect of urban population distribution. For this, he produced evidences regardless of time or place, as the population density within the city conformed to single empirically derived expression :

$$D = De^{-bx}$$

Where *D* is the population density and distance
X from the city centre
De is the Central density
b is the gradient (The slope factors).

In this way we may say that the contributions of these scholars have placed varying emphasis on the role of different factors in urban development. Urban growth is influenced by several socio-economic, demographic, cultural and technological factors. It is, therefore, essential to study the impact of these factors so as to accelerate the tempo of urban development.

The study area of the present chapter relates to the Munger division where there are 12 urban centres. These urban centres possess pronounced rural character with a few urban amenities. The main objective of this chapter is to highlight the different size class of towns and their density growth and economic functions in the Munger division.

Methodology

There are two customary indices to study the level and trends of urbanization :

(i) ***Degree of urbanization*** **:** It is the percentage of total population living in urban areas. The formula for computing is as follows :

$$L = \frac{U}{T} \times 100$$

where L = Degree of urbanization,
U = Urban population,
T = Total population.

In accordance with the international comparison, this index is further defined in two ways :

(a) Instead of total urban population living in cities with 1,00,000 and more inhabitants is considered as urban population.
(b) In place of total urban population the population living in urban places at least 20,000 inhabitants is considered as urban population.

(ii) ***Rate of speed of Urbanization*** **:** It is a simple arithmetic growth rate of degree of urbanization between two or more censuses. It is calculated as :

$$\text{Rate of urbanization} = \frac{L_2 - L_1}{L_1} \times 100$$

Where L_1 and L_2 are the degrees of urbanization at two point of time.

There are two ways of measuring the urban growth. One is the instantaneous method which ascertains the population in all urban categories at each and every census, tracing the changes in each class regardless of the cities that make it up. The other is the continuous method which begins with particular city and traces the subsequent extension of these groups. These two methods have been used to study urban population growth by different categories of towns considering 11 towns of 1951, one of which was declassified in 1971 census.

The concept of national growth rate method has been adopted to study the level of migration in cities and towns. The method describes that if PT and P_T be national population at the beginning and end of the intercensal period respectively and $P_1 P_2 P_3......P_n$. be population of the geographic subdivisions at the beginning of the period. Then the estimated migration rate in m_1, for area 1 is given by :

where K is taken as 100 or 1000.

In this concept, the difference between actual growth rate of the town has been considered as the net migration rate of the town. The net migration was multiplied by the base year population in order to get the volume of migration for each town. These were summed up and divided by the base year total population of the towns of different size classes and multiplied by 100 in order to get the net migration rate for each category of towns and cities.

In search of a new concept, this chapter is concerned with the tertiary and secondary population of different anchals which indicate the level of regional development through urbanization process. Hence, this study aims to highlight the relationship between location coefficient of secondary and tertiary population with the location coefficient of service facilities available in different anchals in the Munger division. In this way, cumulative frequency and percentage of both these values and their plotting on Lorenz in bringing out an inference to the problem an integrated relationship of density of population, functional system and growth patterns in every urban centres of the Munger division have been studied. Towns are classified into six classes according to their population size (Table 8.1)

Table 8.1 : Number and Size Class of Towns in the Munger division (2001)

Size class	Population range	Name and number of towns
I	1,00,000 and above	Munger
II	50,000-99,999	Jamalpur
		Jamui
		Lakhisarai
III	20,000-49,999	Kharagpur
		Barahiya
		Barbigha
		Sheikhpura
		Gogri
		Jhajha
		Khagaria
IV	10,000-19,999	Nil
V	5,000- 9,999	Asarganj
VI	Below 5,000	Nil

Source : Census of India, 2001.

Urbanization and its Characteristics

Human settlement is the expression of social philosophy, economic characteristics and administrative organisations of a given society. System of settlements in terms of rank-size relationship, hierarchy, spatial ordering and centrality of functions differ from region to region. A town where agricultural activity constitutes the principal sector of economy has to develop a system in which agricultural villages occupy the lowest order and the market based urban settlements at the highest order. The system in other towns having industrial economy, of necessity has to be altogether different bias for large size urban settlements accommodating larger proportion of the population in comparison with samaller rural settlements.

According to 2001 census, the total population of Munger division was 6,96,471 distributed in 3,460 villages and 12 urban centres. All the villages and urban centres together constitute the settlement system in the district in which the urban centres form a sub-set in the main system, where a major segment of

44.48 millions people lived in the rural area, about 7 millions people or 14.08% population lived in the urban centres

From Table 8.2, it is evident that the level of urbanization in Munger division is very slow. It is about 3.45% in 1901 which has gone up to 13% in 1971, 14.05% in 1981, 14.06% in 1991 and 14.08% in 2001 showing thereby a rather slow average shift only which was slightly in a decade.

Table 8.2 : Percentage of Urban Population (1901-2001)

Census years	*No. of towns*	*Rate of variationin urban population*	*% of urban population*
1901	4	NA	3.45
1911	4	24.97	4.18
1921	6	26.40	5.56
1931	10	11.66	5.51
1941	11	64.43	8.08
1951	11	28.92	9.38
1961	11	40.56	11.08
1971	12	54.62	13.00
1981	12	58.78	14.05
1991	12	62.96	14.06
2001	12		14.08

Source : District Census of India, 2001.

The pattern of urbanization has undergone a dramatic reversal in terms of percentage, growth of towns in the Munger division.

Table 8.3 shows the percentage growth of towns in the Munger division from 1941-51, 1951-61, 1961-71, 1971-81, 1981-91 and 1991-2001 censuses. From 1941-51 decade may be called a moderate growth of urban centres as for example Sheikhpura shows a 3.3% growth, Khagaria shows 35% growth whereas in Barbigha it is 7.5% and the maximum 19.01% in Jhajha. Similarly, in Jamalpur it is 29.1% for 1951-61 but is just 8.1% in 1961-71. Kharagpur shows highest growth rate in 1981 i.e., 100% whereas the lowest growth rate in urban centre in 1981 is 10.22% in Barahiya and similar is the position in almost all urban centres in the Munger division.

Table 8.3: Percentage Growth of Towns in the Munger Division (1951-2001)

Name of the town	*2001*	*1991*	*1981*	*1971*	*1961*	*1951*
Munger	96.98	97.08	96.07	14.1	20.7	17.7
Jamalpur	45.01	39.45	26.88	8.1	29.1	12.1
Bariarpur	-	-	59.93	-	Village	Village
Sheikhpura	44.97	43.05	41.76	26.1	6.1	3.3
Jamui	25.79	24.26	22.11	22.3	26.1	8.5
Khagaria	49.35	48.25	47.17	24.9	36.6	35.4
Barbigha	54.12	53.07	51.14	13.2	33.8	*7.5*
Lakhisarai	50.68	49.86	48.09	43.9	12.4	17.5
Jhajha	43.21	40.36	39.04	28.7	35.2	19.0
Kharagpur	100.00	100.00	100.00	25.4	30.1	10.6
Barahiya	14.25	12.75	10.22	27.0	1.2	12.5
Gogri	22.61	20.35	16.18	-	Village	Village
Asarganj	48.59	Village	Village	Village	Village	Village

Source : Census of India, 2001, Final Population Table.

Urbanization is the system of settlements in terms of rank-size relationship. Table 8.4 shows the change in the rank of population of urban centres in the district of Munger. It is clearly evident that in most cases only a few towns have gone to the higher rank from its position in 1951 census as for example Barbigha and Khagaria, other urban centres have decreased in rank to lower rate of population growth. Some urban centres are maintaining their status quo from 1951 to 2001 census as for example Munger, Kharagpur and Jhajha. This shows that the pace of urbanization and regional development are not so sound in the Munger division.

The total area covered by urban centres of the region is 16,532 square kilometres (2001) and the urban population of the region is 466,285. The average density of urban population of the region is 2820 persons per square kilometre. The density of population varies from one town to the other depending upon their physical, economic and cultural characteristics. So far as the density of urban population is concerned Table 8.5 shows the density of rural population. There is a sharp contrast between urban and rural population density in 2001 census in different urban and rural areas of the district. As for example in Munger urban centre per sq. km. the density of urban population is 5,693 whereas it is just 318 persons per square kilometre in rural area and similar is the situation in most of the urban centres of the division. As for example in Gogri anchal the population density was 421 in 1971 but it was 1805 persons in Gogri urban centre.

Table 8.4 : Change in the Rank of Towns (1951-2001)

Name of the town	*1951*	*1961*	*1971*	*1981*	*1991*	*2001*
Munger	II	I	I	I	I	I
Jamalpur	III	II	II	II	II	II
Bariarpur	VI	V	IV	III	III	III
Kharagpur	V	IV	IV	IV	III	III
Lakhisarai	IV	III	III	III	II	II
Sheikhpura	IV	III	III	III	III	III
Barbigha	IV	IV	IV	IV	III	III
Barahiya	III	III	III	III	III	III
Jamui	IV	III	III	III	II	II
Jhajha	IV	III	IV	IV	III	III
Khagaria	IV	IV	III	IV	III	III
Gogri	IV	IV	IV	IV	IV	III
Asarganj	—	—	—	—	—	V

Source : District Census Handbook of Munger, 1971 and Census of India, 2001.

Figure 8.1 shows the hierarchy of towns according to the size of population in 2001. It is quite evident that southern Munger is divided of urban settlements including the Kosi belt in the north but on both sides of the Ganga the number of towns are located such as Barahiya, Lakhisarai, Khagaria, Gogri, Bariarpur, Munger, etc. According to population size the largest urban centre is Munger and the smallest one is Bariarpur whose population is 102,000 and 5,000 (in 1971) respectively. Other urban centres are of medium population size.

Functions of urban centres and their spatial linkages constitute important elements in the urban centres distribution pattern. The pattern of spacing and functional mix of urban settlements depend on the character of the space economy. Figure 8.2 and Table 8.6 show the functional characteristics of urban places of Munger division according to 2001 census. For

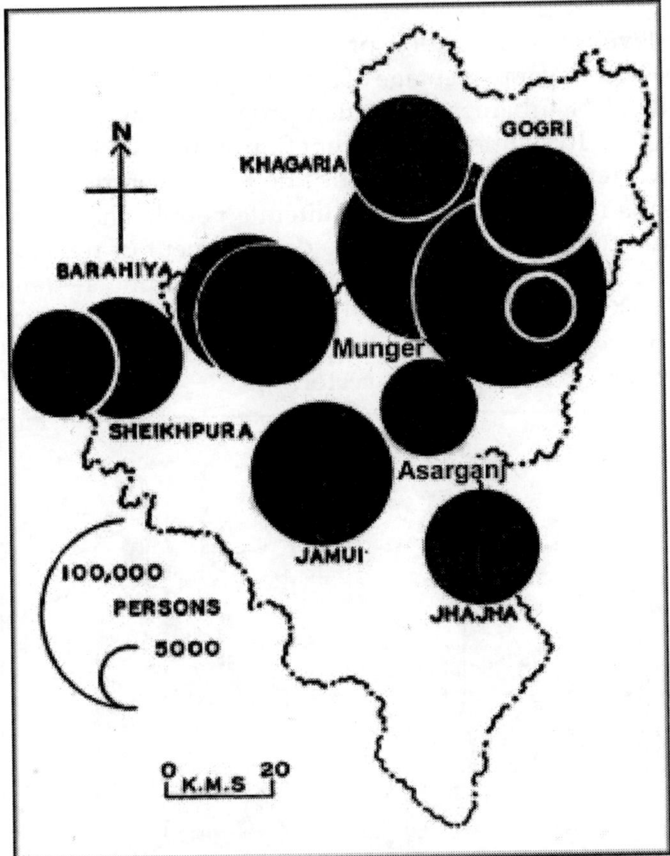

Fig. 8.1 : Munger Division : Hierachy of Town.

this study, the workers engaged in primary, secondary and tertiary sectors are considered for each and every urban centre. It has been found that in northern part of Munger division the towns have higher percentage of workers engaged in tertiary sector making the high status of towns, whereas in smaller town more and more people and workers are engaged in primary sector rather than the tertiary sector. As for example in North Munger—Khagaria 60% people are engaged in tertiary sector while in Gogri only 24% people are engaged in tertiary sector. The predominance of workers in tertiary sector indicates the higher level of urbanization process going on in that place. Whereas the lower percentage of tertiary workers indicates the lower level of urbanization which progress slowly towards higher level. It is quite evident from figures that southern and south-western part of Munger has no urban centre because the area is rugged tract, mostly illiterate people and the bad economic conditions retarded the development of transport, industry, trade and commerce. In southern Munger—Jamui as well as Kharagpur, Barahiya, Lakhisarai and Barbigha have lower percentage of people engaged in tertiary sectors in comparison with the primary sector.

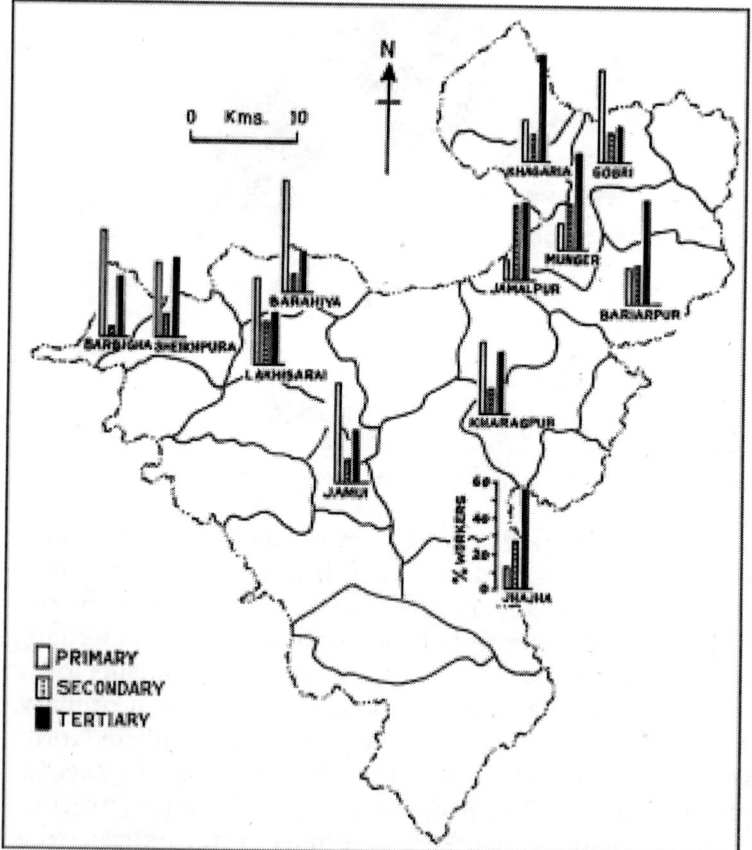

Fig. 8.2 : Munger Division : Functional Characteristics of Urban Centres.

So far as the secondary sector is concerned almost all towns come as a trough between primary and secondary sectors (2001). In the Munger division, Bariarpur, Jamalpur, Munger, Sheikhpura are only to mention a few which have higher level of urbanization in comparison with the other centres of the district.

Table 8.5 shows the force of urbanization in regional development in the Munger division. In this study mainly economic factors have been emphasised through graphical picture. Here *x* axis represents the average composite in —*dx* for six variables—educational facilities, medical facilities, cultural, transport and communication lines and bank as financial institution on anchal basis, whereas *y* axis represents the percentage of secondary and tertiary population because in any place the relationship between secondary and tertiary populations and the availability of different services indicate the economics of regional development because the localization of tertiary and secondary workers and the service facilities in an anchal are the true indicators of the regional development. Hence the product moment correlation coefficient *(r)* has been calculated for two sets of data on anchal level and it has been found that the *r* value is 0.88 which really shows high correlation, Similarly, the regression value comes to Y=1.09+.72X and this is the best-fit line. In this way it can be concluded that the percentage of secondary and tertiary workers and the availability of different service facilities in different anchals are the true sign of development of urbanization in the region.

Problem

From Table 8.5, it may be seen that out of some of the problems of urban centres, they received scant attention in the Five Year Plans of India. Here the vital question is the mode of linkages between urbanization and economic growth and the elements of planning. Problems in close connection with the location, size, distribution, concentration and scatter of population, etc. did not receive serious attention. The crucial role of urbanization is an important task for social and economic

Table 8.5 : Blockwise Abstract of Educational, Cultural, Medical, Communication and Financial Amenities in the Munger Division (2000)

Name of Anchals	*Educational*	*Medical*	*Cultural*	*Communication*	*Financial*	*Total*	*Average*	*% of workers engaged in secondary and tertiary sector*
Munger	72	26	22	115	42	259	56	75
Bariarpur	14	3	9	22	9	46	12	11
Jamalpur	56	16	20	98	18	210	38	43
Dharhara	22	17	15	34	22	110	22	22
Kharagpur	16	4	11	30	10	78	14	13
Asarganj	14	5	8	23	11	55	9	9
Tarapur	40	10	12	60	18	151	29	13
Tetiha Bambor	8	3	4	14	7	22	9	5
Sangrampur	20	7	11	29	9	81	14	9
Barahiya	42	29	27	41	26	157	34	26
Pipariya	4	9	3	8	4	24	3	3
Surajgarha	9	19	8	20	11	71	9	9
Lakhisarai	16	28	14	26	18	96	24	26
Ramgarh Chowk	2	2	0	3	3	12	2	1
Halsi	5	4	1	7	8	36	6	4
Barbigha	20	15	13	29	18	101	18	15
Sheikhopur Sarai	8	6	4	10	8	36	7	6
Sheikhpura	13	23	11	21	17	85	17	20
Ghat Kusumbha	5	11	4	9	7	26	8	9
Chewara	3	8	3	7	6	26	4	3
Ariari	11	21	11	19	16	81	14	11
Islamnagar Aliganj	6	3	4	11	5	27	5	3
Sikandra	21	8	12	35	15	86	18	10
Jamui	28	26	20	34	31	124	29	29
Barhat	2	2	3	1	2	11	2	1
Lakshmipur	5	5	12	9	12	39	8	6
Jhajha	41	26	26	41	22	157	34	40
Gidhaur	3	3	5	4	5	12	4	3
Khaira	25	15	14	26	21	97	25	14
Sono	45	16	24	26	38	126	29	27
Chakai	21	19	24	19	15	89	19	8
Alauli	21	26	29	19	41	115	29	34
Khagaria	34	29	24	21	46	131	31	39
Mansi	4	6	7	3	9	35	9	5
Chautham	9	14	16	7	19	64	19	9
Beldaur	5	9	8	12	15	55	11	15
Gogri	15	19	19	24	29	98	26	15
Parbatta	21	19	24	29	16	98	25	16

development has not been adequately realized. The plan documents do not contain any reference to the positive characteristics of urbanization and the dynamics by which it could create development. At last, urbanization has been treated passive factor in economic development.

At present there are wide disparities in urbanization levels in different parts of the districts, the most urbanised parts are Munger, Jamalpur and Khagaria having a proportion of urban total population in the range of 30% or above. Yet another important aspect of urbanization is the heavy concentration of urban population in class IV towns, i.e., Kharagpur, Barbigha, Jhajha, Khagaria and Gogri which have approximately half of the total urban population of the Munger division.

Policy on Problems of Urbanization and their Development

Planned urban growth is a necessary component of the infrastructure of economic development as town provides inter alia, a variety of centralised services for the surrounding rural areas such as marketing of agricultural surplus and products of village and cottage industries and supply to rural areas of a variety of goods, e.g., fertilisers, engineering products like pumps, and pipes, manufactured essential commodities like bicycles, kerosene, soft coke, building materials and medicines and last but not least, specialist skill in a wide variety of situations required for rural development :

(i) The regional plans through which the desired national economic plans are to be achieved should reflect on the following points : Study on natural and geographical conditions made in order to determine the suitability of the land for various development purposes.

(ii) Population distribution and their structure.

(iii) An analysis of these findings, aimed at determining the location of basic facilities, socio-economic phenomenon and land use in order to ascertain the compactability of existing land use with natural and geographic conditions and functional connection.

(iv) Proposal of investment programmes, housing development, communication network, agriculture and allied industrial development.

Spatial planning has not yet developed in a systematic way in different urban centres in the Munger division. Unless spatial dimension is made for development programmes of towns, the objectives and goals of these plans can hardly be achieved. Priority should be given to formulate perspective regional planes at block level. Sheikhpura, Barbigha, Barahiya, Jhajha and Gogri towns in the Munger division are overgrown rural service centres. These towns show more agglomeration of population with hardly any change in economic character. Munger, Jamalpur and Khagaria towns have tremendous changes in growth whereas Barbigha, Jhajha and Bariarpur are stagnant ones in terms of employment as well as population. Hence, our planners should have a duty to suggest the reduction of disparities of different urban centres of the division in order to help planned development of the region on sound economic footings. The Fifth Five Year Plan of the country has set forth various policies and programmes and earmarked funds to different departments through sectoral allocations. The lacuna has been that these plans when transformed into their physical components do not fulfil the desired objectives in terms of equitable growth of the region. It is obvious that any economic plan should interrelate agricultural development, industrial development and others with each other within the same set of geographical coordinates.

Different agencies function in the same area and implement their programmes but these agencies tend to be administratively remote from one another and do not interrelate their functions. The result of these extra-ordinary functional fragmentation has caused a setback in the balanced development of urban centres. Hence, sectoral approach of the national plans requires modification by the introduction of the horizontal system of coordination through regional planning approach. Appropriate organisational set up at different levels is necessary for horizontal as well as vertical integration of various development programmes at block level.

Industrial Development

Location of industry is dependant on transportation linkages, costs incurred in the assembly of raw materials and delivery of finished products to the market. In case of consumer goods, there are number of finished products which are available in the same price throughout the country leaving cost of transportation of the materials to the production point as the major variable in material costs, labour supplies and regional variations in wage rates have been largely discounted as locational factors.

The first point of locational analysis has been concerned with the study of existing locational aspects of the site, situation, availability of raw materials, labour, market, capital and the facility of transportation of goods. In this section of the chapter, an attempt has been made to show the relationship between urban centres and industrial location in the Munger division.

Munger division has an old tradition of industrial development including the skill of artisans, the excellence of their work and the facility of river transport :

1. ***Indigo*** : According to statistical account of Bengal published in 1877 by Sir William Hunter, the industry connected with the preparation of indigo has now for long time taken the first place amongst manufacturers in this division.
2. ***Slate Quarrying Industry*** : In 1864, Messers Amblar and Company, set up a slate factory firstly to manufacture excellent roofing slate and by 1914 for manufacture of switchboards, fuse bases, knife, switches and school slates.

Due to change in building habits, this industry also gradually declined.

Railway Workshop at Jamalpur

This was set up in 1862 for the repair and construction of rolling stock and plant connected with the railway. In course of time constituent shops were added and at present the complex is spread over an area of 538 hectares. Another important function of this workshop has been to impart training in various branches of trades as also to train young men to hold responsible superior posts in the railways and tobacco manufacturing industry, etc.

In 1907, Munger was selected as a site for a cigarette factory by the Peninsular Tobacco Company Limited, and with the view to use the Ganga river as a means of transport, the factory was set-up at Basudeopur on the bank of the river. The factory continues to flourish even today and manufactures many well-known brands of cigaratte as well as smoking tobacco, like Birmingham during the 19th century. Thus the district has had an industrial climate from the ancient past. Most of these old time industries centred round delicate and superb craftsmanship in the making of gold or silver ornaments, jewellery, stone images and furniture of high order. These small-scale industries led to the development of other minor industries like manufacture of paint, leather goods, palanquin and agricultural implements.

It would be worthwhile mentioning, about the traditional matmakers of Gogri, the "Dabgars" who made leather bags for holding juice of sugarcane, molasses, etc. as the potters of Munger town who made water bottles of clay and images, etc. and stone cutters who manufactured a large number of "Lingas" of the temples of Shiva were renowned.

Important industries which developed during Muslim period were manufactured fire arms, furniture, tables, cane chairs, sofas, footstools, gold or silver inlaid guns and pistols of superb workmanship. Some of the important visitors to Munger who have presented vivid account of industries of the district were Mr. Twining (1794), Walter Hamilton (1820), Francis Buchanan (1800-11), W.W. Hunter (1877) and others. According to them the following industries flourished in the district, e.g., manufacture of double barrel guns, rifles, single barrel fowling pieces, muskets, blunder busses, ordinary match-locks, carved match-locks, single barrel pistols, tea-kettles, fish-kettle, iron ovens, sauce pans, frying-pans, chafing irons, chamber stoves or grates, kitchen stoves, ramrods, swords, spears, table knives and forks, scissors, padlocks, chest locks, door lock, horse auxiliaries, agricultural implements and superior type of household furniture.

Boat building, manufacture of fire hearths and colouring materials, manufacture of coarse soap, tallow candles, torches, leather goods, tailoring machines, etc. were also flourishing trade during this period. Due to abundant and cheap supply of raw materials, the inhabitants of Munger gave more stress in the manufacture of iron utensils of almost every description but the products were scarcely inferior to the European models after which they were made.

Even in 1824, when B.R. Heber visited Munger, he found that these industries were still flourishing. "I was surprised", he wrote in his journal, "at the neatness of the kettles, tea-stays, guns, pistols, toasting forks, cutlery and other things of the sort, which may be procured in this tiny Birmingham."

Many of the industries of the Muslim period continued to survive and flourish during the British period, and to them were added large and important industries such as :

1. Indigo cultivation and factories;
2. State quarrying and tile manufacture;
3. Railway Workshop at Jamalpur set up in 1862;
4. Tobacco manufacturing industry set up in 1902;
5. Biri-making, distillery and a number of cottage industries.

Biri-Making, etc.

Biri-making developed as a very important small-scale industry during this period and still continues to flourish at Jhajha, Munger, Jamalpur and Sheikhpura. The products are presently sent to West Bengal and other neighbouring states.

There also developed a distillery at Mankatha for the manufacture of spirit and country made liquor. The produce is consumed locally. In addition a large number of small scale industries based on agricultural resources (rice, dal, wheat, etc.)

and demand based like paints, furniture, etc. also developed during this period mainly at Khagaria and Lakhisarai.

Present Position and Brief Account of Jamalpur Workshop

At present Jamalpur Railway Workshop has covered an area of 2 lakh 35 thousand square metres and equipped with some 3000 items of plants and machinery and has nearly 13,925 staff on its pay roll. It consumes 12,000 tonnes of steel and 600 tonnes of non-ferrous metal in a year and needs 6000 KVA power for its repair, overhaul and manufaturing processes. It is the only shop having the rolling mill. The Workshop does not build locomotive but manufactures spare parts to the extent of 4842 standard stock items and approximately 2500 standard stock items are purchased from trade by Divisional Controller of Store, Eastern Railway, Jamalpur.

As electrification and dieselisation of engine began gradual reduction in the fleet of steam locomotives in which Jamalpur embarked diversification. In 1961, the Workshop took on the manufacture of ticket printing machinery and travelling cranes which were hitherto imported.

The major activities of Jamalpur Workshop are as follows :

1. Periodical overhaul, intermediate overhaul and special repairs to all steam locomotives based on the Eastern Railway.
2. Manufacture of locomotive spares for supply to shops and sheds of the Eastern Railway and North Frontier Railway Sheds (broad gauge).
3. Manufacture of steam and diesel travelling cranes.
4. Manufacture of wagons like heavy oil and petrol wagons.
5. Manufacture of bolts, nuts and rivets for Eastern Railway.
6. Rolling of Steel sections for railways and wagon builders.
7. Manufacture of permanent raw materials.
8. Periodical overhaul of steam travelling cranes.
9. Special repairs of diesel locomotives.

10. Manufacture of ticket printing machines.
11. Manufacture and supply of diesel locomotive spares.
12. Manufacture and supply of steel casting to all Indian Railways.

The above activities enabled the Eastern Railway to recruit about 700 additional men. The strength of supervisory staff and assistant staff of the Jamalpur Workshop from 2001 is given in Table 8.6.

Table 8.6 : Staff strength of Jamalpur Workshop (1991-2001)

Years interval	*Supervisory staff*	*Assistant staff*	*Total*
1971	890	8,765	9,655
1972	905	8,905	9,810
1973	902	8,888	9,810
1974	904	8,968	9,872
1975	886	8,832	10,081
1976	725	9,790	10,525
1977	868	9,430	10,298
1978	816	9,565	10,381
1979	824	9,602	10,426
1980	796	9,701	10,697
1991	704	9,785	10,784
2001	624	9,841	10,912

Source : Received with due permission from S.M.W., Jamalpur Workshop, 2001.

It is evident from Table 8.8 that Jamalpur Workshop runs since long with a slight fluctuation of staff strength. It is hoped to increase the strength of staff by another 2000 men by the end of 6th Plan. According to 6th Plan Jamalpur Workshop will continue to be responsible for the major overhaul of components for the Indian Railways. In addition to this, Jamalpur has been entrusted with a large number of new lines of production, including the manufacture of steam crane, hydraulic jacks, barrels for tank, wagons, scragging machines and ticket printing machines. A fully equipped modern steel foundry has also been set up for meeting the increasing demands of steel castings for the railways. It has also been decided to undertake periodical overhaul of diesel shunting locomotives of Eastern and South Eastern Railways in Jamalpur Workshops.

The question of providing adequate workload for Jamalpur Workshop is constantly under consideration and there need be no anxiety that the capacity of Jamalpur Workshop will not be fully employed. There is no proposal at present to set up a new steam locomotive repair workshop of the Indian Railways. Capacity of manufacture of diesel and electric locomotives has been developed in Chittaranjan Locomotive Works and Diesel Locomotive Works Varanasi to meet the requirements of Indian Railways. There is, therefore, no need of developing similar facilities in Jamalpur Workshop, where the existing capacity available is being fully utilised with the existing load for repairs to steam locomotives and new lines of production.

During various Five Year Plans of the country, development of industry received continued importance. Infrastructure in the shape of road, railway, electric power, generation, financial institutions, etc. have gradually been developed.

Munger division also continued to receive help but some of the traditional industries have declined.

1. Industrial Directory of Bihar, 2001 depicts the number and types of following industries in the Munger division :

Name of the Industry	*Number of Units*
1. Large and medium industries	3
2. Small-scale industrial units registered with Directorate of Industries, Bihar	
(a) Food product	3
(b) Furnitures and fixtures	2
(c) Paper and paper products	4
(d) Printing industries	13
(e) Chemicals and chemical products	16
(f) Non-metallic mineral products except petroleum and coal	3
(g) Metal products	36
(h) Electric machinery apparatus, appliances and supplies	4
(i) Transport industries	1
(j) Miscellaneous manufacturing industries	12

Present Position and Brief Account of Tobacco Manufacturing Industry

The cigarette factory is equipped with modern and up-to-date machinery. One cigarette making machine is capable of producing 1300 cigarettes per minute. The total strength of workers employed is approximately 25,000. Cigarette factory of Munger has its own printing press for the printing of all materials used in connection with the packing of goods. In addition to these a large saw mill has been erected which fabricates all types of packing cases from the log to the finished articles.

Many well known brands of cigarette are manufactured. Raw materials used in the manufacture of cigarette and packets come from the following sources :

Materials	*Sources of Supply*
Unmanufactured leaf tobacco	Andhra Pradesh
Cigarette paper and wrapping materials	West Bengal
Board for packets	Bihar
Timber for cases	Bihar and Nepal Tarai
Tin plates	Bihar

Factory is working on full double shifts. The following are some of the amenities given to the workers of Munger Cigarette Factory :

(i) A fully equipped Labour Welfare Institute complete with stage, furniture, fans and cinema projector,
(ii) A large canteen with modern equipment,
(iii) A housing colony with houses of one or two rooms,
(iv) A hospital building on the bank of the Ganges, in addition to the ambulance room and creche within the factory premises under qualified staff,
(v) Sports facilities
(vi) Two hundred and fifty school scholarships and fifty seats at the Basudeopur High School are provided for the children of workmen, in addition to the twelve college scholarships awarded each year.

Gun Factory

The gun manufacturing industry of Munger is an old industry of this district in particular and India in general. The able artisans are not only masters of gun making and trade but they can also manufacture revolvers and pistols. They are very eager to start this industry provided the Government gives them licence. The price of revolver and pistol ranges between Rs. 6,000 and 10,000 but the actual expenditure in preparing these goods are Rs. 700 to Rs. 1,000. It is thus saving a huge amount of foreign exchange annually. Nowadays there are 36 licence holders of gun manufacturing with a total capital investment of Rupees one crore only. The annual production of guns which comes to open market for sale is about 8,000 only (1957-95).

The quality of Munger manufactured guns are rated high from the foreign guns because guns manufactured in Munger are heavy in weight which is essential for accurate range of firing and bears the capacity of L.G. long range firings. The very remarkable feature of these local manufactured guns is that it bears the test of 350 ton pressure, whereas foreign made guns can bear the test of only one ton air pressure. The hand made guns are cheaper in rate and superior in quality than foreign machine manufactured guns. Besides this various kinds of guns are manufactured here.

This industry if promoted and fully financed by the Government can earn world reputation as it has a bright future especially in Munger due to the availability of skilled labour. The gun manufacturing industry has brought Munger on the map of India.

Biri-making is one of the most important small-scale industry of the division. Jhajha, Munger, Jamalpur and Sheikhpura are the main centres for this industry. There are quite a large number of factories employing more than 1,000 workers per day at Jhajha. Tobacco for this purpose is imported from Gujarat, and Kendu leaves and other raw materials are available in plenty in the areas around Jhajha. The average production of biri at Jhajha area alone would be about 16 lakh per day. A considerable quantity of biri is exported to Calcutta from Jhajha by several concerns who have a number of branches at different places within the district. It is estimated that a total capital of about Rs. 9,00,000 has been vested in this industry in the district and it gives employment to about 10,000 workers. The rate of wages to a worker is normally Rs. 3 per thousand biris.

Distribution of Industry

The distribution of small-scale and large-scale industries in the Munger division presents a haphazard pattern of industrial development and urbanization. Although there are high sounding words about the industrial development of the district of Munger but the location of industries shows that only few areas are more industrialised whereas others are lagging behind than their urban counterpart. In the Munger division the highest 7.81% industrial development is found in Munger anchals whereas it is 7.45 in Kharagpur, 7.32 in Lakshmipur, 5.86 in Tarapur and others. In Anchal level the lowest percentage of industries is 0.68 in Khaira, 1.37 in Chakai, 1.5 in Halsi. Such an imbalance shows that the extreme southern part of the district is less industrialised and hence lesser degree of urbanization has taken place in those areas.

In urban areas the highest number of industries (481) is found in the city of Munger whereas in other urban centres this figure is very low as in Bariarpur (209), Khagaria (168), Kharagpur (159), Barahiya (140) and Barbigha (523). The lowest number of industries is 25 in Gogri, 73 in Lakhisarai, 67 in Sheikhpura and 90 in Jhajha. This shows that the bigger urban centres have greater units of industrial location due to the availability of cheap skilled labours, administrative convenience, accessibility through roads, railways and waterways, besides a large market at hand. In case of smaller urban centres the most of the service facility are lacking in comparison with bigger urban centres which leads to lesser industrial development but whatever development has taken place is the outcome of the planned effort of the Government and the enterpreneurs too (Table 8.7).

Table 8.7 : Distribution and Concentration of Industries in the Munger division (1991-2001)

Name of Anchals	*No. of industrial units in rural area*	*No. of industrial units in urban area*	*Total No. of industrial units*	*% of industries*	*% of area*	*LC value*
Munger	176	481	657	7.81	3.06	2.55
Bariarpur	110	—	110	1.31	1.68	0.80
Jamalpur	70	209	279	3.32	2.58	1.29
Dharhara	152	—	152	1.81	3.02	0.60
Kharagpur	468	159	627	7.45	4.79	1.55
Asarganj	242	11	253	3.01	2.11	1.43
Tarapur	251	—	251	2.99	1.05	2.85
Tetiha Bambor	168	—	168	2.48	1.73	1.43
Sangrampur	295	—	295	3.51	1.81	1.94
Barahiya	88	140	228	2.71	3.31	0.81
Pipariya	139	—	139	1.65	2.73	0.60
Surajgarha	211	—	211	2.51	5.30	0.47
Lakhisarai	385	73	458	5.44	4.38	1.24
Ramgarh Chowk	39	—	39	0.46	1.49	0.31
Halsi	88	—	88	1.05	2.78	0.38
Barbigha	168	123	291	3.46	1.88	1.84
Sheikhopur Sarai	80	—	80	0.95	0.98	0.97
Sheikhpura	270	68	237	2.82	2.76	1.02
Ghat Kusumbha	111	—	111	1.32	1.57	0.84
Chewara	92	—	92	1.10	1.26	0.87
Ariari	175	—	175	2.08	2.15	0.96
Islamnagar Aliganj	131	—	131	1.56	3.17	0.09
Sikandra	272	—	272	3.24	2.78	1.17
Jamui	76	105	181	2.15	2.20	0.97
Barhat	141	—	141	1.68	3.86	0.44
Lakshmipur	323	—	323	3.84	5.27	0.73
Jhajha	206	90	296	3.52	5.42	0.65
Gidhaur	152	—	152	1.81	4.64	0.39
Khaira	58	—	58	0.68	5.43	0.12
Sono	206	—	206	2.44	4.97	0.49
Chakai	116	—	116	1.37	9.81	0.13
Alauli	225	—	225	2.67	3.48	0.76
Khagaria	102	168	270	3.21	3.32	0.84
Mansi	90	—	90	1.07	2.78	0.61
Chautham	131	—	131	1.55	1.98	0.78
Beldaur	253	—	253	3.10	2.83	1.09
Gogri	377	25	392	4.66	2.59	1.79
Parbatta	324	—	324	3.85	3.64	1.05

The locational coefficient of Table 8.7 gives the level of industrial concentration in the Munger division which has been calculated on the basis of the following farmula :

$$Lc = \frac{N_i}{A}$$

Where Lc is location coefficient, N_i is percentage of industries and A is the percentage of area contained by an anchal. With the division of percentage of industries with the percentage of areas we get the value location coefficient which has been regarded here as the index of industrial concentration. In case it is more than one, this shows that the concentration of industries is increasing whereas the value less than 1.00 shows the sparse location of industries.

In the Munger division, the highest industrial concentration of 3.55 comes for Tarapur anchal, 3.03 for Sangrampur, 2.55 for Munger and 2.34 for Barbigha. The sparse location comes for Chakai (0.13), Khaira (0.12), Halsi (0.43) and Sono (0.49). The former anchals are mostly urban centres, where well planned industrial development has taken place. The lesser concentration areas are extreme rural anchals where only smaller household industries have developed in their crudest form. The distribution and concentration of industries can be seen in Table 8.9.

Industrial Potential

The concept of 'potential' as developed by Warntz (1960) and 'gravity model' developed by Reilley (1931) appear in many forms in social science literature. But here we are only concerned with distance formulation and potentiality of industrial development in urban areas of Munger division.

The potentiality at any point is defined as $\frac{N_i}{r}$ where N_i is number of industries and r is the distance between a particular point to a representative point in each urban centre. Here an example is given in Table 8.8 to show the method of linear distance calculation for urban industrial potential for the city of Munger.

Table 8.8 : Calculation of Industrial Potential for the City of Munger

Name of the town	*Distance in kilometres*	*Industrial Potentials*
Munger	0	481
Jamalpur	8	60
Asarganj	30	2
Sheikhpura	77	1
Jamui	80	1
Khagaria	14	12
Barbigha	91	1
Lakhisarai	58	1
Jhajha	85	1
Kharagpur	35	5
Barahiya	74	2
Gogri	43	1

Source : Calculated by the author.

In this way, the industrial potential for all the other eleven urban centres have been calculated and the result is given in Table 8.9.

Table 8.9 : Urban Industrial Potentiality in the Munger division

Name of the town	*Industrial Potentiality*
Munger	568
Jamalpur	293
Asarganj	30
Sheikhpura	96
Jamui	128
Khagaria	223
Barbigha	142
Lakhisarai	103
Jhajha	113
Kharagpur	192
Barahiya	165
Gogri	56

Source : Calculated by the author from Munger Industrial Directory.

According to Table 8.9, the highest industrial potential comes for the city of Munger (568), the second position is that of (293) for Jamalpur and the lowest urban Industrial potential is (56) for Gogri (Figure 8.3).

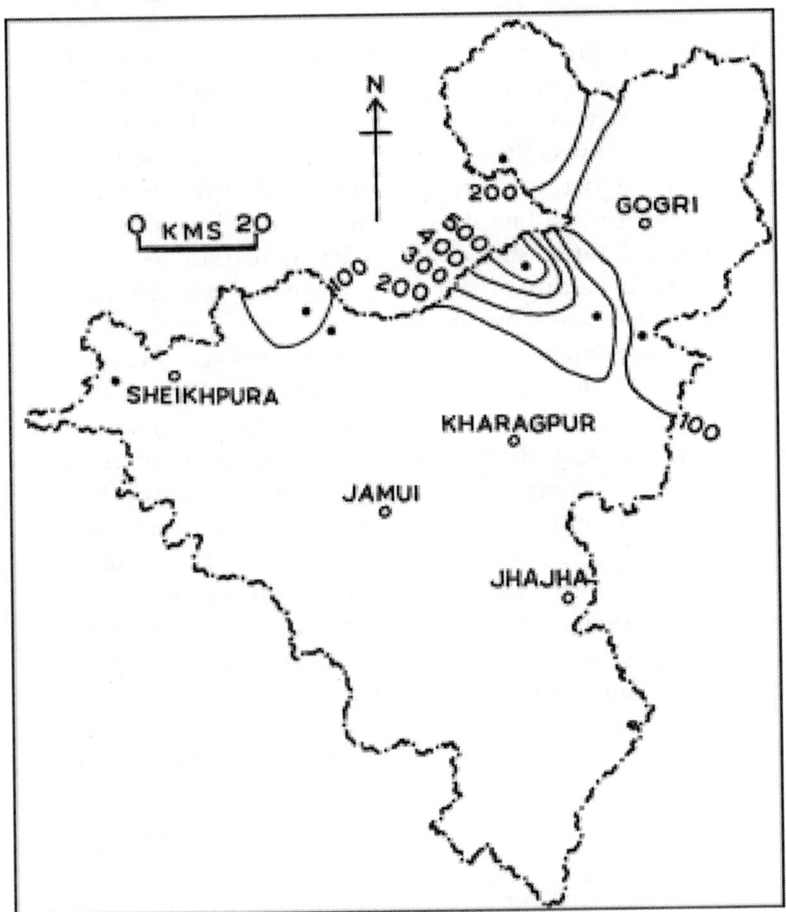

Fig. 8.3 : Munger Division : Industrial Potential.

Prospects

The economic investigation team of the Ministry of Commerce and Industry, Government of India surveyed Munger district and recommended two small-scale units of electrical and porcelain industries to be established at Jhajha.

Regarding slate industry it was suggested that slate slabs would be properly tested with a view to find whether the same is suitable for the manufacture of school slates. According to it, there is good prospect for a small aluminium utensils unit in Sheikhpura area and for the setting up of small tanneries in this Munger division. Regarding gun manufacture, the recommendation was that arrangements may be made for the supply of graded raw materials and suitable machinery to enable the gun manufacture to ensure durability resistance, precision and accuracy and qualities. It was also suggested that the feasibility of setting up a testing centre for guns at Munger will also be technically examined.

The Government has decided to set-up a few industrial establishment of lesser magnitude in the near future in the district. They are plastic, rubber, synthetics, rubber-fibres, sulphur, industrial chemicals, polythene, poultry feeds, agricultural implements, cold storage, one industrial mining school at Jamui, milk products industry in Khagaria district, industrial training school at Munger, paper manufacturing unit under Khadi Board scheme in Jamui sub-division and Industrial Workshop facility for the gun manufacturers of Munger. These will automatically bring an area of regional preferment in the district as a whole.

Table 8.10 : Relationship Between Industrial Workers and the Number of Industrial Units (1999-2000)

Name of the town	*No. of industrial Units (x)*	*No. of industrial worker Sin O' (y)*	X^2	Y^2	*XY*
Munger	481	570	2,31,361	3,24,900	2,74,170
Jamalpur	209	594	43,681	3,52,836	1,24,146
Asarganj	21	16	441	256	336
Sheikhpura	67	79	4,489	6,241	5,293
Jamui	105	121	11,026	14,641	12,705
Khagaria	159	39	25,281	1,521	6,201
Barbigha	123	72	15,129	5,184	8,156
Lakhisarai	43	119	5,329	14,161	8,687
Jhajha	90	123	8,100	15,129	11,070
Kharagpur	159	39	25,281	1,521	6,201
Barahiya	140	60	19,600	3,600	8,400
Gogri	25	75	625	5,625	1,875

Arithmetic average of x

$$\text{or, } AX = \frac{\Sigma x}{N}$$

$$= 138.4$$

Arithmetic average of x

$$\text{or, } AX =$$

$$= \frac{1936}{12}$$

$$= 161.3$$

$$r = \frac{\Sigma xy.N - (\Sigma x)(\Sigma y)}{\sqrt{\Sigma x^1.N - (x)^2}\sqrt{\Sigma y^2.N - (y)^2}}$$

$$= \frac{473163.12 - (1661)(1936)}{\sqrt{392285.12 - (1636)^2}\sqrt{748718.12 - (1936)^2}}$$

$$= \frac{5677956 - 3215696}{\sqrt{4707420 - 2676496} - \sqrt{8984616 - 3748096}}$$

$$= \frac{2462250}{\sqrt{2030924} - \sqrt{5236520}}$$

$$= \frac{2462260}{1425 \times 2288}$$

$$= \frac{2462260}{3260400} = 0.75$$

$r = 0.75$

Regression Coefficient of y on x

$$= \frac{\Sigma xy.N - (\Sigma x)(\Sigma y)}{\Sigma x^2.N - (\Sigma x)^2}$$

$$= \frac{2462260}{2030924} = 1.2$$

Hence, b = 1.2

$a = x\text{-}b.y$

$= 161.3 - 1.2 \times 138.4$

$= 161.3 - 166.0.8 = -4.7$

$a = -4.7$

$y = 1.2x - 4.7$ Sey

Regression Coefficient of y on x

$$= \frac{\sum xy.N - (\sum x)(\sum y)}{\sum x^2.N - (\sum x)^2}$$

$$= \frac{2462260}{5236520} = 0.47$$

Hence, b = 0.47

a = x-b.y

= 138.4 × 0.47 × 161.3

= 138.4-75.8=62.6

x = 62.64+0.47y

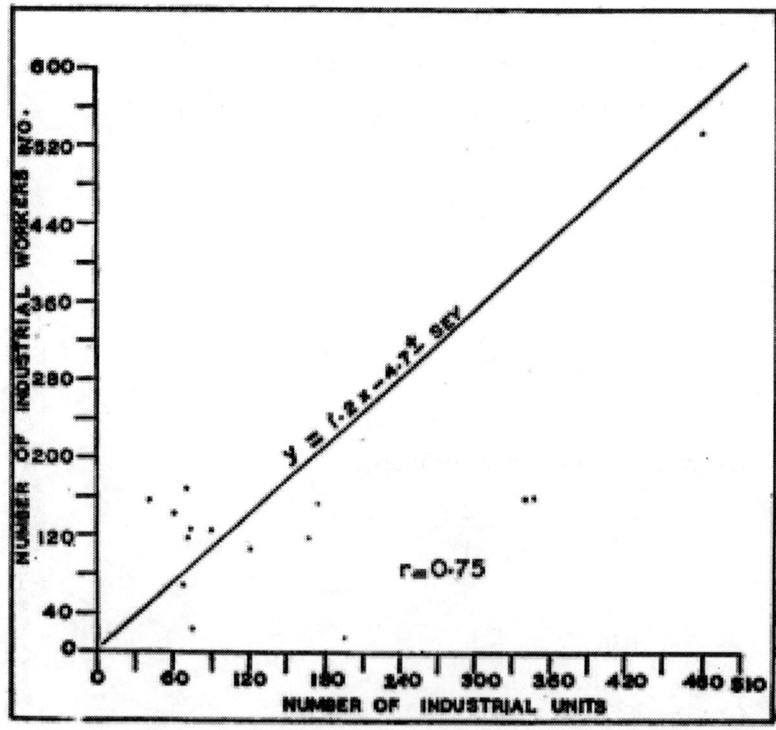

Fig. 8.4 : Munger Division : Relationship between Industrial Workers and the Number of Industrial Units.

The relationship between number of Industrial Units (x) and the number of workers of urban centres engaged in large and small scale industries (y) which gives a result r=0.75 (Table 8.12). This shows that the development of small-scale industries and large-scale industries in urban centres has good impact on regional development. The highest number of industrial units are 481 in the city of Munger and the lowest number is 21 for Bariarpur town, where industrial workers are engaged in these industries are 5,700 and 160 respectively (Figure 8.4).

So far as the regression of x on y is concerned it comes as y= 1.2-4.7 ± Sey.

These values show that industrialization and regional development has a close relationship and hence industrialization of a region is essential for economic development and better utilization of resources.

REFERENCES

Chand Mahesh and R.K. Vishwakarma, "Urbanization as a Strategy of Rural Development", Urban System and Rural Development, The Director, Prasaranga, University of Mysore, Mysore-6, 1972, p. 141.

Chandrasekhar, C.S., G.D. Mathur, and K.V. Sundaram, "Growth Foci in Regional Strategy", The Director, Prasaranga, University of Mysore, 1972, p. 55.

Christaller, W., Central Places in Southern Germany, Englewood Cliff, 1965, Translated from the Original German Version Published in June, 1933.

Clark, Collin, "Urban Population Densities", Journal of Statistical Society (a), Vol. 1, 1951, pp. 492-596.

Draft Five Year Plan: 1978-83, Government of India, Planning Commission, 1978, p. 241.

Gowda, K.S. Rame: "Regional Planning for Integrated Rural Urban Development in Mysore State": Urban Systems and Rural Development, p. 154

Losch, August, Die Raumliche and Nug Der Wirfschaft, 2nd Edition, 1944, Translated into English as Economics of Location, Yale University Press, 1954.

Mallick, U.C., "Development of Small Towns for Balanced Urbanisation and Economic Growth", Place of Small Towns in India, 1979, p. 55.

Reily, R.C., "Industrial Geography", London, 1973, p. 10.

Sastry, S., "Location of Crops", The Indian Geographical Journal, Vol. XVIII, 1942, pp. 110-114.

Sundaram, K.V., "Growth of the Urban Economy of Small and Medium Size Towns", Urban and Regional Planning in India, Vikas Publishing House Pvt. Ltd., 1972, New Delhi, p. 213.

Thunen, J.H. Von, "Wirtschaft and Nationale Isolated State: Peter Hall", Pergamon Press, 1966, pp. 72-73.

9

Transportation and Sustainable Economic Development

Means of transport and communication are very essential for the smooth working and the further development of the economic life of a country. It has been rightly observed that "if agriculture and industry are the body and bones of national organism, communications are its nerves". Kipling goes further and says, "Transport is civilization".

Transportation contributes to the economic, industrial and cultural development of any country. Transportation is vital for the economic development of any region since every commodity produced whether it is food, clothing or medicine needs transport at production and distribution stages. In the production stage, transportation is required for carrying raw materials like seeds, manure, cotton, sugarcane, coal, steel, etc. In the distribution stage, transport is required from production centres, viz., farms and factories to the marketing centres for distribution to every household. The adequacy of transportation system of a country indicates its economic and social development.

The above analysis shows that transportation lines are the life blood of regional economic development, because they are immensely helpful in moving the money, men and material up to the consumers or required places. In this way transport network is the channel through which production—economic differentiation of areas into the various specialized types of land uses and, indeed, the existence of cities themselves, in the modern sense would be impossible. Beyond this transportational lines have multilateral effects and represent "positive feed-back mechanism" on the overall growth of regional economic development. The development of transport keeps the population mobile and also results in the development of service centres including their population size." It is, therefore, felt that the provision of road linkage is a primary requirement for all development. Roads are also economical for shorthands as compared to waterways, which are mainly suited to long hands.

Different Modes of Transportation

Human being has always remained with a curious nature of invention since human being is surrounded by three basic mediums, viz., land, water and air, the modes of transport are also connected with these three mediums for the movements. Land has given scope for development of road and rail transport. Water and air have developed waterways and airways respectively. The roads or the highways not only include the modern high-way system but also the city streets, feeder roads and village roads catering to a wide range of road vehicles and the pedestrians. Railways have been developed both for long distance transportation and for urban travel. Waterways include oceans, rivers, canals and lakes for the movement of ships and boats. The aircrafts and helicopters use the airways. Apart from these, pipelines are used for the transportation of water and some other fluids, elevators belt conveyors, cable cars and aerial tramways are also functioning as minor transportation system.

Science that covers the building of roads and providing the adequate facilities to the roadways and the safety of road vehicles may be defined as road or highway geography. The science which deals with laying of railway track for the movement of heavy locomotives and rail vehicles including the control of their movement with station yard facilities may be called railway geography. Similarly, the techniques used in development of docks and harbour on sea or river shores for the departure and arrival of ships, encircle the river and harbour geography. The knowledge of development of an area for safe landing and take-off of aircrafts, i.e., providing facilities of runways, taxiway, hangers, terminal buildings and control devices, etc. may be included under airport geography. There are, therefore, four different modes of transportation which may be enumerated as :

(i) Roadways;
(ii) Railways;
(iii) Waterways; and
(iv) Airways.

On studying the transport characteristics of each of these listed modes, it may be broadly said that road and rail transport are easy and economical for internal movements.

Characteristics of Road Transport

It has been very well said that of all communications, road communication is the nearest to man. Man has to go to railway, or a waterway or an airway, but roadway comes to him.

The characteristics of road transport are briefly listed here :

(i) Roads are used by various means of transport, i.e., road vehicles, like animal drawn carts and carriages, cycle rickshaw, passenger cars, buses and trucks ;
(ii) Road transport requires a small investment from the government side. Motor vehicles are comparatively cheaper than other carriers like railways, waterways and airways. Road construction and maintenance is also cheaper ;
(iii) Road transport offers a complete freedom to road users to transfer the vehicle from one lane to another and from one road to another according to the need and convenience. This flexibility is not very much available to other modes of transport ;
(iv) In particular, for short distance travel, road transport saves time; and
(v) Road transport is the only means of transport that offers itself to the whole community alike.

Railways

The Railways are clearly the most efficient and economical form of transport for bulk traffic and long distance passenger movement. For meeting the requirements of this class of traffic, the growth of the railway station is both necessary and economically sound. The Indian Railways system has been and will continue to occupy a key position in the strategy of building up the basic infrastructure required for a growing and diversified economy. The main objective in the past has been to carry the projected quantum of passenger and goods traffic with a view to ensuring smooth traffic movement so as to avoid bottlenecks in the production process. However, there has always been a special objective of ensuring an efficient rail transport system. In the First Five Year Plan, the special objective was the rehabilitation and replacement of the overaged assets, in the Second Five Year Plan, particular emphasis was being laid to prepare the Railways for carrying the traffic generated by the new steel plants and the increased production of coal. In the Third Five Year Plan, attention was given to build up additional capacity so as to be ahead of the traffic demand which could otherwise act as a restraint or as bottlenecks in the economic development; and in the Fourth Five Year Plan, the focus was on modernization of the system to improve efficiency of operations. In the Fifth Plan cognizance was taken of the need for better utilisation of the existing track and the rolling stock capacity and higher operational efficiency through maximising of movement in block rakes and reducing turn-round time. As a result, there has been a significant improvement in the operational efficiency of the Railways as measured by the indices of utilisation of railway rolling stock in terms of wagon turn-round time, wagon, vehicle, engine, kilometres per day and net tonne kilometres moved per wagon, engine, day etc. Since independence, however, the direction and extent of road planning has been much influenced by the need for promoting the economic development. The construction of existing roads with bridges were taken up almost immediately after the achievement of independence in

1947 linking Munger and its important market centres with the neighbouring districts.

Present Position of Transportation Linkages

Waterways

Water transport is useful for the carriage of bulky and heavy commodities of low grade coal, timber, raw ores, etc. The cost of transporting goods through waterways is very low, but water transport suffers from the drawback of slow speed. Configuration of the terrain also affects water transport as the rivers of South Munger are shallow and rain-fed where the year around navigation on permanent basis is never possible whereas in North Munger the rivers are perennial and this favours year round navigation on permanent basis especially in the Ganga, the Burhi-Gandak and the Kosi.

Early History of Transportation Linkages

Reliable information is not available regarding the ancient trade routes and highways until later part of the Muslim Period in the Munger division. In the beginning of 19th century, the chief highways of the division were connected with the other parts of India. At the time of the rule of Shershah, the Muslim armies quite often passed through the district of Munger towards West Bengal and North Bihar. The Ganga, however, appears to have been used by several travellers such as Huien Tsang, Tavnier and Francis Buchanan.

The first rail track was laid in India by the East India Railway Company in the year 1853. It was extended to Munger in the year 1862 and since then several other lines have been constructed. The first rail line was constructed up to Khana junction in the north which follow the course of the river Ganga and later on after the increase of traffic, chord line was constructed from Lakhisarai in the west to Khana junction in Munger.

In 1900 A.D., the Bengal and North-Western Railway extended their system in Khagaria sub-division of the district to Katihar and since then several branch lines from Mansi in Purnea district to Supaul, Bachwara, Samastipur, Sahebpur Kamal to Munger Ghat and from Barauni junction to Simaria Ghat were constructed.

In pre-independence period the road systems have grown up under the British rule especially near strategic and administrative locations.

Since independence, however, the direction and extent of road planning has been much influenced by the need for promoting the economic development. The construction of existing roads with bridges were taken up almost immediately after the achievement of independence in 1947 linking Munger and its important market centres with the neighbouring districts.

Present Position of Transportation Linkages

In the Munger division both metalled and unmetalled roads are found. Metalled roads are mainly State Highways and National Highways. The roads which connect different villages are also metalled owing to the planned development of roads by the Rural Engineering Organization (REO), Government of Bihar. National Highways have a breadth of 12 metres in which 7 metres is metalled. State Highways have a breadth of 7 metres in which only 4 metres is metalled. The other roads have a breadth of just 3 metres.

The landscape features of the Munger division is both plain and mountainous. The total area of the district is 7,884 square kilometres out of which the middle southern plain covers 2,499 square kilometres, mountainous area in the south covers 3,888 square kilometres and northern portion extends over 1,497 square kilometres. The main types of road are PWD, REO and District Board (Table 9.1).

The northern part of the district has a well metalled 10 metres wide National Highway Number 31. It has a length of 33 kilometres in North Munger. The National Highway

Number 31 passes through Khagaria, Chautham, Gogri and Parbatta anchals.

The southern upland plain have a total number of 1068 rural settlements besides 7 towns. This zone consists of only 163 kilometres of PWD road, 244 kilometres of REO road and 232 kilometres of District Board road.

The middle part of the district is the heartland from where one can move either north or south properly due to its great accessibility either through roadways, railways and waterways. When we glance at the number of transportation facilities it seems troublesome to cope with its population and agricultural produce. The road should be so planned that each and every village must be interlinked with any means of transport as we are desirous for more and more development. The ratio of present population and the length of road would be most insufficient to cope with the burden of coming generation. Seeing the prospect of development, it is essential to increase the length of road at least double from the existing roads based on scientific lines the road planning form development has been framed in the Chief Engineers Conference held in 1943 in Nagpur. The Nagpur Plan classified the highway in five categories according to their location and financial responsibilities. They laid down recommendation on all India basis according to inherently constructed, geometric standard of highway and bridge specification on different categories of landscape. The second road development plan (20 years plan 1961-81) was finalised in 1959 at the meeting of Chief Engineers and the same has been adopted by the Central Government of India. The total length of road would be almost double in comparison with Nagpur Plan of the development of roads.

According to the decision of the Bihar Government, the villages having at least a population of 500 should be linked with the metalled roads. In this connection there are at least 1284 villages in the Munger division which have more than 500 of population. On this basis the total requirement of road is 2,600 kilometres. But the caution has been made that each and every village must be provided with at least 2 kilometres in total and the plan should be framed in such a way that this objective must be fulfilled.

Table 9.1 : Length of Different Types of Roads and Railways in the Munger Division (2008-09)

(in Kilometres)

Name of Blocks	*R.E.O.*	*P.W.D.*	*District Board*	*Railway*
Munger	9.40	12.00	24.40	6.40
Bariarpur	14.90	8.60	15.20	3.00
Jamalpur	13.30	9.40	17.60	13.00
Dharhara	19.10	7.00	8.70	8.30
Kharagpur	15.40	28.12	28.00	-
Asarganj	8.70	7.40	4.10	-
Tarapur	9.50	5.60	4.30	-
Tetiha Bambor	6.30	8.57	5.10	-
Sangrampur	4.20	14.31	6.10	-
Barahiya	22.20	18.00	12.70	14.08
Pipariya	8.70	8.09	11.95	-
Surajgarha	13.70	16.91	28.25	17.00
Lakhisarai	35.10	12.00	20.60	29.52
Ramgarh Chowk	19.70	2.30	1.70	-
Halsi	23.40	3.70	2.60	-
Barbigha	16.80	12.60	9.90	-
Sheikhopur Sarai	11.80	7.40	7.70	-
Sheikhpura	24.60	14.20	52.50	8.32
Ghat Kusumbha	7.20	9.80	38.90	-
Chewara	2.60	13.50	4.80	-
Ariari	20.40	2.86	7.60	14.08
Islamnagar Aliganj	7.50	5.13	3.82	-
Sikandra	13.90	10.87	6.78	-
Jamui	19.10	22.00	89.60	11.50
Barhat	10.90	12.40	6.54	4.90
Lakshmipur	10.50	10.60	6.76	-
Jhajha	57.40	8.64	47.60	25.0
Gidhaur	4.50	7.00	3.70	4.70
Khaira	17.00	30.72	49.20	-
Sono	12.60	13.10	5.40	-
Chakai	39.40	29.44	56.20	-
Alauli	8.00	31.60	36.80	-
Khagaria	9.50	8.80	10.10	9.00
Mansi	16.50	10.40	17.70	-
Chautham	8.50	9.80	10.10	-
Beldaur	17.50	14.00	59.20	-
Gogri	18.20	14.00	13.60	12.00
Parbatta	16.80	38.40	17.60	3.00

Source : Compiled by the Author.

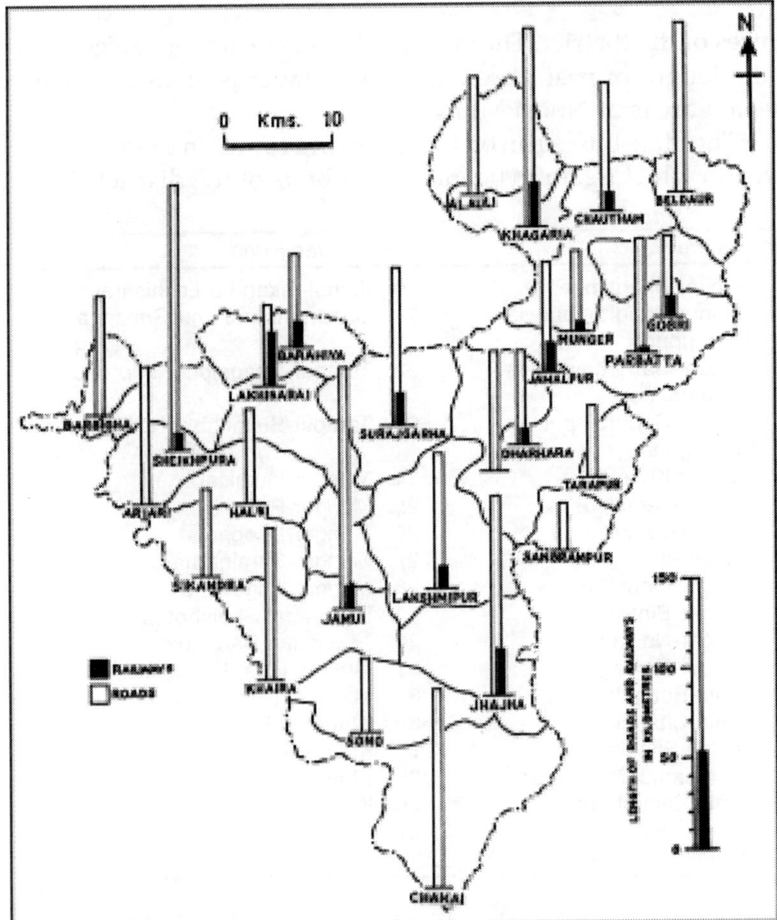

Fig. 9.1 : Munger Division : Length of Roads and Railways.

Figure 9.1 and Table 9.1 show the length of roads and railways in the Munger division (2000). It has been found that the total length of road is 1935.06 kilometres and the railways is 209.40 kilometres. The highest length of road is found in Jamui Anchal (137.25) and the lowest length of road is about 19.5 kilometres in Tarapur. In between these two extremes, the high length of road is found in Khagaria, Chakai, Jhajha and Jamui anchals, all of which have important nodal transport centres of the district. The anchal which have lower categories of the length of roads are either mountainous areas or flood affected areas of North Munger.

The State buses run on the following routes in the Munger division which seem to be main road links of the district :

Name of the Road	Name of the Road
1. Munger-Bhagalpur	20. Jamui-Sikandra-Lakhisarai
2. Munger-Sangrampur via Kharagpur	21. Sangrampur-Jamui Sultanganj
3. Munger-Sangrampur via Gangata mor	22. Tarapur-Bhagalpur Kharagpur
4. Munger-Sangrampur via Sultanganj	23. Tarapur-Sultanganj Gangata mor
5. Munger-Kharagpur	24. Munger-Lakhisarai
6. Munger-Jamalpur	25. Munger-Ranchi
7. Munger-Jamui	26. Munger-Deoghar
8. Munger-Kiul	27. Munger-Surajgarha
9. Munger-Gourabdih	28. Munger-Begusarai
10. Munger-Biharsharif	29. Lakhisarai-Sheikhpura
11. Munger-Nawadah	30. Lakhisarai-Barbigha
12. Munger-Nawagarhi	31. Munger Ghat-Begusarai
13. Jamui-Bhagalpur	32. Munger Ghat-Khagaria
14. Jamui-Chakai	33. Khagaria-Begusarai
15. Chakai-Simultala	34. Khagaria-Gogri
16. Ariari-Jamui-Mallehpur	35. Khagaria-Chautham
17. Jamui-Biharsharif	36. Khagaria-Beldaur
18. Chakai-Deoghar	37. Khagaria-Parbatta
19. Jamui-Nawadah	

Table 9.2 shows the density of roads in the Munger division. The division as a whole has 0.241 kilometres road density per square kilometre of area. The anchals which have higher value of road density of 0.241 are Parbatta, Gogri, Beldaur, Mansi, Chautham, Khagaria, Jhajha, Jamui, Ariari, Barbigha, Sheikhpura, Tarapur and Jamalpur. This shows that these anchals have advantageous position in terms of road in the Munger division, whereas the anchals of Dharhara, Kharagpur, Lakhisarai, Lakshmipur, Khaira, Sono and Halsi have very poor density of road such as 0.10 in Jamui per square kilometre. This shows the underdevelopment of the area in terms of road development which is the result of physiographic condition only.

Table 9.2 : Per Square Kilometre Density of Roads and Railways in Munger division

Name of Blocks	*Geographical Area km²*	*Roads*	*Railway*
Munger	242.00	0.1892561	0.026
Bariarpur	160.52	0.241914528	0.018
Jamalpur	85.45	0.4716208	0.152
Dharhara	282.85	0.1424783	0.029
Kharagpur	306.79	0.17575	-
Asarganj	59.48	0.33961	-
Tarapur	71.04	0.2730856	-
Tetiha Bambor	101.38	0.2634642	-
Sangrampur	85.94	0.2195718	-
Barahiya	241.06	0.20229	0.053
Pipariya	58.39	0.492207569	-
Surajgarha	389.66	0.099420007	0.040
Lakhisarai	283.30	0.2077656195	0.085
Ramgarh Chowk	104.27	0.2272945238	-
Halsi	151.06	0.1966106183	-
Barbigha	94.36	0.416900382	-
Sheikhopur Sarai	55.12	0.4877606528	-
Sheikhpura	185.18	0.4960338049	0.033
Ghat Kusumbha	92.59	0.6037369046	-
Chewara	115.06	0.2424821832	-
Ariari	146.65	0.3331742243	0.068
Islamnagar Aliganj	172.89	0.09514720342	-
Sikandra	184.01	0.1714580729	-
Jamui	173.91	0.748	0.066
Barhat	232.16	0.1716057891	0.021
Lakshmipur	251.77	0.1090678	-
Jhajha	427.39	0.257	0.058
Gidhaur	71.11	0.2137533399	0.066
Khaira	418.71	0.194	-
Sono	392.27	0.104	-
Chakai	774.04	0.146	-
Alauli	274.47	0.233	-
Khagaria	262.36	0.453	0.977
Mansi	70.01	0.637	0.038
Chautham	163.97	0.5263	-
Beldaur	223.56	0.405635	-
Gogri	250.31	0.273	0.058
Parbatta	241.04	0.28087	0.10

Note : Length of road in comparison to total geographical area in km².

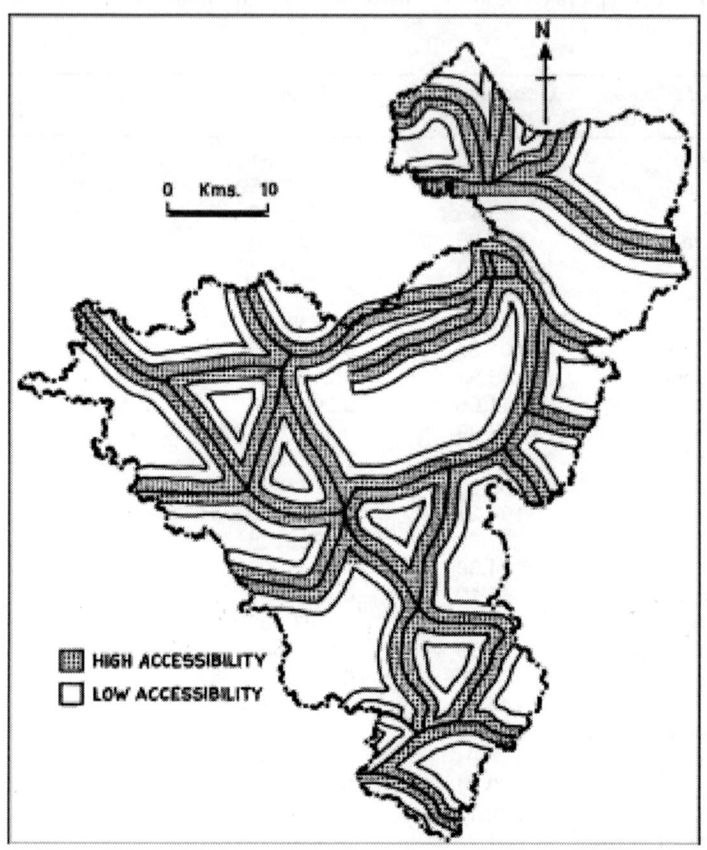

Fig. 9.2 : Munger Division : Accessibility of Road.

Figure 9.2 shows the accessibility of National and State highways in the Munger division, where a distance of 2 kilometres from road is considered as highly accessible area whereas the distance of 2.4 kilometres from the road is considered as low accessible areas and the areas beyond these two limits are considered as inaccessible one. In the district of Munger, there are three pockets of inaccessible areas such as the Kosito belt of North Munger, Kharagpur hills, Barahiya tal and the mountainous terrain of Khaira, Sono, Jhajha, Chakai and Halsi anchals in the extreme South. The zone of highly accessible areas are mainly found around the urban areas such as Munger, Jamalpur, Jhajha, Jamui, Kharagpur, Bariarpur, Barahiya, Lakhisarai, Sheikhpura, Barbigha, Gogri, Khagaria, etc. The low accessibility areas are found as transition zone in between the highly accessible areas.

So far as the per capita distance of road is concerned (Table 9.3) the Munger division has 0.00069 kilometres of road. The anchals which have a higher figure than this average values are Chautham, Chakai, Jhajha, Khaira, Jamui, Ariari, Sheikhpura and Dharhara anchals. This shows that mostly in anchals of low population density per capita length of road is higher, whereas in anchals of high population density such as in Munger per capita lowest length of road is found (0.00026).

Table 9.3 : Per Capita Length of Roads and Railways in Munger division

Name of Blocks	*Population*	*Length of road*	*Per capita roads*	*Lenth of railway*	*Per capita railway*
Munger	1,73,931	45.80	0.0002633	6.40	0.0000367
Jamalpur	1,46,636	62.20	0.0004241	16.00	0.0001091
Dharhara	62,836	51.60	0.0008211	8.30	0.0001320
Kharagpur	1,19,116	66.40	0.0005574	-	-
Tarapur	83,811	39.60	0.0004724	-	-
Sangrampur	71,060	33.70	0.0004742	-	-
Lakhisarai	1,40,492	45.10	0.0003210	29.52	0.0002101
Surajgarha	1,36,757	87.60	0.0006405	17.00	0.0001243
Barahiya	96,547	52.90	0.0005479	14.08	0.0001458
Sheikhpura	1,06,983	147.20	0.0013759	8.32	0.0000777
Barbigha	1,00,766	66.20	0.0006569	-	
Ariari	68,362	76.76	0.0011228	14.08	0.0002059
Jamui	83,368	130.70	0.0015310	11.53	0.0001347
Lakshmipur	1,11,676	72.50	0.0006491	9.60	0.0000851
Khaira	96,225	83.20	0.0008646	-	
Sono	86,103	41.00	0.0004761	-	
Sikandra	1,26,830	48.00	0.0003784	-	
Halsi	84,247	53.40	0.0006338		
Jhajha	1,10,846	110.00	0.0009923	25.00	0.002255
Chakai	95,386	113.00	0.0011846	-	
Khagaria	1,37,534	119.00	0.0008652	25.60	. 0001861
Alauli	1,14,272	64.00	0.0005600	-	
Chautham	91,063	70.00	0.0007686	9.00	0.0000988
Beldaur	78,296	90.70	0.0011584	-	
Gogri	98,755	55.80	0.0005650	12.00	0001215
Parbatta	1,21,282	72.00	0.0005936	3.00	0000247
District	27,45,180	180.36	0.000691	209.40	0.000076

Source : Compiled by the Author.

Screening Test for Viability of Road

International Development Authority has derived the road viability formula to compare cost-benefit ratio.

The formula is as follows :

Maintenance cost (Cm)

+Cost of capital at the rate of 12% interest (Cc)

(Construction cost or improvement Cost)

= Cm+0.12 Cc

Annual benefit per kilometre where there is already a road carrying traffic-365 × No. of vehicles per day (q) × Average saving in vehicle operating cost per km(s)

=365 x qa.

If 365 qa Cm+0.12 Cc, Road is viable where there is no road. Area of influence in Hectare (ha), Length of Road (w), Proportion of area that is arable (Pa)

Average incremental producer surplus per hectare due to irrigation (ba) =W × Pi × Pa × ba

If W x Pi x Pa x ba>Cm+0.12 Cc, road is viable.

Cm is REO estimate

Cc is REO estimate

Case of Karmatia to Agahara via Balendih-Dubedih Road

Ha = 2200

Km = 8.00

$$\frac{\text{Ha}}{\text{km}} = \text{w}$$

$$\text{W} = \frac{2200}{8} = 275 \text{ (length of road)}$$

Average incremental producer surplus per hectare due to irrigation (ba) = Rs. 400

Proportion of arable area (Pa) = 0.90

Proportion of arable land that will be irrigated (Pi) = 0.87

Hence, W × Pi × Pa × ba

= 275 × 0.87 × 0.90 × 400 = 6.013

Cm = Rs. 10,000

Cc = Rs. 2.50

CM = 0.12 Cc = 0.40

Hence, this road is viable, data collected and analysed the viability test. The tested result figure is 0.40. Thus road should be constructed.

Railways

The Munger division is well-served by broad gauge railways in South Munger and metre guage in North Munger. South of the Ganga it is known as Eastern Railway and north of the Ganga it is known as North-Eastern Railway. The total length of railway in the plain land of South Munger is 89.04 kilometres mountainous tract of the south, 70.40 kilometres while North Munger has 49.60 kilometres only.

The highest length of railway is about 30 kilometres found in Lakhisarai anchal. The lowest length of railway is about 3 kilometres in Parbatta anchal, whereas in other anchals railway facility is very poor which are not conducive for regional economic development.

Following are the Railway Stations on different rail lines of the division :

1. **In between Kiul to Lahaban**
   Kiul, Bansipur, Mallehpur, Bhalui halt, Jamui, Chaura, Gidhaur, Deopurhalt, Jhajha.
2. **In between Jamalpur to Munger**
   Jamalpur, Safiasarai, Purabsarai and Munger.
3. **In between Kiul-Gaya Section**
   Lakhisarai, Garsanda, Sirari and Sheikhpura.
4. **In between Kiul to Kalyanpur**
   Kiul, Dhanauri, Urain, Kajra, Abhaipur, Masoodan, Dharhara, Jamalpur, Ratanpur and Bariarpur.
5. **Kiul-Mokameh Section**
   Lakhisarai, Mankatha, Barahiya, and Rampur-Dumra.
6. **Katihar-Samastipur Section**
   Pasraha, Maheshkuth, Mansi, Khagaria, Olapur and Imli.

According to Table 9.2, the Munger division has several anchals where the railways are not found such as Beldaur,

Alauli, Halsi, Sikandra, Sono, Khaira, Barbigha, Sangrampur, Tarapur and Kharagpur. The average railway density in the division is 0.027 kilometres per square kilometre. The anchals which have more than average density of railways are Gogri, Chautham, Khagaria, Jhajha, Jamui, Ariari, Sheikhpura, Barahiya, Surajgarha, Lakhisarai and Jamalpur. The lowest railway density is 0.010 in Parbatta anchal. So far as the per capita length of railway is concerned the average condition is 0.000076 kilometres in the Munger division. In comparison with this average figure in almost all anchals which have railways have higher amount of the length of railways except Parbatta and Munger anchals. This regional variation is the result of the unproportionate coverage of railways in different anchals of the district.

Nowadays river route is not found to be suitable for long hauls in the Munger division except a few ferry services.

Waterways

The total stretch of river Ganges is about 88 kilometres inside the Munger division from Barahiya to Binda Diara having sufficient width and depth of water even in summer season also. The river is perennial so development of waterways should be considered by the assistance of State and Central Governments. In its planning it may be suggested that from Commissionary headquarter of Bhagalpur to the State capital Patna, should be connected with a water route of the Ganga.

Airways

There is an air base (aerodrome) at Safiasarai about 6 kilometres west of Munger where planes reach casually.

Problems of Transport

The problems of transport development are both physical and cultural. The physical problems are flood, undulating ground etc., and the cultural problems are mainly associated with the number and quality of population of the area. Some of the problems of different modes of transport are as follows :

Problems of Road Transport

1. Poor maintenance of road due to carelessness of the public as well as the government.
2. The road becomes muddy in the rainy season and this accelerates erosion.
3. The encroachment of settlement on the roadside.
4. In flood affected areas submergence of road is the main problem especially in North Munger and along the riparian tracts of the Ganga.
5. Most of the roads are unmetalled.
6. Most of the rural settlements are not connected with the metalled road.
7. Due to undulating ground in the hilly area there is a problem of tunnel construction, whereas in the plain land there is a problem of bridge construction on several smaller rivers.
8. Overcrowded buses have made the travelling most inconvenient.
9. There is a dearth of road in forest areas, which hampers the efficient movement of forest produce.
10. According to the increase of population the provision of buses, tamtam and other means of transport have not increased proportionately. Hence the lack of buses, etc. are felt almost in all parts of the district of Munger.
11. Recently the acuteness of the problem is felt due to non-availability of petrol, diesel and inflated freight rates, etc.
12. Most of the rural areas are still unaccessible parts of the district.

Problems of Railways

1. Inefficiency increased due to single rail line.
2. Burglary in trains.

3. Overcrowding in almost all bogies.
4. Passengers travelling without ticket.
5. Unbooked load of goods on passenger and express trains.
6. Even Railway Police Protection Force collects money from the traders who have not booked their goods.
7. Thus the lack of coal, electricity and even diesel are the main hurdles of timely running of trains.
8. Among other problems strike, lack of a good number of trains in proportion to the increase of population, regional physical problems and quite often the inflated freight rate are some of the problems which should be tackled immediately.

Problems Related to Waterways

1. Accumulation of vast amount of water as flood especially in the rainy season.
2. Formation of Diara in river bed.
3. The stretch of shallow water due to deposition of immense quantity of sand in the river piracy.
4. Prevalence of river piracy.
5. Dearth of boat, steamer and bridge in most of the rivers.

Suggestions for Transport Planning

Roadways

1. Resources should be incurred for the proper maintenance of old dilapidated road.
2. Alkatra should be given over the road each and every year for preventing the road to be muddy in rainy season.
3. The encroachment on the side of road should be immediately removed after making a housing colony, because this will reduce the accidents which are frequent especially in areas where encroachment on roads are found.

4. The unmetalled road should be immediately converted into a metalled one.
5. The metalled road should be extended to the unaccessible areas also.
6. The present narrow road should be made wider in order to facilitate the both way traffic.
7. Road should be extended in forest area also so that the forest produce may move up to other parts of the State also.
8. According to the increase of population the bus traffic and the number of buses should be increased.
9. The ensured availability of petrol, diesel should be arranged for efficient running of buses.
10. Almost all the rural settlements should be linked with the metalled road at least.

For the planned development of the road in the Munger division, the abstract of the programme which has submitted to the government is as follows :

1. To connect the villages and towns having 2,000 of population.
2. To connect all the industrial, religious and historical places by road.
3. To keep all the villages up to a maximum distance of four kilometres from the road.

Railways

1. Both way rail line should be constructed to facilitate the unhindered movement of trains.
2. Efficient police protection should be provided in each and every train to prevent burglary.
3. With the provision of number of quick running trains the overcrowding may be removed easily.
4. With the intensive ticket checking drive, the problem of mass passengers travelling without ticket may be checked in trains.

5. The movement of goods for trade should be strictly made prohibited in passenger and express trains.
6. The police force found to be collecting money in trains should be immediately prosecuted and suspended from the service.
7. Sufficient stock of coal and diesel should be made especially in railway junctions so that the cancellation of trains due to lack of coal and diesel may not be possible.
8. In each and every region the provision of trains should be made according to concentration of population and the strike by workers should be completely banned.

Waterways

1. In order to check the flood, on underground reservoirs, river embankment and mini dams a caution level should be made.
2. With the timely dredging of important rivers the formation of sandy Diara and shallowness of the river bed may be checked.
3. With the provision of efficient administrative machinery the prevalence of river piracy may be checked.
4. The bridge over the Ganga near Kastharni ghat of Munger should be constructed.
5. In spite of bridge for the time being the provision for continuous plying of boat steamer services should be introduced in important points in almost all rivers of the district.

Airways

Munger should have a dry port in North-Eastern India in order to export forest produce, mica, gun, cigarette, spare parts, etc. so that Munger may develop as a nodal centre in the district.

REFERENCES

Mayer and Kohen, Readings in Urban Geography, Central Book Depot, Allahabad, 1969.

Munsi, S.; "Railway Network Growth in Eastern India 1854-1919", Centre for Studies in Social Sciences, Calcutta, No. 3, 1975.

Sdasyuk, G., "Transport and Formation of Economic Regions in India", quoted from T.P. Patil, "Transport in Sholapur District : A Geographical Survey", The Deccan Geographer, Vol. X, No. 2, 1972, p. 1.

Singh D.N., "Road Planning in North Bihar", Uttar Bharat Bhoogol Patrika, Vol. V, No. I, 1969, pp. 33-35.

Singh, J., "Research Methodology in Spatial Functional Interaction", The Deccan Geographer, Vol. XII, No. 2, 1974, p. 119. Mandal, R.B., "Role of Transportational Linkages in Rural Community.

10

Sustainable Resource Systems

The available endowment on the earth surface for the well-being of mankind is known as resource. It may be natural, biotic, human, economic, social, political and atmospheric and lithospheric in its dimensions. The system on the other hand, may be defined as "any set of objects and the linkages between them can be called a system". Men and their artifacts make up a manmade system. Natural phenomena and the processes which link them can be thought of as natural system. Hence the system shows how resources exert their integrated influence on man—the supra resource of the earth surface and how in configuration together they react on the earth by virtue of their functional structure in the environment. In any system, the process of feedback machanism is always at work, for example, if man needs a resource to be raised for his betterment and if it does, then it is man who developed the resources for human consumption. Thus the systems of resources are always interacting with each other in between the welfare of man and the investigation, growth, production, planning and conservation of resource at the same time.

Objective of the Study

The purpose of the paper is to highlight some of the natural resources and their systems in relation to man who plays a vital role in shaping and utilising different kinds of resources, as spatial linkages on economic, social and political grounds. The production and destruction of resources are the two extreme points, and in between these two the same takes several forms in order to serve mankind—a region under study. This study has been designed to throw light on the forest and agricultural resources in the Munger division

Review of the Literature

Resource utilization is the greatest function of the human society among which, according to Leonard Zobler, the resources can be grouped under three stages. Firstly, when technological development is in the initial stage with poor technology and unskilled labour. At the second stage, the cultural advancement and technological change, abundant and proper use of resources are found. But at the third stage, with the exhaustion of resources again scarcity prevailed for most of the resources of the world. Water is an exception which is available ubiquitously. The complex system of acute scarcity may be expressed in Table 10.1.

Table 10.1 shows that the Munger division has been faced with a vicious circle of poverty. This challenge has to be met with the economic development in the country. This major process requires mobilization of natural resources on a huge scale. All the frontiers of natural resources have been opened during the period of planned development in the district. Utmost precaution regarding resource-use policies should be taken. The resource-use policies should be so planned where no stone should be left unturned to utilise optimally the existing natural resources.

Secondly, the study of Duncan shows that with the opening of vital resource fronts their best utilization is possible. But only a few areas in the world have sufficient resources in a viable economy. This condition is available in the Mediterranean countries and China. According to Morgan and Munton, "Agricultural system cannot be meaningful unless it is associated with the production system along with the forest, soil, irrigation and systems of human resources." Similar attempt has been made by Simmons who had successfully analysed the interdependence of resources in close connection with the ecological conditions of the region.

Importance of Resource System

It has been observed that resources act as factors of production when these are used in the economic process. Entrepreneurial ability, labour capital and natural resources counted as factors of production are known as human resources and the others are called property resources. Technology teaches us to make effective use of scarce resources in different ways. The other capital resources machinery, plants and the like have to be adopted and adjusted to other uses. Sometimes, due to technological advancements, it has been observed that some resources are still unused. The four factors of prodution, e.g., land, labour, capital and entrepreneurship become scarce because they have alternative uses. If natural resource like land is used for a new golf course, the society provides space for an industrial park or a residential development. Similarly, capital resource and material resource building can be used to produce something new in resources. Thus the most scarce factors of production rotate behind choice and pricing.

Population as resource is the great creator, fashioner and destroyer of resources. In this way, systems of resources always revolve round a mechanism of integrated push-up in raising the production for the functioning of society. A man needs nearly twenty two hundred calories of food for sustenance in a healthy manner in the Indian environment. The gain of these calories may be possible with the eating of prepared food whose preparation requires coal, oil, edible items, utensils, home, defence, amenable social stratification, amicable environment, energy, water and several other products of nature. It proves that only those items come under resource system which are useful for man.

The function of a system may be exemplified with the help of the production of food item. A man produces foodgrains to eat. The foodgrains give energy in the form of proteins and calories to mankind, and on the other hand, the same is used as seed to produce foodgrains. But here the system is not a meaningful one, because once a particular foodgrain is consumed, the same cannot be replenished, time and again whether one used water in a steam engine or in a hydraulic turbine or the same is operated in the air, but ultimately after doing its job the water again comes into the form of water on the earth surface. Thus water balance in a man, in a field, in a machine and environment is a complete system of its regulation, and cannot be the food item. The production of food items in this sense gives an incomplete system wherever the replenishable capacity of water gives a full-fledged and complete system.

Nature

The materials used by man comes from the physical, biological, social and economic components of the earth. The interest and the same is known as ecosphere in which a complex collectivity, alternation of quantity and the behaviour of elements and environmental components are followed by the adjustment of one with the other. The satellite imageries taken from the space craft tell us that the earth has sub-planet and limited dimensions. The rate of the flow of solar energy on the earth surface is at the rate of 3,400 K cal/m^2 day and the same becomes the basis of light and energy of the biotic resources on the ground. The importance of solar energy lies in the proportion which is observed by the plants and its conversion into an organic matter to make it available to the other organism in the form of food through insolation. This process is known as photosynthesis and the world over yearly 2% of the solar energy is converted into the organic matter in the form of food. Thus unless and untill this proportion is received by technological means, it may not form a limit in raising the resource level of an organism in this world.

There is a series of stages by which the energy flows through a system of human and biological organism (Figure 10.1).

Table 10.1 : Conditions of Resource Use : A Conceptual Framework

Limited production in terms of agriculture, industry and trade	*Drain of resources agricultural and industrial*	*Drain of capital through defective system of trade and agriculture*	*Ruin of industrial skill*	*Population explosion*
1	2	3	4	5
		Stage of Scarcity		
More than under industrial-cum-commercial crops.	Limited per capita agricultural production.	Negligence of scientific and technological training.	—	Misuse
Export of food and industrial raw materials.	Lesser capital formation and investment and economic activities.	Limited industrialization.	—	Misuse
Limited improvement in agricultural efficiency.	—	Limited trained hand.	—	Wastage
More mouths of feed.	Lesser capital for mobilization of resources.	Scarcity of skilled labour.	—	Agricultural scarcity of agricultural raw materials.
Per capita scarcity of all resources.	Scarcity of capital.	—	—	—

In the system shown in Fig. 10.1 at each transfer of energy, there is a loss of heat back to the atmosphere. Such systems are also in need of the factors of inorganic elements which are available to the plant from the soil. Animals derive them from plants in case the same is herbivore and by animal prey if it is carnivore. In nature these elements are recycled within the biological system which is susceptible to loss as far as it is possible. Such systems of living resources and their inanimate habits which affect man's general characteristics is known as ecosystem. Here we are interested in discussing only one or

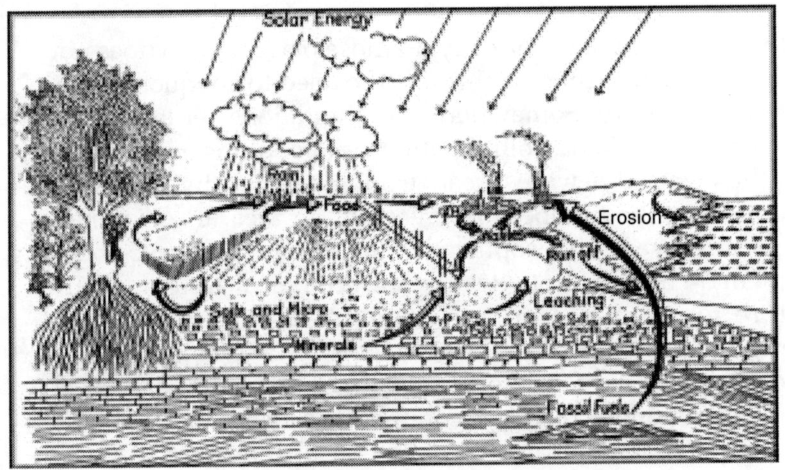

Fig. 10.1 : Resource System (After Kolars, et al.).

two components of ecosystems. In the domestication of wild animals the number of energy path can be reduced for the preparation of the structure of a system. Hence, per unit area of crop may be greater under cultivation than under natural conditions, but the general level of organic matter in the form of crop and the rate of energy flow for the greater production of resources is lower. At the time of harvest of all the crops, the essential elements of the system can be removed and the same can be reinstated artificially if the cropping is to be maintained. This is one of the ecological reasons for the use of fertiliser in the farm. Ultimately, man's modification of ecosystem removes their natural barriers and balances, and makes them prone to instability as in the case of the soil erosion which suddenly changed the patterns of cropping.

Culture

Among all the resources, man's idea is the greatest storehouse of resources because unless a man perceives a thing as a resource potential, the same can never be one. In the western world, the cultural attitudes are dominating over nature due to their intellectual superiority. In those places a man does not care about nature and for this fact the need to conquer nature is not a necessity. Somewhere, the limitations set by nature may change the behavioural patterns of people as for example Malthus sensed a widespread trouble created by nature in maintaining the number of persons of an area. In Asia, man and nature were philosophically viewed as one because both of them do not set out to destroy the ecological systems set by nature for the rapid growth of population. Recently man viewed himself as a part of the total community of the planet. The people of western world said that in the composition of world community, a series of commodities emerge as balancing forces. Recently cultural attitudes are dominated by the economic principle of demand and supply and the system of price mechanism. In nature, the demand is set by consumers for environmental resources, spatial arrangement of wilderness and pure air for breathing, but the price of industry quite often restricts, the supply of natural resources in bulk to the inhabitants. In this sense, it is desirable to develop a reformed economics in which all external diseconomies are fully taken but studies of environmental preception show the objective knowledge given by science through imagination, experience and prejudice, all of which play a vital role in resource use.

Resource-use can also be measured on a scale of rational to individual and immediate or unthinking selection of an action. In most cultures, only coercive modes of behaviour persist which includes the modification of ecosphere in garnering resources. The idea of duty to posterity is strong in the western world especially in the present use of the planetary systems which may not be set to endanger future generations and resources as well. In religious writings, the concept of stewardship of man as the temporary guardians is also set in motion to make this planet as a useful storehouse of resources for the moral and perceptual well-being of mankind.

System Analysis

In a resource system man considers natural system because it is also the resource producing system. Every system has to receive energy for movement from its natural state according to the use set by man because man is dominant in the system. The system is also the response of economic and behavioural aspects of man, cultural values but an ecological aspect of the system, man and society have no power to limit nature.

System of Forest Resources

The Munger division is rich in forest resources where forest lands are found in Sono, Chakai, Jhajha Kharagpur, Sheikhpura, Jamalpur, Lakshmipur and other anchals of the district. In system analysis forest resources come under the close system with special reference to the topographic, climatic, hydraulic and general environmental conditions. In forest environment rugged topography, higher amount of rainfall, lesser density of population, unproductive ground, the menace of wild animals, etc. are the rule. On the other hand, from forests as the resource, mankind derives wood by forest cutting, wildlife with the prey of animals for meat, tending of trees for glue and several other forest products as raw materials for paper industry and so on. Thus forest as the marginal ground of human resource provides a firm base to the economy of the region, directly with the forest products and indirectly with the industrial production giving employment to the people. Cultural advancement to the society and the modification of environmental conditions it brings rain, protect soil from leaching, increasing the fertility of land, excretion of leaves as nitrogenous element and humus content to the soil including the construction of houses, highways, bridges, industries, railway wagons, etc. On forest resource, the primary function of human society lies with the forest cutting, secondary function of human is the manufacturing of goods from wood, tertiary function with the sale of manufactured goods by the shopkeepers, and building in urban areas. Quarternary function is provided by the forest rangers, etc. who teach the society the ways of preservation of forest resource and the virtues of resources for shaping mankind, whereas the quinary function is provided by the chairman, and ministers of the Forest Department who direct other persons on telephone for the protection, conservation, planning and extensions of forest resources. Its regeneration is not possible because once the wood drawn from the forest resource in itself cannot be a full-fledged system but the incentive of man to grow and protect forest provides incentive for raising the forest resource including their growth and wood is an example of fund resource. Wood itself is the form of manufactured resource material. The presence of tree in nature is a natural resource. The utilisation of wood in the construction of house acts as a basic resource. Thus the wood of forest resource and the desire of man fulfills and in turn acts as a catalyst for the development and conservation of forest resources on the earth surface.

In resource system with the utilization of solar energy and moisture by rain, the growth of plants on the ground takes place where micro organisms, minerals, leaching of soil, erosion by water run-off, and industrial waste are the ecological cycle which runs from one resource to the other. The growth of plant, production of food and agricultural development system in between the food production and raw materials along with fossil fuels, of soil fertility be leaching, the growth of grass diet for animals, deposition of alluvial soil by meandering courses of the river Ganga and again the evolution of the initial cycle of landscape on newer land are the various cycle of natural and biotic resources which form the base for economic development, cultural advancement, and strategic position over different parts of the Munger division.

In the Munger division, the distribution of forest resources, net sown areas and production of foodgrains are given in Table 10.2.

Table 10.2 : Distribution of Forest, Net Sown Area and Production of Foodgrain (1991-2000)

(Area in Square Kilometres)

Name of anchals	*Percentage of forest*	*No. of registered - craftsman industries*	*No. of workers engaged in craftsman in '000*	*Net area in '000 acre*	*Total production '000 MT*
Munger	—	209	41	27	148
Bariarpur	160.52	55	23	18	112
Jamalpur	85.45	159	26	15	121
Dharhara	282.85	45	28	14	95
Kharagpur	306.79	175	38	43	240
Asarganj	—	78	17	15	85
Tarapur	—	65	12	14	55
Tetiha Bambor	101.38	22	11	9	42
Sangrampur	85.94	52	17	14	62
Barahiya	—	54	27	47	195
Pipariya	58.39	39	11	21	78
Surajgarha	389.66	85	16	61	241
Lakhisarai	283.30	141	78	65	224
Ramgarh Chowk	—	11	8	9	211
Halsi	—	18	23	22	232
Barbigha	—	15	15	16	201
Sheikhopur Sarai	—	8	12	17	168
Sheikhpura	—	32	29	53	205
Ghat Kusumbha	—	5	7	15	209
Chewara	—	11	6	18	106
Ariari	—	22	14	15	109
Islamnagar Aliganj	172.89	9	11	23	187
Sikandra	184.01	17	14	31	201
Jamui	—	105	19	18	102
Barhat	232.16	12	18	32	58
Lakshmipur	251.77	41	23	47	66
Jhajha	427.39	48	15	19	76
Gidhaur	71.11	15	8	18	72
Khaira	418.71	27	19	31	98
Sono	392.27	55	16	16	99
Chakai	774.04	26	35	32	72
Alauli	—	35	25	44	107
Khagaria	262.36	103	14	49	132
Mansi	—	22	14	32	104
Chautham	—	45	18	36	105
Beldaur	—	41	16	39	103
Gogri	—	23	14	36	95
Parbatta	—	21	15	41	111

Source : District Statistical Office, Munger.

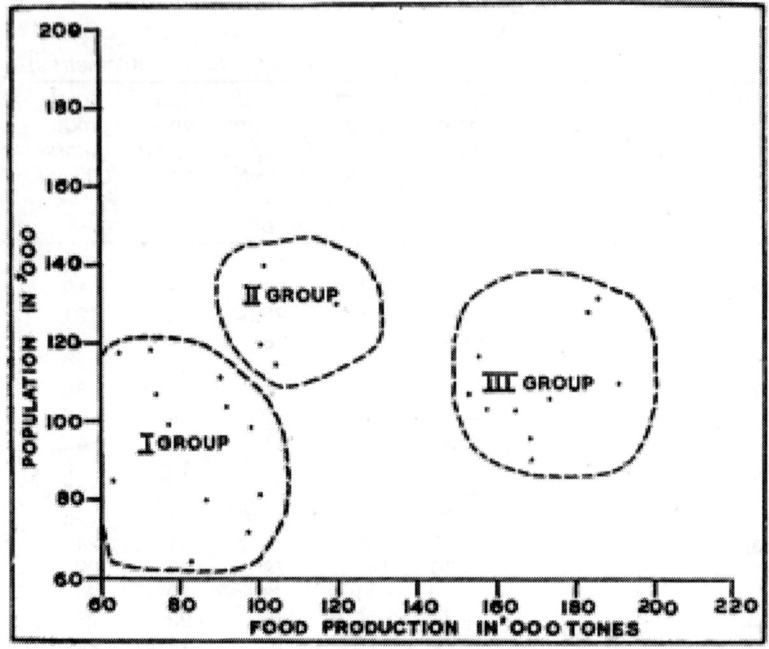

Fig. 10.2 : Munger Division : Relation Between Population and Food Production.

Agricultural System

Agricultural system provides the sustenance almost 80% people in the Munger division. It is a traditional society of agricultural production where intensive subsistence farming is the rule due to larger number of people engaged in agricultural production and tiny plots of land owned by the farmers under permanent settlement of land tenurial system. For agricultural system, the concept runs through land, labour, capital, suitable environmental condition and the incentive of local people for the success of such a system. In the Munger division only 50% land is available for cultivation where the density of population is about 400 persons per square kilometre and the per capita income is only Rs. 200 in the area. The level of resource development has been achieved by the production of food-crops in four different seasons of Aghani, Garma, Bhadai and

Rabi. Most of the years the failure of natural rain, flood havoc, the spell of heat, cold waves, dissipation of crops by frost in the winter and the presence of innumerable natural hazards are responsible for present agricultural system of the Munger division. Agriculture is a vital resource from which local people derive food, capital materials of personal amusement, industrial raw materials, and in turn agriculture provides employment, energy and shelter for living to man. In any region, the mechanisation of agriculture, use of higher number of agricultural labourers are examples of employment avenues. In the Munger division, the relationship between the net sown area and agricultural labourer is very nice. In terms of the availability of agricultural labourers for almost all net sown area is relatively more than 800 per anchal.

Similarly, the correlation between total population and production of food is also high but it is curvi-linear in the graph (Figure 10.2). This shows that some areas are overpopulated with lesser production of food, while some areas are highly productive with lesser number of population in group III. Some areas have both lesser population and lesser food production in group I. In the Munger division, the total net sown area and production of foodgrains is given in Table 10.2.

Thus in the study of systems of resources, it has been observed that various resources available on the earth surface to their usability for man, it has been found that the resources, e.g., forest, water, mineral, agriculture, population, livestock, etc. are interlinked for the well-being of man and on the other hand, modified the usability of a particular resource of different places according to their need and hence act as catalyst for resource development in a systematic manner. This is determined by the man himself. In this study with justification of close relationship between forest resource and cultural and economic advancement of people shows the existence of a variable resource system. Similarly, the close relationship between total population of anchals and food production with net sown area and agricultural labour justified the wholeness of the system.

Panchayati Raj and Self-Employment Programme

Panchayati Raj have traditionally been used as a word of politics and power. Whatever meagre authority they have is frittered away in political game. The priority was contracting and implementing this tendency left panchayat in many states with a bad reputation.

The time has come when Panchayati Raj institutions have to be involved actively and purposefully in economic activities including self-employment priorities. Due to neglect of Panchayati Raj Institutions in overall development initiatives of their area a huge gap has been created between expectations and fulfilment. It is here that the issues of self-employment policies—programmes lending to reduction of dependence on States becomes important. There are Panchayats which have offered banks security of advances and loan provided they are involved. This means transparency and viability. These are the keywords when it comes to economic programmes on decentralised basis getting success in the rural areas. Panchayati Raj has to understand the mindset of bureaucracy and thus prepare itself to facing challenges. It is not an easy transformation. There are areas required in entrepreneurship, policy, clarification, coordination, transparency and overall viability. Glamourisation of self-employment with better policy packages is possible.

Panchayati Raj has to understand the challenges of time specially in states like Uttar Pradesh, Bihar Madhya Pradesh, Orissa, Tamil Nadu where they have hardly ventured into these new areas.

Plan of Action

1. Any Plan of Action has to be area-based say a village or a community.
2. The felt needs of the people specially from demand angle needs proper appreciation. Hence whatever is going out from the village must be stored, value added and marketed at the best prices fetching at that time.

On the other hand, whatever comes from outside in terms of industrial service or business initiative should be locally developed where possible.

3. There is need to involve all local leaders and initiatives in the action projections. Hence they are leased and the contribution they can make needs codification.
4. SWOT analysis and Hidden Investible Potential Power of Organizations (HIPPOO), Known Investible Potential Power of Organisation (KIPPOO) and Threatening Investible Potential Power of Organizations (TIPPOO) of the area would make the job much easier for analysis and action.
5. Planning, action research codification, extension, loaning has to be the base. Hence cooperation and coordination become crucial with critical areas for monitoring.
6. Training has to be on practical basis. Organisership and entrepreneurship will require two different states of attitudes funding, monitoring and training capsule. Philosophy, Organisership management, Marketing, Production, Training and Technology, Extension and Codification (POMPTEC) would have to be the centre of training for Organisership. While for self-employment marketing led production should be the essential of training.
7. Monitoring system with regular interaction with multi-disciplinary team including banks, NGOs for Panchayati Raj is must.
8. Emerging areas of investment in Tourism, Trekking, Health, Food, Sanitation, Nurseries, Transport, Servicing, Education, Computers, Guides, Packaging etc. must be regularly watched.
9. Cost-effective systems development need periodical identification of cost-escalative activities with better local raw material use and increasing intake should make costs lesser and lesser.
10. Comparative SWOT analysis should be carefully done periodically. The aim should be to develop model so

that replicability could be ensured. Before the above is done there is need to exchange notes and experiences with all the role players for avoiding repetition of mistakes. A re-visiting success type of activity would definitely help in identifying reasons for poor performance of earlier initiatives. This would help in better project formulation and implementation.

11. In the world of materialistic greed led growth there is need to plan fruit eating versus root eating. In the plan of action systems for income with honesty, transparency must be in-built. The programmes must have lesser gestation period within in-built possibility and propensity to grow. Traditionally herbs, plants, pilgrim centres, heritage and tourism will be other areas of stable income provided better marketing systems could be developed.

REFERENCES

Duncan, C., "Resource Utilization and Conservation Concept of Economic Geography", Vol. XXXVI11, 1962, p. 39.

Kolars, F. John and Nystuen, D. John, "Ways to Weigh the World : A Preface to Man Environment Systems Geography, The Study of Location, Culture and Environment", 1964, p. 247.

Morgan and R.J.C., Munton, "Agricultural Activity and Physical Environment", Agricultural Geography, 1974, p. 39

Ronald, A., "Introductory Economics", Harper and Row Publishers, New York, 1971, p. 15.

Simons, C., "Resource System : Aspect of Geography", 1974, p. 7.

Verma, P., "History of Resource Use, Applied Geography", 1968, pp. 135-36.

Zobler, L., "An Economic Historical View on Natural Resource Use and Conservation", Economic Geography, Vol. XXXV1II, 1962, p. 191.

11

Integrated Development of Sustainable Resources

Integrated rural development implies development of all sectors of the rural economy and society. Rationale of the development of all sectors of rural economy is that they are inextricably interlinked with each other. Hence agriculture is closely linked with the allied activities like irrigational facilities, forestry, industry, road, railways, dairy, poultry, fishery, piggery, sheep farming, etc. Agriculture itself covers a vast field in various dimensions. Development of irrigation, rural electrification, plant protection, distribution of improved seeds, fertilizers and pesticides, health, education and sanitation in rural areas are several aspects which are interlinked with each other. Integrated rural development connotes suitable programmes in all these fields.

The rationale for development adversely affect all sections of the rural society, modernisation of agriculture and increasing agricultural productivity and prosperity which nowadays are confined to the big landholders. In such a situation small and marginal farmers and agricultural labourers will not be the beneficiaries of the green revolution. That is why special programmes like Small Farmer Development Agency (SFDA), Marginal Farmers and Agricultural Labour Agency (MFAL), Drought Prone Area Programme (DPAP), Rural Works Programme (RWP) and Antodaya Programmes (AD) have been designed to deal with the less privileged sections of the rural society.

Purpose of the Study

A Comprehensive Programme of integrated rural development is required to deal with the rural programmes covering all sections of rural society. Hence the content, means, method of study and agencies of such transformation are designed to analyse herein brief.

The purpose of this study is to highlight the regional imbalance of resource development in terms of forest, agriculture, population, irrigation, percentage of cultivable area, soil, water and industrial resources along with road and railways networks as available in the region.

History

Programmes of rural development in India have been going on since 1952 when the National Extension Service and Community Development Programmes were initiated. They were followed by various specialised programmes of rural development like IADP, DPAP, SFDA, MFAL, IRD, etc. For reaching these programmes to the remotest corners of this vast country, a number of techniques were employed with different degree of success.

Concept of Integrated Area Development

After going through the weakness of the existing rural development programmes let us now examine the alternatives. At present our country is going through a process of green revolution in the field of agricultural development which means enhanced prospects of self-sufficiency in food and savings. The savings, if properly harnessed and utilized, it may lead towards greater urbanization of Indian villages. If the improved economy of the village is not properly managed it may lead to rural unemployment, increased migration to the already congested towns and cities, social tension, political unrest and a real imbalance in economic growth. Moreover, if agricultural sector is not properly managed, the green revolution may die before making the country green. If this happens the green revolution will certainly turn into red revolution.

What is then needed is to provide not only the agricultural inputs on increasing scale but also employment to the people and the facilities to distribute, operate and serve them. The provision of these facilities cannot be made in each village. Thus it has to be centralised and so distributed that each village have an easy reach at a reasonable cost on each part of the country.

Availability and cost factors have to go together. The reduction of cost takes place only if there is economy of scale of both horizontal and vertical linkages among different facilities. The village as a unit cannot support more than a couple of grocery stores. We have then to rethink the principles on which villages can be grouped together to form viable communities for socio-economic development.

The integrated area concept helps us in making these groupings. The individual villages are linked with one another. These linkages do not extend too far in the present context but as developmental activities proceed the linkages are mostly functional and well reflected by the frequency of trips of people to other villages or towns. If we study the trip pattern of each village we will find they are more frequent within a radius of 15 kilometres. This is mainly because of the poor transportation and communication facilities. We further notice that various villages have a tendency to cluster around a common centre for a particular function. If we study the frequency of trips made by individual villages to different places for varied functions we will be able to see clearly the emotional as well as the physical linkages of each village to certain centres.

A further study of the relationship between centres of varying sizes will reveal that the primary centres to which villages are attached are functionally linked with larger centres, e.g., town or cities. If we can identify a common centre which caters to the varied needs of a large number of villages through subsidiary centres, we can easily identify a micro-region. It is thus, micro-region, which in our opinion, should be the area for planning and development at the lowest level.

The block as a planning unit has not been thought of in these functional towns. Hence it has failed to generate economic development as it was expected. The present block boundaries are located at points which cannot be said to form the functional centre of the nature explained above.

In order to meet the increased needs of the rural areas we have to develop infrastructure facilities as well.

Let us then propose that the development of physical and cultural regions that will meet the development needs of the coming year require the development needs of structure networks or systems or places or towns forming a hierarchy based on functions. Such towns have to be distributed in such a manner that they lead to the most efficient flow of goods and services and indeed the ideas. Unless we include the programme of infrastructure development, the development of towns on the line suggested above, we will not be able to develop our villages fast enough at the least possible cost. And if we fail to do this the reversal will begin and the whole process of development create chaos in the society.

Level of Regional Development

In the Munger division various types of resources are available according to the speciality of ecological conditions and patterns of agricultural growth. The area north of the Ganga is mainly flood prone belt susceptible to flood, still it is one of the biggest granary and densely populated part of North Bihar owing to its agricultural superority. On the other hand, the southern plains of Munger is the theatre ground of historical events, population concentration in large and medium urban-industrial centres, availability of fertile plains with higher percentage of net sown area, etc.

The areas around Kharagpur hill and Sono, Chakai, Jhajha, Khaira, Sheikhpura, Lakshmipur, etc. present diversified picture of ecological components in terms of resource development. Here rugged terrain conditions and dense forests retarded the progress of agricultural, industrial, human and other sorts of resource development in comparison with the forests and biotic resources.

Methodology

This study attempts to devise a new method for measuring the integrated development of resources with special reference to the Munger division. The measures will be the percentage value of different variable of resources with the fininding out of the rank order to each and every item of each and every anchal and the summation value for all the percentages have been calculated according to their ranks. The summation value of these ranks orders by the number of variables available, a final index of integrated rural development emerge in assessing the level of regional development.

In the Munger division for measuring the integrated development of resources, percentage of seven variables have been added together, had a composite picture of resource development is seen with percentage data of all variable, e.g, percentages of cultivable area, forest cover, irrigated area, production of foodgrains in thousand metric tonnes including percentage of population, number of industrial establishments and length of roads (Table 11.1).

So far as the percentage of cultivable area is concerned it has been seen that the highest area of cultivable land is found in Ariari, Tarapur, Sheikhpura, Jamui, Sangrampur, Gogri, Barbigha and Lakhisarai anchals whereas the lowest percentage of cultivable area is found in Jhajha, Sono, Sikandra, Dharhara, Munger and Kharagpur anchals of mountainous forest covered areas, whereas the lowest percentage of 50% is found in flood affected Chautham anchal of North Munger.

The percentage of forest gives a varying picture. The forest cover is almost abstained from Munger, Jamalpur, Tarapur, Sheikhpura, Barbigha, Ariari, Halsi, Gogri, Parbatta, Chautham and Beldaur anchals whereas lowest percentage of forest is found in Khagaria, Jamui, Barahiya and Surajgarha anchals. The higher percentages of forest cover which ranges between 10% and 53% are found in Sono, Jhajha, Chakai, Lakhisarai, Lakshmipur, Khaira, Kharagpur and Dharhara anchals of South Munger.

Table 11.1 : Level of Regional Development of Resources in Munger Division (1991-2000)

Name of Anchals	*% of Culti-vable area*	*% of total forest*	*% of irri-gated area*	*Total pro-duc-tion in '000 metric tons*	*% of popu-lation*	*No. of indus-try*	*Leng-th of road*	*Total rank*	*Ave-rage of all ranks*
Munger	6	-	2	14	7	23	10	52	9
Bariarpur	8	6	4	7	12	14	8	51	8
Jamalpur	6	-	2	5	1	16	4	43	6
Dharhara	5	9	7	5	19	9	11	65	9
Kharagpur	7	8	3	22	17	22	12	91	13
Asarganj	21	-	10	13	12	18	7	65	10
Tarapur	19	-	12	12	14	16	6	68	10
Tetiha Bambor	14	4	8	6	3	10	3	56	8
Sangrampur	18	4	9	6	4	10	5	56	8
Barahiya	9	-	8	17	12	13	7	66	11
Pipariya	9	2	15	18	12	18	6	88	13
Surajgarha	10	3	15	20	14	18	8	88	13
Lakhisarai	16	5	10	21	11	20	15	98	14
Ramgarh Chowk	9	-	9	15	14	1	13	65	11
Halsi	11	-	10	17	16	3	16	73	12
Barbigha	17	-	17	19	1	1	2	57	10
Sheikhopur Sarai	14	-	12	13	1	1	1	37	13
Sheikhpura	20	-	9	15	5	7	9	65	11
Ghat Kusumbha	16	-	4	11	3	4	6	55	13
Chewara	20	-	11	8	10	3	1	65	10
Ariari	22	-	13	11	13	5	3	67	10
Islamnagar Aliganj	2	4	3	14	13	2	17	56	9
Sikandra	2	4	5	16	15	2	18	62	9
Jamui	19	2	6	1	8	19	17	72	10
Barhat	2	5	14	4	18	10	8	55	12
Lakshmipur	3	7	18	6	20	12	11	77	11
Jhajha	1	8	6	3	19	14	14	65	9
Gidhaur	2	3	11	2	13	6	5	42	13
Khaira	15	16	16	18	18	6	6	90	13
Sono	1	6	1	4	20	15	6	53	8
Chakai	5	10	12	2	21	5	13	68	10
Alauli	10	-	14	10	-	9	1	52	9
Khagaria	12	1	20	13	6	17	4	73	11
Mansi	4	-	12	6	-	11	2	54	10
Chautham	8	-	18	7	-	12	2	55	10
Beldaur	13	-	18	10	-	5	-	53	9
Gogri	16	-	14	9	5	4	4	44	7
Parbatta	14	-	16	9	10	5	6	60	10

Source : Compiled by Author.

So far as the percentage of irrigated area is concerned the lowest 2% is found in Sono, 3.5% in Munger and Jamalpur, 5% in Gogri, 3.8% in Kharagpur anchal whereas 7.4% is found in Khagaria, 50% in Chautham, 42% in Beldaur, 38% in Barbigha, 33% in Parbatta and Khaira and so on.

These variations in north and south is mainly dominated by the private tubewells in North Munger whereas in South the irrigated area is mainly found in the command irrigated areas of Badua Canal, Morwe Canal, Lower and Upper Kiul reservoir, Anjan reservoir and others.

The percentage of total food production is lowest in mountainous forest cover areas of Jhajha, Sono, Chakai, Lakshmipur, Jamui, Dharhara and Jamalpur anchals in South Munger. In flood affected areas north of the Ganga and the Burhi Gandak the lowest percentage is 16.60% in Gogri, Parbatta and Chautham anchals.

The highest percentage of food production is noted in Kharagpur, Lakhisarai, Sheikhpura, Barbigha, Barahiya, Khaira and Halsi anchals of South Munger whereas more and more land is devoted in food production even if the area is susceptible to annual flood especially in Barahiya tal areas (Figure 11.1).

According to 2001 Census, the lowest 32% to 61% of population is found in Barbigha, Jamalpur, Tarapur, Sangrampur, Gogri, Khagaria, Beldaur, Alauli and Jamui anchals of North and South Munger. In plain area although the density of population is high but the concentration of lowest percentage of population is caused by the smaller size of area in the North. In southern Munger the lowest population gives somewhat different picture where the lowest percentage is coupled with large size anchals.

So far as the industrial concentration is concerned the total number of registered craftsman industry along with large industries are around 1,152 which are distributed in an uneven fashion in different anchals in the Munger division. The highest industrial concentration is found around Munger, Kharagpur, Jamalpur, Lakhisarai, Jamui and Khagaria anchals whereas the lowest number of industries are 10 in Barbigha, 12 in Sikandra,

13 in Halsi, 14 in Beldaur and 16 in Ariari. These numbers show the lack of interest among entrepreneurs in the above-mentioned area where the rural surrounding hinders the industrial development due to lack of capital, raw materials and lack of entrepreneurial interest.

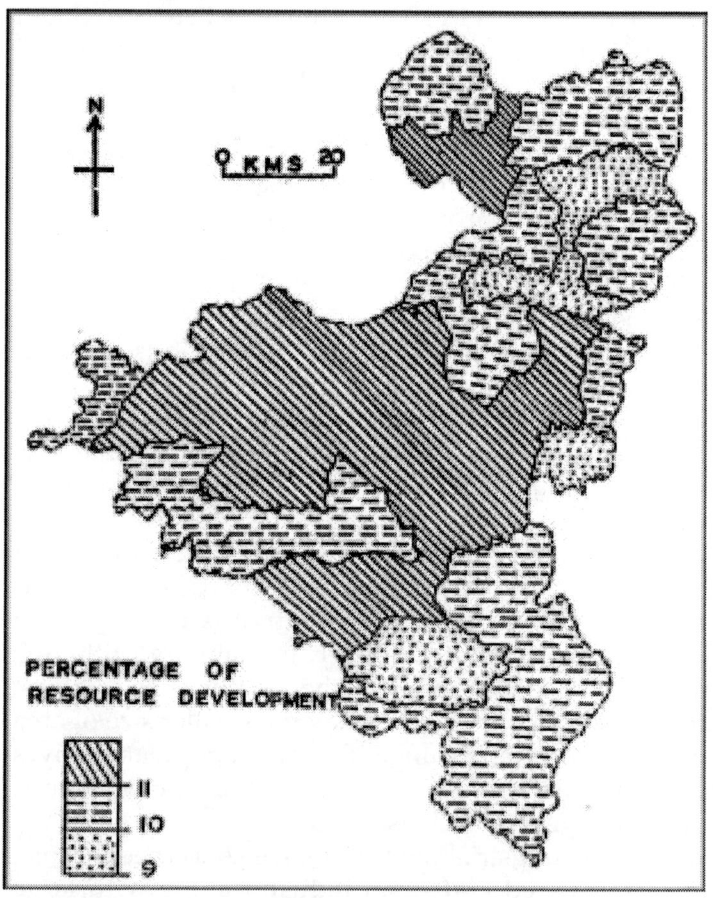

Fig. 11.1 : Munger Division : Level of Integrated Development of Resources.

So far as the length of road is concerned the lowest rank is for the flood ridden anchals of Alauli, Chautham, Khagaria and in the mountainous tracks of Tarapur and Jamalpur. The lengh of road is not so high in Sikandra, Halsi, Jamui, Lakhisarai and Kharagpur anchals. So far as the railways are concerned the highest rank is obtained by Lakhisarai followed by Parbatta, Chakai, Gogri, Chautham, Barahiya, Barbigha and others in decreasing fashion.

Finally, the integrated picture of all the other resources show that the highest average rank is achieved by Lakhisarai anchal and the same is followed by Khaira, Kharagpur, Tarapur, Jamui, Lakshmipur, Halsi and Khagaria anchals. From the lowest side the lowest rank of 7.2 is achieved by Jamalpur whereas Sangrampur has 8.0, Sono and Alauli 8.1, Gogri 7.6, Sikandra 9.0, Ariari and Barbigha 9.1 and others.

These represent the importance of Canal irrigation in southern Munger. North Munger show a lowest average rank due to ravages of flood each and every year.

Correlation Matrix

Table 11.2 shows the correlation matrix and their graphical representation for the different variables of planned development of resources in the Munger division, such as percentage of net sown area as cultivable land, forest irrigated land, agricultural production, population, number of industries and the length of road.

Table 11.2 : Correlation Matrix for Different Variables Selected for Integrated Development

	Culti-vable area	*Total Forest*	*Irrigated area*	*Total produc-tion in '000 hectares*	*% of popu-lation*	*No. of indus-tries*	*Length of road*
	X_1	X_2	X_3	X_4	X_5	X_6	X_7
X_1	1.00	-8.1	0.81	0.40	0.91	0.47	-2.1
X_2		1.00	0.42	.52	0.14	-0.12	0.36
X_3			1.00	0.94	-0.27	-0.41	0.67
X_4				1.00	-0.14	0.97	0.47
X_5					1.00	0.67	0.94
X_6						1.00	0.47
X_7							1.00

The per cent of agricultural land has a positive and a high correlation of 0.91 with the population and 0.81 with the irrigated land, whereas negative correlation of minus 0.81 is reported with the forest and the connection of roads in the entire Munger division. This shows that forest environment restrict planned development of agriculture due to poor road development.

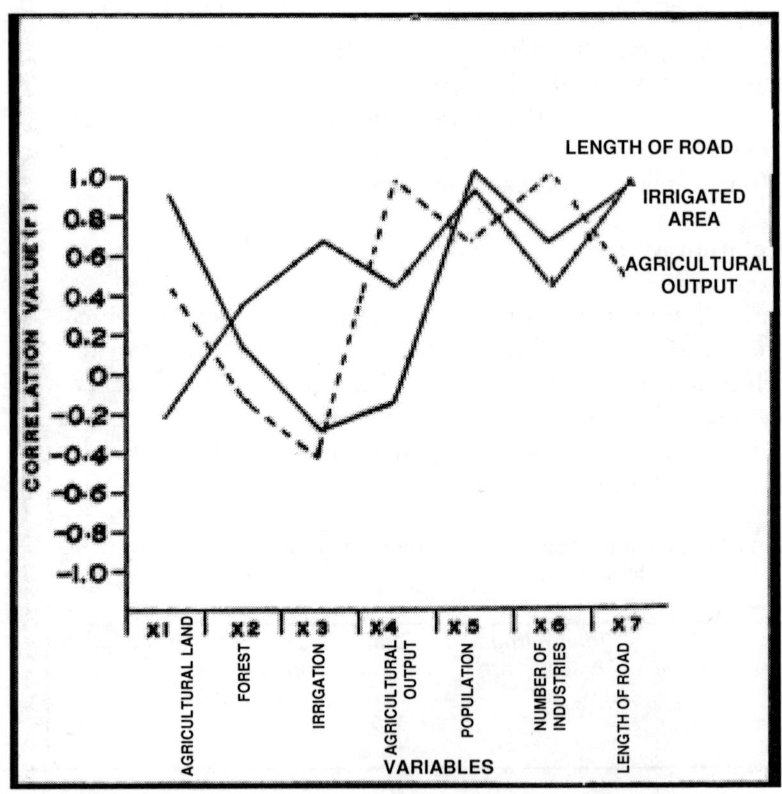

Fig. 11.2(a) : Munger Division : Correlation Matrix for Different Variables.

The forest resource shows a high correlation of 0.52 with the agricultural production and 0.42 with the irrigation. This means that in forest environment with the help of irrigation higher agricultural production may be obtained with certainty in any planned development of forest and agricultural resources in the Munger division. A negative correlation 0.12 is reported with the number of industries in forest areas which shows that as yet no sincere effort has been made to develop forest-based industries in the forest covered areas.

Irrigation, the third variable has a high correlation of 0.94 with the agricultural production and 0.67 with the development of road.

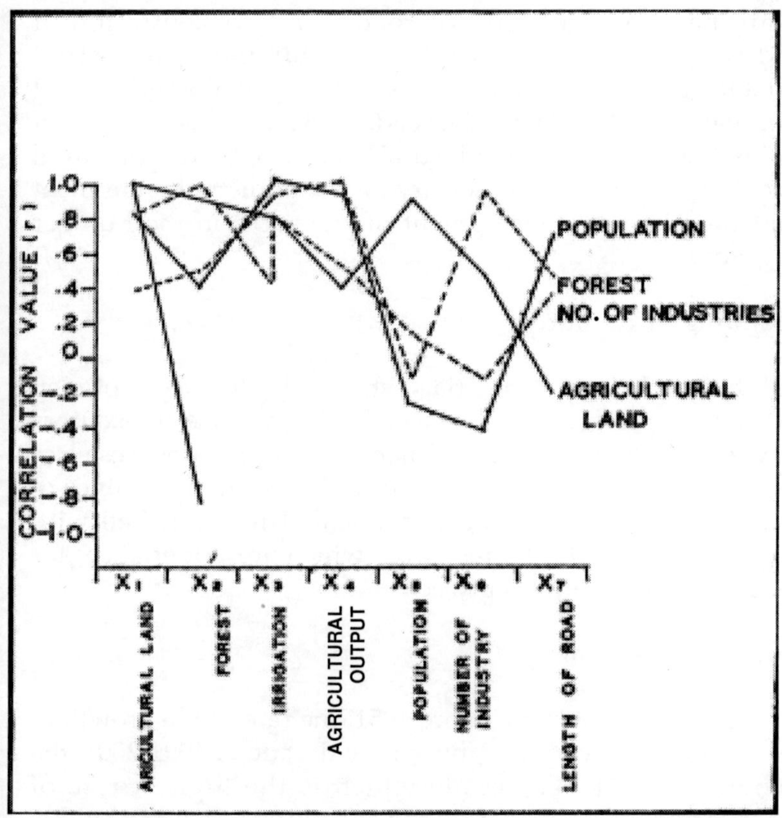

Fig. 11.2(b) : Munger Division : Correlation Matrix for Different Variables.

This shows that irrigation is highly needed for better agricultural production as well as for the development of road which provide structure facilities for the development of road in any area.

The fourth variable agricultural production has high correlation of 0.97 with the number of industries and 0.47 with the road. This clearly shows that agro-based industry is prevalent in the entire Munger division on the road side. Population has a better correlation of 0.94 with the road network, 0.67 with the industrial development and 0.91 with the agricultural land. This clearly shows that areas of fertile plain have high density of roads, and concentration of industries in the Munger division. Similarly, industrial establishment has high correlation of 0.97 with the agricultural population and 0.47 with the road. It means better network of roads, higher number of industrial labourers, good agricultural production are almost necessary for the industrial preferment and the planned development of resources in the Munger division.

Planning of Resources

All the goods or raw materials which fulfil the needs of men are known as resources. Hence the systematic use of resources have started with the origin of man which is the super resource among all the resources of world. In this section, the lines of planning process have been delineated in a nutshell with special reference to the resources which are undertaken, for detailed analysis in this project.

Population Planning

In the Munger division before 1951, the rate of the growth of population was very low, but in the decade of 1991-2001, the growth was 21 per decade which is the highest rate of population growth in comparison with all the previous censuses. In 2001 Census, the Munger division has 51.44 lakhs of population and the number of occupied residential houses were 4.04 lakhs. In case we consider an occupancy rate of 5 persons per room, according to 2001 Census, there should be 6.62 lakhs of houses in the Munger division and hence we may say that there was a dearth of 2.58 lakh houses in 2001 Census itself. On this basis in 2001 Census, there should be an extra houses of 4,30,507 over 2001 Census. In this way with the growth of population the requirements of the number of houses can be seen in Table 11.3.

Table 11.3 : Growth of Population and Requirements of Houses in the Munger division (2001-2021)

Census year	*Actual and estimated population*	*Shortage of houses Total number of over 1981 Census*	*Total number of house required*
2001	33,14,806	25,896	6,40,993
2011	37,31,665	2,42,333	7,14,776
2021	41,54,535	4,30,507	8,80,516

In the Munger division the average population density was 650 persons per square kilometre in 2001. There are anchals like Kharagpur, Dharhara, Surjgarha, Ariari, Khaira, Halsi, Lakshmipur, Jhajha, Chakai, and Sono which have low density of population in comparison with the division average. Hence, in case in the abovementioned anchals with the provision of roads, irrigational facilities and multiple cropping we may suggest the lines of migration of population from densely settled area to the area of low population density. In this sense there should be a planned and systematic population migration from Munger and Jamalpur to Surajgarha area, from Tarapur and Sangrampur to Kharagpur anchal, from Lakhisarai and Sheikhpura to Dharhara anchal, from Barbigha and Barahiya to Ariari anchal, from Jamui and Sikandra to Khaira anchal, and almost all anchals of North Munger should send their extra population to the southern hilly margins such as Sono, Chakai, Jhajha and Lakshmipur anchals.

According to the estimated population from 2001 to 2021, it can roughly be said that up to 2001 there will be a treble number of population in the Munger division if the present rate of population growth continues.

In order to overcome this issue, the author suggests to introduce rigorous birth-control measures in almost all inhabited villages not only in the Munger division but allover the State in Bihar and India, so that an environment may be created for planned growth of national economy on static population growth. If this dangerous problem of population growth is not checked immediately then it may jeopardy our efforts of Five Year Plans and would nullify the achievements of planning.

Forest Planning

In the Munger division, 16.5% of land has forest cover. According to the world standard, there should be 32% land under forest in order to maintain ecological balance between natural resources and men. In India there has been 23% land under forest cover. Hence, in the Munger division there is a very low amount of forest in comparison with the national and international standards. Besides, the forests are sparse one and not protected, from the onslaughts of ranchers, woodcutters and occasional break out of fire especially by the mahua peakers. At present there is no restriction on the prey of wild animals. Hence, the Government should first of all give protection so the forest resource, forest conservation and its systematic utilization with the establishment of sanctuary and suiting blocks. In forest of South Munger there is a lack of good approach road and this should be constructed urgently in order to facilitate the export of forest products of the division which are highly required in North Munger and North Bihar Plain including the industrial belt of Calcutta-Hooghly region. Based on the forest products of the Munger division the Government should establish Paper Mills, Biri Industry, including the extraction of wood for good furniture as required in all urban and industrial centres of the State. About the industrial potentiality based on forest products the establishment of medium-scale industries and wood wastes will not only increase production of various industrial goods but also increase the employment opportunities for unskilled and skilled labourers as well as for the educated unemployed youths. (Table 11.4).

Table 11.4 : Industrial Planning Based on Forest Products in the Munger Division (2001)

Name of Industry	*Category and Investment*	*Direct Employment (man days) per annum*	*Indirect Employment (man days) per annum*	*Approximate value of production per annum*	*Subsidiary Industry on the basis of original Industry*
1. Particle Board	Medium (Rs. 130) (lakhs)	75,000	150,000 (Logging transport, etc.)	Rs. 100 lakh	Furniture and low housing industries.
2. Manufactur of Mechanism	Small scale (Rs. 3.25) lakhs	7,500	15,000 (Logging transport etc. Fishers and rowers)	Rs. 7 lakh	Railway carriage industry, fishing, net fishing manufacture, fish canning
3. Manufacture of black lead pencil	Small scale (Rs. 805) lakh	6,000	12,000 (Logging transport, sales, etc.)	Rs. 23.4 lakh	Lead slip manufacture
4. Manufacture of Oxalic acid from saw dust	Medium (Rs. 11) lakh	8,000	10,000 (Collection and transport and sales, etc.)	Rs. 15 lakh	Polishing industry
5. Manufacture of Tartaric acid from Tamarind leaves and fruits	Small scale (Rs. 5.20) lakh	6,000	12,000 (Collection and transport)	Rs. 7.50 lakh	Manufacture of fruit jelly and bleaching powder
6. Industry for manufacture of oil based on non-edible oil seeds	Small scale (Rs. 3.03) lakh	5,000	15,000 (Collection and transport)	Rs. 1.75 lakh	Soap industry
Total	**Rs. 160.20 lakh**	**10,750**	**214,000**	**Rs. 154.65 lakh**	

Many subsidiary industries will also be established on the basis of the products of the aforesaid industries. The employment and production position will be as follows :

Immense benefit will, therefore, accrue to the division by establishment of these six feasible forest-based industries at a total estimated cost of Rs. 1.60 crores. These industries will produce goods worth Rs. 1.55 crores annually and generate employment of about 322,000 men per day. Moreover, at least ten more industries can be established in the division based on the industrial products of aforesaid forest-based industries. These ten industries will further generate additional employment of about 700,000 men per day.

Agriculture Planning

Besides irrigation and electricity, the provision of other infrastructure facilities is most essential for agricultural development.

1. The Use of Multiple Crop

In the management of multiple cropping 1.25 lakh hectares irrigated land is involved in Rabi crops, whereas 30,000 hectares irrigated land is under Kharif crops. Altogether these two crops are sown in 2.25 lakh hectares of land. In the Munger division about 13,000 hectares land is involved in treble cropping. In case multiple cropping has been adopted the production of about 75,000 tonnes of foodgrains can be increased without much investment. In some areas the provision of irrigation is essential to increase the coverage of irrigated area by sinking 450 tubewells only. In areas of assured irrigation facilities the farmers should necessarily know the technical know-how of scientific cultivation so that the proper utilization of land and water resources could be possible for sustaining the life of the people.

2. The Use of High Yielding Varieties of Seeds

In the Munger division high yielding varieties of rice has been cultivated in 1.96 hectares of land, wheat in 0.96 lakh hectares and maize in 0.86 lakh hectares. Among all these high yielding varieties of seeds have been used for paddy, maize and wheat and over 60% of total area engaged in these crops. According to the increase in irrigational facilities and the use of Chemical fertilizers it is necessary to enhance the technical knowledge of farmers regarding the use of high yielding varieties of seeds for planned agricultural development. In coming years if the Government will achieve this target of producing high varieties of maize in 80,000 hectares and wheat in 50,000 hectares then the production of food crops of the State would increase to 1.10 lakh metric tonnes.

3. The Adoption of New Techniques of Agriculture

In the Munger division especially in drought and flood affected areas the Government has not a definite plan to develop an agriculture on scientific lines. The agricultural land of Jamui is suitable for producing crop but due to lack of irrigation it looks barren as a stony waste. With the help of irrigation, the cultivation of soyabean, groundnut, etc. are possible if the technical knowledge of the production of these crops would be given to the farmers before the time of their sowing in the field.

4. Cultivation of Cash Crops

Along with the production of food crops the development of cash crops like cotton, chilli, etc. are essential because the Munger division has enough potentialities to produce these crops if the technical knowledge of the production of potato, chilli, onion, sugarcane and cotton are given to farmers then the Munger division will be a leading producer in these crops in future.

5. Required Use of Fertilizer

In order to increase the production of food crops besides irrigation and use of hybrid varieties of seeds, it is essential to use compost, green manure and chemical fertilizers in the field because these help in increasing the production of crops from 2 to 3 times. Without proper testing of soil and without knowing the suitability of any fertilizer in any field, the excess use of fertilizer may hamper the productive capacity of the field. The use of fertilizer depends upon the characteristics of soil, crops grown and the climatic conditions of the area, because the need of fertilizer varies from crop to crop and field to field.

6. The Use of Insecticides and Pesticides

The use of insecticides and pesticides along with fertilizer is essential to weed out certain destructive Kajra caterpillar and termites in the field which are very dangerous for crops in tender age. Each and every insecticide will be of higher price and hence the farmers generally use the same after knowing the problems of particular crop in the planned development of agriculture. It is essential that farmer should be given insecticides and pesticides including the knowledge of their utilization in each and every season with the provision of loans from the financial institution like hank.

7. The Establishment of Agriculture Science Centre

In order to increase investigation and dissemination of new practices in the field of agriculture, it is essential to give training to the farmers of such scientific knowledge with the help of demonstration farm. Hence it is requested to the Government that the proposed agricultural science centre should be established without any further delay on the basis of thorough investigation and proper infrastructure facilities in Munger so that the farmers of the division will get proper training for increasing the production of crops.

Drought and Flood Control

In the Munger division the physical problems of flood and drought are always felt since times immemorial. On the one hand in Diara, Tal area and northern part of North Munger the flood havoc is seen each and every year. But on the other hand, Jhajha, Chakai and Jamui are the anchals in South Munger where due to lack of water the drought hampers agricultural development each and every year. Hence, it is essential to control flood in North Munger with the construction of river embankments, digging of river channel, dredging the course of rivers and the construction of reservoirs and mini dams to store water for the summer months. On the other hand, it is essential to obstruct free flow of water through the construction of reservoirs, mini dams and regulation of irrigation water through canals in South Munger.

In the riparian tracts of the Ganga Diara area lie under 10 anchals which have a total expanse over 59,242 hectares in which 26,609 is cultivable and the rest 3,980 hectares is under irrigation. During rainy season most parts of Diaras are submerged from the knee-deep to neck-deep water and hence no irrigation schemes would be managed permanently over the Diara land. With the provision of electricity, bamboo boring may be feasible in Diara land on the basis of which the production of Parwal, number of vegetables, cucumber, watermelons, chenna, etc. are possible. In the Diara tract of Munger division with the help of Central Government Operational Research Project was adopted long ago. But as yet the problem is unsolved and the area is annually susceptible to flood and drought in rainy and summer seasons respectively.

The problem of Tal area is similar to Diara belt. The total cultivable area is 104,366 hectares out of which only 1,730 hectares have the facility of irrigation and in these areas Rabi crops are mainly grown. In 1979, the Government has adopted the project of Tal area reclamation but as yet no progress has been made in this direction. In the Tal area the farmers should give a proper training to rear dairy cattle and fowls because these flourish in the Tal belt.

In drought affected areas of Jamui, Chakai, and Jhajha with the help of technical assistance of Evard, the State Government has adopted a scheme to eradicate the problems of drought in South Munger. Under this scheme deep boring have been established at 313 places along with the energising of 300 tubewells. Besides this, loan has been given for the purchase of tractors, etc. It is hoped that sinking of 300 tubewells will not be going to solve the problem of irrigation under DPAP scheme. Hence there is need for coordination between the activities of farmers and the Government especially in the field of irrigation, livestock-rearing, construction of road and providing the facility of electricity as these will help in integrated development of the area.

Dairy and Livestock Rearing

The last five year plans have witnessed several schemes of dairy development. At present in Lakhisarai centre there are no dearth of the buffaloes of Murra breed and cows of Haryana breed. Among the cattle wealth of 2.74 lakh of the division, 1.24 lakh come under the controlled rearing scheme. The launching of crash programme of livestock development is essential for the integrated development of dairy cattle rearing in and around Munger.

The development of chilling centre is also essential under cooperative sector so that the supply of cold milk can be ensured with special reference to the urban centres of Munger and Jamalpur. In rural areas where most of the cattle wealth of the division are found, it is essential to establish a well organized veterinary hospital in almost all panchayats of the division.

Irrigational Planning

In order to increase agricultural production, it is essential to develop irrigational facilities in the division in a planned way. Out of total cultivated area only 20% is irrigated in the Munger division, whereas rest of the area is unirrigated. In anchals of Jhajha, Chakai, Khaira, Jamui and Lakshmipur most of the areas are stony, waste and drought affected and hence the provision of irrigation is essential in these areas.

The following irrigation schemes have been undertaken in 2001-02 in the field of irrigation in the Munger division :

(a) Barnar Irrigation Scheme;
(b) Anjan Irrigation Scheme;
(c) Dakranala Irrigation Scheme;
(d) Surajgarha Pump Canal Scheme;
(e) Belbarna and Badua Reservoir Project; and
(f) Upper Kiul Reservoir Project.

But due to mismanaged execution and lack of coordination between objective and achievement, these schemes are as yet not completed. Hence in order to provide irrigation in southern part of the division it is essential to complete all these schemes as early as possible. In plain areas of the division tubewells can be easily utilized for irrigation, which is under active consideration of the government. In North Munger bamboo boring works efficiently especially in Alauli, Beldaur and Parbatta anchals. It is a cheap source of irrigation. Hence the government should adopt this scheme right earnestly for irrigating the land under planned agricultural development.

Geo-Thermal Power Plant

As yet there is not even a single geo-thermal power plant in India, hence, the development of the Bhimbandh area does not exclude the future development of geo-thermal power plant at different locations. On the contrary, a permanent power supply will facilitate the super project. For a geo-thermal power plant a temperature of 300-400 °C is desirable. The test drilling or sensoring for temperature increase per each 10 m drilling is most necessary. Geo-thermal power, once developed, will provide an equally permanent power supply all the year round, as it is not dependent on the supply of water from the thermal spring.

Another advantage is the absence of air and water pollution. As there is a fault line from Bhimbandh to Rishikund in a distance over 40 kilometres there is ample scope for a number of geo-thermal power plants, with a big enough total output to possibly benefit power to develop new industries or by selling the surplus power to the railways or big industries in Bihar and West Bengal.

Underground reservoirs of streams and hot water have long been tapped in Iceland, Italy, Japan and Newzealand. The figures of output given by the Sonoma Country California is 800 kilowatt of electricity, which is about 40% of San Franscisco city's total power requirements. In case the plant will be established then the hot springs of Munger may generate electricity in excess from the total requirements of the division.

It is observed from published literature that geo-thermal power station is feasible for thermal springs having temperature of 200°C under 3 to 4 atmospheric pressure. It is also seen from Table 11.5 that geo-thermal power station has been feasible even under the following pressure and temperature at turbine-point :

Table 11.5 : Geo-Thermal Power Station

Nation's station	*Tempera-ture °C*	*Pressure*	*Kind of Stream*
1. Italy			
A. Lardorello No. 2	120	2	Hot exchanger stream
B. Caste lonuro	120	2	Hot exchanger stream
2. USSR : Pauzhetsk	120	1.2	Wet stream

Source : Geohydrological Report of Thermal Spring, Munger.

It is interesting to note that geo-thermal power station is one of the most economical process of generating electricity. The capital cost and the average working expenditure has been worked out for a 30 MW Power Station.

Table 11.6 : Capital cost and working expenditure of Power Station

Mode of generation	*Average generating cost/kwh*	*Capital cost*
Geo-Thermal	2.63 Milis	61,30,000
Coal fired	7.10 Mills	74,40,000
Nuclear	7.20 Mills	1,43,40,000

Source : Geohydrological Report of Thermal Spring, Munger.

It is also seen in the most developed geo-thermal area of the world, namely Italy, Newzealand and California, U.S.A. indicate proven field capacity running into hundreds of MWs. Hence, it will be advisable to undertake the detailed geophysical and geohydrological investigation of thermal springs in the country.

Bhimbundh hot spring is located in a picturesque surrounding with hills all-round covered with fairly dense mixed jungle. Of all the seven hot springs of Munger division, Bhimbandh and Sitakund are readily accessible owing to the copious discharge of hot water in the cluster of hot springs the existence of perennial fresh water stream. Hence this area can be developed into a beautiful tourist centre. Since the terrain is sloping, pebble bedded hills and the hot bathing enclosure can be constructed.

At the meeting point of the hot water with the cold water, bathing ghats can be constructed where hundreds of people can take bath in the following tepid water. Since the stream at this point flow with the moderate velocity there is scouring effect of its southern bank. It will, therefore, be advisable to build in a retaining wall along the bend. In between the hot springs and the fresh water stream there is a level ground where public bath with a canteen and picnic spots could be developed.

The watch tower which is located on the hilltop could be developed in such a way wherein tourists can watch the beautiful sceneries especially at the time of sunrise and sunset. There should be facility of a Cafeteria where tourists can relax and have refreshments.

The State Department of Tourism and Development should do adequate advertisement to attract tourists from the Eastern region of the country. As the climate is tropical, it may not attract much foreign tourists. Students from different Universities particularly of Eastern region may be encouraged to establish holiday camps on the similar lines as done by the students of Viswa Bharati University, Shanti Niketan.

Lastly, camping and conveyance facilities should be provided at the site to attract tourists.

Urban and Industrial Planning

In the Munger division there is a high density of population in urban areas which result into the development of slums, congested housing conditions, lack of civic facilities and the spread of vice of the under worlds. In order to make planned urban development there should be a regulation in urban areas that housing should not be developed in congested parts of the city. In order to eradicate slums there should be provision of separate housing colony so that the congestion may be reduced from the slum areas. The roads, communication lines, condition of housing, facility of drinking water, arrangement of street lights should be provided in slums so that a sense of cleanliness may help the slum dwellers for ameliorating the ill-health of slum conditions.

In most of the urban areas there is a problem of unemployment but with the establishment of new industries we may overcome these difficulties. In this way industrial planning is as follows :

Industrial Estate : Industrial estate for small scale industries must be built in a phased manner either on hire-purchase basis or rent basis in Munger to help the entrepreneurs. An industrial estate with tool room facilities, adequate power and other physical amenities such as water, canteen, etc. will help small investors as they need not spend their scarce resources on non-productive items.

There is vast scope for organizing small-scale-cum-cottage industries if adequate technical training is imparted to ambitious youngmen with adequate capital to invest. In this respect Industries Department of Munger can play a vital role. I am listing below a few industries that could provide huge employment opportunities for unskilled labour, if only few technical personnel can be trained to organise in cooperative or private sector :

1. Handloom industry to make bed sheets and towels.
2. Hoisery industry.
3. Cycle manufacturing industry.

4. Safety match industry.
5. Dairy industry.
6. Leather industry.
7. Bidi/cheroots industry.
8. Leather industry including tannery.
9. Food processing industry for canning fruits, etc.
10. Furniture making industries :
    a. Steel.
    b. Wood.
11. Plastic goods industries.

The above industries are not capital oriented, but are highly labour oriented which are ideally suited for a developing region. There is lack of technical know-how in this area. If industry department can identify young men in the age group of 25-35 to enter in this field, they should be given necessary training facilities with the help of Regional Engineering College in and around the city.

Scope for Small-Scale Mineral Based Industries

The Munger division with its rich natural resources encompassing some of the important minerals, particularly building stones, bauxite, mica, clays, quartz, quartzites, felspars, kyanite and low grade limestone etc., provide ample scope for the development of new mineral based industries.

The close proximity of some of the workable mineral deposits, to the lines of communication road, railway and river, water supply etc., opens up new avenues for the development of this division.

If that could be done, the rich and glorious past of the division particularly during the Moghul period, had enjoyed a coveted position among the industrial towns of this country, might be revived easily.

Keeping in view the natural resources, industrial backwardness, the economic condition prevailing in this division, the possibility of finding employment to the landless labourer through the setting up of mineral based industries on small scale need be specially explored. That would perhaps provide the much needed base for an all-round integrated development and mass employment of poorer section of the population of this division.

A number of small-scale industries based on the locally available minerals are already flourishing in the division and giving good dividends, a few important ones among them are enumerated below :

(i) Pottery industry at Malaypur (Jamui).
(ii) Limestone Kiln at Jhajha.
(iii) Road metal in Jamalpur-Kajra belt of quartzite.
(iv) Silica brick and foundry sand from white quartzite of Patam area (25° 18' : 86° 32') containing about 98% Silica.
(v) Roofing slate—Dharhara area.
(vi) Brick industry along the entire alluvium tract parallel to the Ganga.
(vii) Mica industry—Balia and Chakai.

The mineral potential of the division, the close proximity of the Jamalpur-Kiul Section of the Eastern Railway within 10 kms of the Kharagpur hill mineral deposits and the ideal topography of the plains having a fall of 400 m. bordering the northern slopes of these hills for hauling the load of minerals by ropeway would go a long way in boosting up the mineral-based industries in this area.

Over and above that, the perennial water supply from the Ganga, the power supply from the Maithan and Panchet Power Projects of D.V.C. and the Kosi Project will prove additional boom to the industrial complex around Kharagpur hill area.

In keeping with the above factors and reservations, the limited resources vis-a-vis demands of the local entrepreneurs, it would perhaps be more appealing if only the following small-scale industries (based on the locally available raw materials) are initiated as a first phase enterprise towards the fuller achievement of our efforts for developing this industry.

Aluminium Industry

Sufficient reserves of bauxite and aluminous laterites are available from Khapra (25° 10' : 86° 27'), Maruk (25° 11' : 86° 33'), Maira (25° 14' : 86° 31') and Thadi (25° 13' : 86° 33'), in the Kharagpur hill range. Preliminary appraisal of these deposits have indicated that the alumina content of the ore is 50-60% and Silica varies between 2-4%, while the titania is fairly low. Iron oxides are rather moderate to high.

A tentative reserve of 1.5 million tonnes of various grades of bauxite has been estimated from these hills.

Detailed investigations will certainly be helpful in delineating similar reserves and assessing their potentiality for finding extensive use in the following industries :

(a) Extraction of Aluminium,
(b) For lining and steel Furnaces,
(c) Manufacture of Cement,
(d) Preparation for abrasive material,
(e) Petroleum industry and for refining kerosene oil and cracked spirit.

Building Materials Industry

The division has unlimited reserves of building materials like quartzite, granite, slate and kankar. Granite and slate of Kharagpur hills are ideally suited for building roof ornamental purposes. The master joints in the granite make the splitting easier and dressing them into ornamental stones and slabs of varying dimensions. Similarly, the cleavage planes in the slate make them vulnerable for roofing purposes.

Kankar is a fairly good quality of building material which is found in village Chain (24° 44' : 86° 25'), Musadih (24° 40' : 86° 17'), Chontari (24° 42' : 86° 32') and Lahahon (24° 37' : 86° 36').

Mica Industry

Mica belt of Munger division forms part of the famous mica belt of Bihar, and there are sufficient workable deposits of ruby mica which can find wide use in electrical industry for the manufacture of insulating materials and allied goods as well as manufacture of Sunmica Sheets from the scrap. There is no dearth of skilled labour for the industry in the division. With proper encouragement and resources, this industry can attain importance in this division.

Lime Industry

Though some lime kilns exist between Jhajha and Simultalla, the sizable deposits of low grade lime stone in this area still provide ample scope for opening such type of lime kilns to meet the local requirements.

Refractory Industry

The good quality of fire-clay from Pirpahar (Kharagpur hill ranges) close to Munger town, and Kyanite deposits of Belbhinda-Bhika area are potential sources of raw materials for the proposed refractory industry in the division.

Transport Planning

In the Munger division there is a lack of road and railway facilities in each and every part of the division. Areas of forest cover mountainous zone, flood prone area and Tal lands have no road and railway facility due to the inherent drawback of physical and economic conditions which hamper transport development in those areas. There are several inaccessible parts of the division which require transport development on priority basis. Per capita roads and railways and their densities suggest that the Munger division is poorest in transport connections in comparison with developed regions of the world. Hence, the proper maintenance of road and railway tracks, the construction of double way wider roads and its extension up to rural areas and the construction of double rail tracks which are necessary for the proper development of the economy of the Munger division. There is an urgent need for the construction of river bridges over the Ganga at Munger near Kastharni Ghat which has commendable stable site for the construction of bridge. There is also an urgent need of starting river borne traffic, especially in the river Ganga from Mokamah and Barauni in the west up to Sultanganj and Bhagalpur in the east.

In southern forest covered areas of Munger the construction of road is necessary for the movement of forest products and mineral resources up to industrial centres of the State.

The systems of resources suggest that integrated resource development is necessary for planned development of resources in the Munger division under open system which has given chance for making linkages of one resource with the other.

Case Study of Binda Diara Development Scheme

Diara can be defined as the unequal deposition of sandy layer which even during the normal floods gets water spills from the river and thereby causing sufferings to the inhabitants.

The main problems of such areas are :

1. Poor returns from agriculture, as only unirrigatcd crops are raised in this area only two crops namely Rabi and Bhadai are grown, out of which there are always chances of the Bhadai crop getting damaged due to early floods.
2. Problem of rehabilitation during the flood season when the villages get inundated.
3. Disruption of communications during floods.
4. Submergence of walls and pollution of drinking water.

Thus essential needs of Diara region are : Protection from floods, provision of assured irrigation and drinking water, and maintenance of communication lines. For flood protection, embankments are generally not favoured. Since they are likely to affect the river regime and their maintenance is also difficult and costly. Moreover, they exclude the fertile soil spilling on the Diara lands. Alternatively, the villages should have raised platforms, where people could take shelter especially during floods. Reduction in damage to crops can be effected by growing such crops which can mature before floods. However, this will require early sowing for which irrigation will have to be provided. This can be accomplished by providing tubewells on high grounds or on raised platforms. These tubewells can also be used for drinking water supplies and with assured irrigation. It will also be possible to introduce high yielding varieties of wheat in Rabi cultivation. Maintenance of communication lines to the marooned villages is possible either by providing adequate number of boats or by constructing roads above high flood level and adequate waterway for spill channels. However, boats are slow moving and at the same time risky. Hence construction of roads for maintenance of communications all the year round is a better solution. The roads will, however, need maintenance of high order. For extending electric conditions to the tubewells and to the villages for domestic consumption, the electric lines can be laid along the roads.

There are several such areas where people are living in villages which lie in the flood plains of the Ganga in the States of Uttar Pradesh, Bihar and West Bengal. In order to provide relief to such areas, investigations have been carried out by the Bihar State Irrigation Department in a typical area.

Location of Binda Diara

The area extends from village Maheshpur to Kalyanpur in Jamalpur block and falls on the north of Soni Nala which runs along Munger-Bhagalpur road. The area falls within 25°15' N to 25°21' N latitude and 86°37' E longitude. The area is approachable through Munger-Bhagalpur P.W.D. Road and lies north of the same. The area is also approachable from Bariarpur Railway Station. There is no road connection in between different villages within the Diara area especially in rainy season. In dry weather Katcha roads having lower levels than the division fields serve as communication line.

The Diara area is in the flood plains of the Ganga. The general ground level of the area proposed to be developed in between river level of 34.13 metres and 35.99 metres. The general slope is from west to east and from north to south. The area is longitudinally traversed by two *nalas.*

The maximum discharge of the Ganga in this region observed at Hathidah 0.76 lakh cusecs which is 58 kilometres upstreams of this area. The maximum width of the Ganga during monsoon is more than 9.666 kilometres. The level of Ganga during last 10 years has not gone below 34.4 metres and the maximum level attained in last 17 years has been 37.33 metres during 1971 (Table 11.7).

Table 11.7 : Frequency of the Ganga Water near Binda Diara, Munger (1955-2001)

Year	*High flood level at Binda Diara*	*High flood level arranged in descending order*	*Rank*	$T = n + \frac{1}{m}$ *n = number year* $\frac{17+1}{n}$
1955	115.50	122.34	1	18.30
1956	117.25	118.40	2	9.06
1957	115.85	117.75	3	6.00
1958	115.75	117.75	4	4.50
1959	118.40	117.25	5	3.60
1960	115.26	126.85	6	3.00
1961	116.86	116.05	7	2.55
1962	115.26	115.85	8	2.25
1963	114.86	115.75	9	2.00
1964	115.16	115.50	10	1.80
1965	112.86	115.26	11	1.83
1966	113.56	115.26	12	1.50
1967	117.75	115.16	13	1.39
1968	115.65	114.86	14	1.30
1969	117.75	113.65	15	1.20
1970	116.05	113.56	16	1.12
1971	122.46	112.86	17	1.10
1981	119.47	112.56	18	1.09
1991	118.32	112.38	19	1.07
2001	122.58	111.59	20	1.05

Soarce : Pilot Scheme of Binda Diara, Bihar, 2002.

Flood observation of the Ganga near Binda Diara have not been made except in 2001 rainy season. High flood levels of Binda Diara for the period 1.8.2001 to 14.9.2001 and the corresponding period have also been collected (Table 11.8).

It is evident from Table 11.8 that the period of submergence of the area is generally between August and September and the duration of submergence above average Diara level of 34.13 metres has been as long as 68 days in 1961 and the minimum period has been only 6 days in 1965. On an average, Diara is submerged for about 45 days per year. In 2001, however, the period of submergence was about 70 days and was maximum during the last 17 years. Depth of flooding in 2001 had been about 3.35 metres while the flooding in most of the years is about 1.22 to 1.52 metres.

Table 11.8 : Statement showing Water Level at Binda Diara for the Period 1-8-2001 to 14-8-2001

Date	*Water level near village Ekasi in Binda Diara (in metres)*	*Date*	*Water level near village Ekasi in Binda Diara (in metres)*
1-8-01	37.21	24-8-01	36.56
2-8-01	37.38	25-8-01	37.27
3-8-01	37.51	26-8-01	36.10
4-8-01	36.84	27-8-01	35.97
5-8-01	36.96	28-8-01	35.82
6-8-01	37.20	29-8-01	35.77
7-8-01	37.31	30-8-01	35.68
8-8-01	37.20	31-8-01	35.65
9-8-01	37.08	1-9-01	35.66
10-8-01	37.03	2-9-01	35.63
11-8-01	37.02	3-9-01	35.62
12-8-01	36.97	4-9-01	36.51
13-8-01	36.91	5-9-01	35.63
14-8-01	36.88	6-9-01	35.66
15-8-01	37.08	7-9-01	35.63
16-8-01	36.56	8-9-01	35.57
17-8-01	36.41	9-9-01	35.24
18-8-01	36.39	10-9-01	35.21
19-8-01	36.39	11-9-01	35.28
20-8-01	36.39	12-9-01	35.39
21-8-01	36.68	13-9-01	35.59
22-8-01	36.71	14-9-01	35.79
23-8-01	36.70		

Source : Pilot Project of Binda Diara, Bihar, 1972.

As discussed above, since the area is flooded almost every year and the depth and duration of flooding varies from year to year the maize crop usually has to be harvested before it is damaged due to standing water. The early variety of maize crop cannot be grown due to lack of irrigation facilities. The only dependable crop is Rabi which also suffers when there is drought. In monsoon season communication between villages and also between the P.W.D. roads and village does not exist.

During the monsoon drinking water facilities are also not available since most of the low lying wells are submerged and their water becomes polluted. These problems exist almost every year.

The inhabitants are anxious that tubewells should be constructed for irrigation facilities. Few private tubewells have already come up in the area. In addition they want schemes for communication, flood protection measures, drinking water supply and extension of electric lines. Hence the present scheme for development of the area has been drawn up.

Proposals

The area remains submerged during the entire monsoon period. Therefore, a proposal of constructing a ring bund as a flood protection measure was examined. For this purpose, it was seen that 3.05 metres to 4.50 metres embankment would have been required. It was felt that maintenance of ring bund in the flood plains of the Ganga and also to provide drainage of the area enclosed would be difficult and maintenance of such bund would also be costly. Besides, area within the ring bund will encroach upon the free flow of the river. Hence such bund was not considered to be feasible. The existing villages in the area get submerged either partly or fully causing untold myseries to the people when the Ganga is in spate. If raised platforms are provided in each village, marooned people can save their lives and properties.

As the communication to the villages pose a problem during floods, some roads could be constructed linking these villages with the P.W.D. road. To ensure free flow of spill water causeway has to be provided across the spill channels. The causeway has to be for adequate waterways. The estimate also provides for construction of tubewells for irrigation as well as drinking water supply purposes. Provision for extension of electric lines for domestic and other purposes has also been made.

The proposals made in the scheme can be summarised as follows :

(A) Communications

It is proposed to extend the facility for approach road connecting the village Ekasi tola, Kalayan tola, Harizan Kalyan tola, Pairumandal tola, village Jharkahwa, Phulkia and Nazira 6.53 kilometres long and 4.87 metres wide road is proposed. This has been done with a view to provide communication facility so that the people of the area can approach the market and school, etc.

The embankment is proposed across the flow. Therefore, possiblity of erosion to the villages closer to the Ganga main flow cannot be ruled out. Hence it is also proposed to spread 15 centimetres layer of coal ash on the top of the embankment so that the top may not become muddy in the rainy season. The coal ash will be available from the Jamalpur Railway yard within 16 kilometres from the site.

There will be water on both sides of the embankment. Stability analysis has been done and section has been provided accordingly. The embankment will be constructed from local sandy silt soil. In even normal years there will be water on both sides of the embankment for a period of about 45 days, hence, maintenance of the embankment has necessarily to be of high order.

(B) Flood Protection

Almost all the villages of Binda Diara have been affected by floods. The villages are located below high flood land of 37.33 metres and the lowest flood land of 35.43 metres. It is proposed to construct raised platforms at the level of road, or 3.8 metres in villages for the use of villagers during the flood.

(C) Irrigation and Water Supply

Altogether only 647 hectares of cultivated area has been irrigated in Binda Diara with the help of other facilities. Rabi can be grown in 750 hectares, early Maize in 400 hectares and paddy in 240 hectares. Hence, shallow tubewells and turbine pumps will be provided to fulfil the water requirements for irrigational purposes.

(D) Electric Installations

The KVA transmission line extends along Bariarpur road. This line close to the area is proposed to be developed. For extending electrical lines L.S. provision should have been made and transformers should be established.

Benefits **:** Development of the area is the long felt demand of the people and locality. The scheme after execution will improve the communication facilities of the area. Land proposed to be irrigated by tubewells will yield at least two crops per year. The provision of tubewells would also be useful to supply water for drinking purposes to the villages of the area. The extension of electricity would raise the standard of living of the inhabitants of the area and will also open avenues for starting small scale industries. The scheme will raise the social and economic status of the population of the area. The total cost of the scheme comes to Rs. 24.89 lakhs and the area to be brought under assured irrigation is 647 hectares. Similar schemes can be drawn up for other Diaras also.

Suggestions for Infrastructure, Development of Diara Belt

In the Ganga riparian tract of the Munger division the area susceptible to flood is highest in percentage (40%) in the State of Bihar. In 1991, 1996, 1998 and 2000 there was a lot of devastation in which lakhs of people were rendered homeless, hundreds of villagers were washed away. Crops of millions of rupees have destroyed, thousands of livestock wealth swallowed away by the destructive flood and the problem is still continuing from Barahiya and Lakhisarai in the west up to Gogri, Parbatta and Bariarpur in the east.

Considering the gravity of abovementioned problem, the Government has accepted in principle, the eradication of the problems of flood prone areas in the Munger division. But as yet nothing is in sight regarding the solution of any problem. On these problems the Central Flood Commission, Irrigation Department of the Government of Bihar, Geological Survey of India, Indian Institute of Technology of Kharagpur and Survey of India have given their opinion along with concrete planning but all are useless if the Government is not taking these problems right earnestly for solution.

Considering the abovementioned facts the author suggests the following points for integrated development of flood prone areas especially on the northern and southern side of the Ganga in the Munger division :

1. Just like the Pilot scheme for 'Binda Diara of Bihar' in all the Diaras of Munger division such pilot scheme should be adopted and in each Panchayat at least five farms of 3047 hectares should be immediately constructed.
2. Near Barahiya, Dharhara, Ratanpur, Bariarpur and Kalyanpur the flow of the water of the Ganga is much speedy and hence flood protection platforms should be constructed near railway lines.
3. Near Surajgarha, Abgil, Amarpur, Sundarpur, Bahachauki, Hemjapur, Shibkund, Sinhia, Farda, Parham, Herudiara road flax should be constructed for the safety of cattle wealth and peasant of the area.
4. There are thousands of villagers whose settlement sites are eroded away by the gushing water of the Ganga. Thus at least the provision of habitable site is necessary because each and every year 1 lakh people of Diara take shelter in Munger, Jamalpur, Nawagarhi and Bariarpur especially in school and college buildings. Hence Government should manage the affairs in such a way so that the problem should be solved immediately and right earnestly.
5. Due to recurrence of flood the inhabitants of flood prone area want financial assistance in the form of

developed seed, fertiliser, pesticides and loan to each and every farmer.

6. Great portion of the land is eroded away and now the river Ganga is flowing from such land and the peasants of that area still giving rent for the same to the Government and hence exemption should be made for such a renting system.
7. The Government of India has sanctioned "Operational Research Project" for Diara development and this Scheme should be immediately implemented for the eradication of problems faced by the Diara people.
8. Approach road should be immediately constructed for the Diara up to the division headquarter.

The integrated development shows that different resources are available in the Munger division but only the co-ordination of one resource with the other will help in integrated development of the division on the basis of which different levels of regional development have been searched out. This study enables the planning of different resources in regional development and economic preferment of the division in an integrated way. In order to test the validity and role of different resources on regional development of Binda Diara has already been given. Thus the resources have played a vital role in planned development of the Munger division due to its commanding position, chequerred historical accounts and nerve centre of cultural advancement as well as economic development in South Bihar Plain.

REFERENCES

Dubhashi, P.R., "Communication and Rural Development", Yojana, Vol. XXIII/10, 19, p. 13.

Dutta, S.K., "Geohydrological Report on Thermal Spring", Rural Electrification Corporation, Government of India, 1973, p. 3.

Jha, Premshankar, "The role of the Mass Media in Promoting Integrated Rural Development", Science and Integrated Rural Development (Eds.) M.S. Swaminathan, P.K. Bose and Archana Sharma, 1978, p. 65.

12

Summary and Conclusions

Resource is the base of all regional and economic development in the world since times immemorial. In ancient period men tapped fewer resources on the earth surface due to their limited technological development but nowadays the dimensions of resources have developed manifold due to cultural advancement, economic development, chequered history and growth of population in almost all parts of the globe. In this way, it is quite discernible that only planned development of resources have brought the world economy into such a pass when we have achieved the miraculous technological advancement, a new vistas of economic development and the systematic growth of not only the human habitations but also the resources which immediately helped in shaping the planned development of resources for the betterment and well-being of human society. At this juncture we are standing on the crossroads of physical, cultural, economical, political, national, international, biotic and environmental modifications through technological advancement of men and economic gains of resource development of the world.

Sustainable rural development hints at the rotational use of scarce sustainable resources with the support of people and economic politics adopting strategic environment management practices. Hence, a multifarious strategy needs to be formulated to solve the rural problems through development process and to protect the environment through community participation.

Environment Quality is an integral part of development. Without environment ethics development is simply undermined. Natural resources are the wealth of nation, stocked in free and better environment. However, at present they are facing environment hazards due to several reasons. Sustainable development is forced on any kind of betterment that should not harm the environment, so that the well-being of future generations is guaranteed, and the harmonious relationship between environment and development is further sustained. In other words, the process of sustainable development tries to build social and economic progress satisfying the needs and values of all social groups without foreclosing future option. The outcome of Rio-Earth Summit (1992) highlights the view that socio-economic development and environment protection are interdependent and mutually reinforcing proceses.

Priorities

The environment problems that a country faces often have a correlation with the stage of its development, the structure of its economy, and its environment policies. Some problems are associated with the lack of economic development. Inadequacy of sanitation and clean water, air pollution due to biomass burning and many types of land degradation have poverty as their root cause. Poverty and ignorance are the allies of environmental degradation, which in turn, has two major damaging effects. First, it harms human health and secondly, it reduces human productivity. Impaired health may lower human productivity. Environmental degradation also reduces the productivity of many resources used directly. Water pollution damages fisheries and waterlogging and salinization of the soil lower crop yield. The third significantly damaging effect is the loss of 'environmental amenities' such as forest reserves, sanctuaries, wild parks, abundance of flora and fauna, all representing an unspoiled environment.

The importance that societies give to different environmental problems evolves, often rapidly, in response to gains in standards of living and to other social changes. In a country like India, these priorities should mainly focus on prevention of water and air pollution, safe disposal of solid and hazardous wastes, and protection of land and habit.

Policies *vis-a-vis* Protection

Broadly, there need to be three sets of development policies that will help in protecting environment.

The first set should include measures that have no net financial coats for the government and will contribute to both their economic efficiency and the environmental protection. These are elimination of subsidies for energy inputs (such as electric power and fossil fuels), pesticides, fertilizers, irrigation water and commercial exploitation of forests. Not only would these measures yield economic dividends, they will be even more beneficial when environmental benefits are considered.

The second set should include public and private investments that have positive net economic benefits even when environmental benefit are not considered. These are (i) investments in water supply and sanitation, (ii) investments in soil conservation, and (iii) improvement in the education of women because world-wide, women indeed play a central role in resource management, yet have much less access to education, credit, extension services and technology. The justification for these measures is that their environmental benefits outweigh their costs.

The third set should include the measures that have a positive net economic benefits when environmental effects are included. These are : (i) taxes on emission and wastes, (ii) regulation of disposal of hazardous wastes, and (iii) fees for commercial exploitation of forest.

Elimination of Subsidies

Subsidies that cause environmental damage by encouraging resource use are common. Both economic and environmental benefits will be achieved by removing subsidies that encourage the use of coal, electricity, pesticides and irrigation water. These reforms will require considerable political will because the subsidies typically benefits the politically influential or are intended to serve such goals as food self-sufficiency and rapid industrialization.

Recognition of the environmental costs of such subsidies would provide a powerful additional reason for removing them. The inefficiencies that these subsidies encourage in the capital-intensive and highly-polluting industries worsen pollution.

Public enterprises must be given greater autonomy and be exposed to competition. If managers of public utilities are made accountable for their performance, they are more likely to maintain price at levels that improve cost recovery and to compare the costs and benefits of investments. Private investments should also be encouraged, particularly where private benefits are high, such as in irrigation, water supply and electricity generation.

Targeted Policies

Environmental damage imposes certain costs that are not easily quantifiable. Comparing the benefits of environmental protection with the costs of remedial action helps policy-makers to make better-informed decisions. In choosing environmental priorities, standards and policies, government places value on different kinds of damages. It is better that these choices be guided by comparisons of the costs and benefits for environmental improvement. While deciding the priority of environmental problems to be addressed, it is inevitable to compare the costs of setting environmental goals, comparisons should be made between the benefits from environmental improvements with the costs of achieving them.

Regulation and Incentive

Since development and environmental policies differ substantially in cost and effectiveness and since a country can ill-afford to waste resource, the policy-makers' choice of the measures should be guided principally by the cost of effective implementation. The cost-effective policy mix depends in general on the characteristics of the environmental problem at hand as well as the capabilities of regulatory institution. The behaviour of the polluters and resource-users can be influenced in two ways; first by stipulating standards and regulations (command-and-control policies) and secondly by pricing additional pollution or additional resource use (incentive-based policies). Although the regulatory approach is more prevalent, interest is now reviving in incentive-based measures also. Notable examples are effluent charges on water pollution, emission charge on sulfur dioxide, surcharges on fuels, automobiles, pesticides and fertilizers.

Regulatory policies are best suited to situations that involve a few public enterprises and non-competitive private firms. This is particularly true when the technologies for controlling pollution or resource use are relatively uniform and can easily be specified by regulators. For example installation of precipitators in highly polluting industries and changing over to low-sulfur fuel, for improving the air quality.

If effectively implemented, policies that use economic incentives will be less costly than regulatory measure in meeting the environmental goals. With such policies, all polluters or resource-users are faced with the same price and must choose their degree of control.

Regulations by contrast, leave these decisions to regulators who are rarely well-informed about the relative costs and benefits faced by the users.

Incentive-based policies that put a price on environmental damage affect all to polluters, in contrast to regulations, which affect only those who fail to comply. This means that incentive-based policies give the right long-term signals to resource users. The polluter or resource-user has an incentive to use whichever technologies that most effectively reduce environmental damage. Regulations that mandate standards give polluters no reason to go further then the standard demands.

Policies that use economic incentives will be effective only to the extent that polluters and resource-users respond to them. Responsiveness depends on three factors : ownership, competition and differences among users. State-owned enterprises are particularly insensitive to policies that use economic incentives because they generally do not care much about costs. Lack of domestic and foreign competition reduces the pressures even on private businesses to minimize costs. Thus, in India, state-owned petroleum refineries showed little interest in passing out leaded petrol. By contrast, a scheme of incentives attracted highly competitive private refineries in the United States to phase out leaded petrol in 1980s.

Environmental degradation can be controlled either by altering the price of environmental resources of by restricting their use. Policies that specify quantities of pollution or resource-use fix the level of environmental damage, whereas those that alter prices fix the cost of controlling environmental damage.

Quantity-based policies are also appropriate when it is extremely important that certain thresholds should not be exceeded, for example, radioactive or toxic wastes. By contrast, the social costs of other types of environmental damage, for example, mining operations do not rise dramatically if standards are exceeded by small margins. In such case, it is more important to avoid spending too much on controlling degradation than to risk a little more environmental damage.

So, quantity-based policies are most appropriate for pollution problems that involve threshold impacts, for example, pollution through hazardous wastes and heavy metals and for natural resources such as unique habitats. Similarly, enforceable zoning laws may be more reliable than differential property taxes in preserving unique habitats such as wetlands, sensitive shorelines and coral reefs.

Among incentive-based policies, the choice between surcharges and quota permits depends partly on the capabilities of regulators. Although permits can be used for control of air and water pollution, they tend to be administratively more demanding than surcharges because the latter can typically be implemented through the fiscal system.

Policy Reforms

In India, there is a widespread preference for regulatory measures, which has put under strain monitoring and enforcements capabilities of the concerned Government departments. As a result, many of these measures cannot be enforced consistently. For example, despite ambitious goals and regulations, air pollution remains a major problem in cities. Policy reform, by implication, will require considerable political will. But the gains from well-designed measures would be enormous. For ensuring sustainable development, policy reform should proceed in four ways.

First, the initial step would be to remove policy distortions that damage the environment and retard growth. These measures must be accompanied by supplementary measures aimed at recognizing the future environmental impact of the reformed policies.

Second, the measures to change the behaviour of the polluters and resource-users should rely on economic incentives. Putting a price on the environmental damage would reduce implementation costs, encourage faster adoption of environment-friendly technologies and supplement public revenue.

Third, imposition of surcharges on inputs and products that breed pollution. Because most environmental problems stem from the actions of numerous and dispersed resource-users, it is often prohibitively costly to enforce direct regulations. Blanket measures would simplify administration and also make enforcement more likely.

Fourth, early action can reduce the costs of implementing effective environmental measures. It is usually possible to take steps such as setting up regulatory institutions, initiating charges, and encouraging adoption of cleaner technologies that can reduce the eventual magnitude of the problem. Any delay in acting until problems become crises will eventually need extreme and costly responses such as closure of plants and factories and restricting vehicle use.

21st Century Needs

Good environment policies are good economic policies and vice versa. Economic growth need not be an enemy of the environment. Environmental protection will most certainly promote sustainable development.

With the rising focus on free market economy and investment in human resource, it is hoped, within the next generation, widespread poverty could be eliminated as well as air and water and adequate sanitation could be made available. Agricultural productivity could continue to grow at the present or even higher rate and also in a manner that would reduce pressure on natural habitat. Industrial output could rise to six time of the present lable, with lower total emissions and wastes.

Such development could be powered by clean fossil fuel technologies and also by renewable energy. Valuable natural habitats could be much better protected than at present. Only everyone would need to accept this as a joint obligation.

The present study aims to investigate the Planned Development of Resources in the Munger division which have analysed within its limitations of the physical as well as cultural resources such as population, forest, agriculture, water transport, urban and industrial development. The integrated systems of resource appraisal and the planning, process of all the resources in different sub-heads have discussed and summary is given below in the following paragraphs.

Specific Analysis of the Developing Region

The First chapter on introduction delineated the quest for sustainable development, meaning and scope, objectives of study, appraisal of various resources, relationship of resource with culture, role of industrial resources in regional economic development, besides classification of resources. The influence of resources on culture has been shown with special reference to various concepts, characteristics, natural principles, conservation, planning for development, statement of problem, methods of study, sources of data and at the end references pertaining to the conceptual background of resources.

The Second chapter highlighted the review of literature, approaches of study, principles of resource development and hypothesis testing. In review of literature, it has been shown that various disciplines such as economics, geology, geography, botany and zoology have direct concern with resource potential study. Hence, in almost all these disciplines we find literature regarding resource appraisal. In this study the author has incorporated the idea of Zobler, Stamp, Fagg, Finch, Latham, Berry, Baker, Haggett, Dayal, Mandal and Singh on resource development. The review of the work of Chatterjee, Hart, Ganguli, Safi, Karnik and Rao's have also found place in resource analysis. This study also throws light on retrogressive, retrospective, prescriptive, eco-system, quantitative and systems approaches of resource analyses. The principles of resource planning centred round the comparative advantage, diminishing returns, variations of economic rent, minimization of input, maximization of out-put, optimization of crops protecting space including quality of produce, environmental integration and the demand as set by the people. In hypothesis formulation, conceptual ideas have been represented regarding selection of problem, statement of problem and the purpose of study. Further analysis have been made for selection or rejection of null and alternative or research hypothesis. It also helps in the determination and collection of evidences, processing of data, analysis of evidence and finally presenting the result. Ultimately, it has been found that the correlation between area of land under different crops and anchalwise population of the Munger division has a close relationship of r = 0.60 the value of Karl Pearson coefficient of correlation and hence the null hypothesis is rejected in favour of research hypothesis.

Chapter Three throws light on location and physical setting in the Munger division so that the geographical conditions of the division may be analysed fully. The southern part is mountainous, forest covered, undulating surface whereas North Munger is a fertile plain land. In this way, these two regions present a varying picture of resource development especially in the field of dynamic natural resources.

The Munger division lies on 24°22′N and 25°49′N latitude and in between 85°36′E and 86°51′E longitude. It has an area of 7,884 square kilometres in total, whereas the rural area is 7716.1 and urban area is 168.3 square kilometres. According to area the largest anchal is Chakai which has an area of 774 square kilometres and the smallest anchal is Tarapur which has an area of about 131 square kilometres. So far as the urban area is concerned the greatest extension is Barahiya, 26.5 square kilometres, whereas the smallest urban area is Khagaria 2.9 square kilometres.

According to population the Munger division has a total population of 3.3 millions in 1981 whereas the rural population was 0.35 millions in 1981. According to total population the biggest anchal is Ariari.

Munger is located in temperate climatic region where the effect of North-east and South-west monsoonal rains are highly felt. Economically, it lies in the transitional belt of Chotanagpur plateau in the south and the riparian flood plain fertile alluvial tracts of North Munger. The division is divided into four sub-divisions namely, Munger Sadar, Jamui, Lakhisarai and Khagaria, having 26 development blocks in the command. In the Munger division, Kharagpur hill is the outlier of Dharwarian systems of rocks whereas the flat alluvial plain is the outcome of the deposition of silt in an unequal manner especially in Pleistocene period. In South Munger especially in Dharwarian systems of rocks, such as bauxite, beryl, fireclay, felspar, kyanite, mica, mineral water and the vast reserves of quartzite are found. But in northern part of the division the deposition of *bhangar* and *khadar* soils are found in the far-off upland plains and near riparian tracts respectively. Kharagpur hill has an area of 776 square kilometres and it has a height of 415 metres. Maruk is the highest point having a height of 488 metres from mean sea level. Besides these Abhainath has an height of 482 metres and many others have height of over 458 metres. The other important hills are Gidheshwar and Satpahari. The climate is monsoonal with four well-marked seasons of summer, rainy, winter and autumn. The average rainfall is around 1,000 millimetres in almost all anchals. Soil is the outcome of destruction of forest leaves, deposition of silt by rivers especially in high flood area. In this way, it can be said that before going through the planned development of resources in the Munger division, it has been felt necessary to discuss the location and physical setting with due attention.

The chapter Four on human resource throws light that man is the greatest resource who creates, consumes, destroys the same after fulfilment of his demand. This shows, the way for future utilisation of resources, their conservation, planned exploitation and development in the region. Per capita resource analysed here with special reference to the demographic attributes, population planning and the availability of food production in comparison with growing population of Munger. The Munger division had the population of more than 33.14 lakhs in 1981. It is distributed in an area of 7,884 square kilometres. The density of population comes to 419 per square kilometre. The total population in 1951 was only 20.55 lakhs, while it was 22.68 lakhs in 1961, and 27.45 lakhs in 1971. This shows that the growth of population is not so alarming but it is moderate in nature. More than 13% people live in urban areas and the rest in rural areas. Before 1889, the Census were conducted by the Collectors only. But since 1891, Census records are available in a recorded form. Since then the population growth seems to be steady but in 1921 due to various diseases, famine conditions and influenza the population growth in that Census was counted very low in almost all parts of India, for which the Munger division was not an exception to the rule. From 1921-51, the population growth was slower but with effect from 1951 Census it is slightly higher due to better medical facility, control over various diseases, good agricultural harvest due to the application of chemical fertilizers, use of improved varieties of seeds, provision of irrigation facilities both from the state and private tubewells have helped a lot in determining the higher growth rate of population because all these have produced a congenial environment for human living.

Chapter Five deals with the present position, problems and prospects of forest resource. Forest is the greatest renewable resource in the Munger division which balances the ecological conditions between man and the nature. Forest is the hidden resource because it provides essential commodities like fuel, medical herbs, fodder, timber, different kinds of forest besides wildlives. During prehistoric period there are several references of forest in the Munger division in Rigveda, Manusamhita, Veda, etc. During the British rule, the Banaili Raj and Darbhanga Raj were ruled over forest areas of Munger. The study of forest resource highlights the problem of the area, distribution of forest, types of forest including its products, wildlives, utilization, conservation, problems and prospects of forest base industry in the Munger division.

Chapter Six highlights the agricultural resources which is essential for sustaining human resource in the division. The agricultural system of division is said to be the gamble in the hands of monsoon. The saying is true due to the uncertainty of the amount of rainfall and hours and days of monsoonal rain.

There are four seasons of agricultural harvest in Munger, i.e., *Aghani, Rabi, Garma* and *Bhadai*. In *Aghani*, the predominance of rice, some varieties of pulses and the sugarcane crushing for the preparation of jaggery are important. Among all these, production and introduction of dwarf varieties of wheat has almost revolutionised the *Rabi* harvesting. It is termed as 'green revolution' by the use of chemical fertilizers, deep harrowing of fields and the use of mechanised irrigation systems and private tubewells, etc. In *Garma* harvest the predominance of moong varieties of grass for livestock rearing and the production of sugarbeet, cheena, etc. are important. In sandy Diaras of river banks the production of water melons, varieties of vegetables and several other items of food crops are important. In *Bhadai* season the production of aunce, corn, ragi, sama, cheena, kodo, etc. are found.

Chapter Seven deals with the water resources of the Munger division which need conservation, effective utilization for the well-being of human society because at present it is not efficiently used for power generation, irrigation of land and recreational purposes. The Munger division has some hot springs besides several rivers such as the Ganga, the Harohar, the Burhi Gandak, the Karch, the Kosi, etc. which are the potential and perennial source of water. This chapter mainly concentrated on analysing the hydrological cycle, hot springs, irrigational use of water, irrigational history of the Munger division, method of irrigation, suitability of irrigation, sources, net and gross irrigated area, Badua Reservoir Project, Kharagpur Lake Scheme, Dakra Nala Project, Surajgarha Pump Canal Scheme, Barnar Reservoir Project, besides the major irrigational problems and suggestions by the author for better crop production.

In chapter Eight discussions have been made regarding the urbanisation, industrial location and regional development in the Munger division. Here about 13% people live in urban areas whereas the rest 87% people live in rural settlements. In 1901 Census only 3.45% people lived in urban areas but in 1981 Census a considerable progress has been made in this direction. The hierarchy of towns shows that southern Munger is devoid of urban settlements including the Kosi belt in the north but on either side of the Ganga the number of towns are located, such as Barahiya, Lakhisarai, Khagaria, Gogri, Bariarpur, Munger, etc. According to population size, the largest urban centre is Munger, having a population of 102,474 and the smallest town is Bariarpur with a population of 8251 in 1981. It has been felt that for regional development urbanisation provides a necessary infrastructure facilities in the region because urban centre functions as the modest transport centres, administrative headquarters, social capitals, economic capitals and functions as the levers of regional development. Industrial development is an important characteristic of urban development because industries are lccated at the point of cross-roads on the side of the river and hence the Munger division has innumerable such sites with water frontier, cross-roads, skilled labour and the huge endowments of the mineral resources which have enabled the localisation of gun manufacturing, railway workshop of Jamalpur, Indian Tobacco Company at Munger and Biri-making industry at Jhajha. Besides, there are enough potentiality to start a number of industry such as food processing, saw mill, paper mill, cardboard factory, etc. At the end of integrated development chapter, planning of industries has been delineated. This chapter highlights the methodology of urban study, urbanisation and its characteristics, size class of towns, percentage of population growth, change in the rank-size, distribution of urban centres, problems and policies of urban development, relationship of urbanisation with industrial development and industrial potentiality of Munger.

Chapter Nine highlights the role of transportation line in economic and regional development of an area. In modes of transport such as roads, railways, waterways and airways are delineated in different sub-heads besides analysing the early history of transportation linkages, present position, different roads links, accessibility through road and per capita length of road. In the sub-heading of railway, different rail routes in the division along with railway station are given with accessibility, per capita density and problem of railway development. Besides, screening test for the viability of road, etc. at the end, the problems and suggestions of transportation networks are delineated.

Chapter Ten throws light on resource systems of the division in order to see the integrated picture of almost all resources and their linkages in the Munger division. The sub-heads of this chapter are objectives of study, review of literature, conceptual framework, importance of resource system, natural and cultural environment, systems of forest resousce, agricultural systems and at the end it has been observed that various resources available on the earth surface have their differential accessibility for man. Almost all resources are interlinked for the well-being of human society and their accessibility vary from place to place.

Chapter Eleven is concerned with the integrated development of resources in the Munger division. It highlights meaning and scope, methodology, level of regional development, correlation between different variables besides planning of population, industrial location and transportation in a nutshell. In the last section of integrated development, case studies of Binda Diara, suggestions for infrastructure of

Diara belt and Dhanari village have been given besides suggestions for tackling the problems of flood, drought, unemployment conditions and increasing the population, etc. at the end.

Suggestions

First problem has been concerned with the location, growth and development of resources in the Munger division. It is a well-known fact that the northern part of Munger has fertile alluvial tract which form the home of agricultural resources of the region whereas in southern part only forest resources are located. So far as the growth and development of resources are concerned, in the Munger division it has been properly analysed whether the same is a physical or man-made resource. Resource has been developed according to the technological advancement of the inhabitants of the region and optimum utilisation of available resources for the cultural and economic well-being of society. So far as the theoretical background is concerned in the present study with a view to chalk out the problems of study, a huge amount of literature has been studied. Besides this, the concept of crop association, hypothesis testing, principles of resource maximisation, optimum utilisation, minimisation of effort and input, maximisation of output find place in study. Besides the various aspects of population growth, urbanisation, industrial location, system analysis and concepts of integrated development have been employed perfectly which provide the theoretical background of resource utilisation in this study.

Secondly, this study highlights the location and physical setting in the Munger division in orber to establish relationship between geographical environment and resources. Discussions have also been made regarding situation, geology and minerals, relief features, drainage, climatic conditions so that the region can properly be developed within the perimetres of geographical and economic setting of the region under study.

Thirdly, the level of resource study is delineated in chapter 2 'Review of Literature and Hypothesis Testing'. Here the author tries to show the early literature on resource study with special reference to Indian conditions. Ultimately, it has been contended that resource development is closely related with the environmental conditions and technological advancement of the people. In the Munger division only eight resources are capable of development as for example duality of population, forest, agriculture, water, urban centres and industrial establishments. All these have been discussed here in different sub-heads.

In Chapter Eleven the level of resource development have been shown and the average ranking value of almost all these variables and their correlation matrix shows that these resources have very close connection with one another.

The present standard of resource development and their enhancement in the Munger division should be properly assessed. The most baffling problem today faced by the study area is the explosion of population which has nullified the achievements of all the Five Year Plans due to high rate of population growth in different parts of the division. Hence, in order to plan our resources, it is our first and foremost duty to check the growth of population by means of several birth control measures. This has been also suggested at the end of the chapter on 'Human Resources'.

The chapter on 'System of Resources' and 'Integrated Development', it has been shown that the interlinkages of resources are possible and the same should be planned in an integrated way. The sectoral and lopsided development would not yield any fruitful result because the same will be infructuous without developing other resources. Hence, the authority should try to develop the resources of the division on planned basis with equal emphasis, either through managerial control, technological assistance, institutional help and local participation whatever may be possible in the region under study.

In this way, the questions are put forward by the author in a sub-head 'Statement of problem' in the first chapter have been properly answered and the author hopes to usher an era of planned and systematic development of sustainable resources as this will generate much heated debate and enthusiasm among the scholars and planners too.

Bibliography

Ackerman, E.A., K.N. Singh and J. Singh, "Bhoogol Ke Mul Tatwa", Tara Publications, Varanasi, 1973.

Adler, R.W. 2002. "Fresh Water. In *Stumbling Toward Sustainability*", edited by J. C. Dernbach. Washington, DC: The Environmental Law Institute.

Agarwal, Virendra, "Integrated Rural Development", *Yojana,* Vol. XXIII, 24, 1980, p. 19.

Aghion, P.; Caroli, E.; Garcia-Penalosa, C., "Inequality and Economic Growth : The Perspective of the New Growth Theories". *Journal of Economic Literature,* 1999, No. 37, pp. 1615–60. [A comprehensive review of the empirical findings on inequality and growth.]

Agosin, M.R. and Bloom, D.E., "Global Integration in Pursuit of Sustainable Human Development : Analytical Perspectives".

Ahmad, K.S., "Environment and the Distribution of Population in India", *Indian Geographical Journal,* 1941.

Ahmad, K.S., "Environment and the Distribution of Population in India", *Indian Geographical Journal,* 1941, p. 16.

Ahmad, Enayat, "Bihar : A Physical, Economic and Regional Geography", Ranchi University, 1965, pp. 77-78.

Amin, S.; Arrighi, G.; Frank A.G.; Wallerstein I. 1982. *Dynamics of Global Crisis*. New York, Monthly Review Press. 248 pp. [An assessment and forecast of global economic change in the perspective of unequal development that proved to be off the mark in many of its conclusions.]

Ananta Padmanabhan, N., "Density of Rural Population and Terrain Types in Madras State", *Bombay Geographical Magazine,* 1957, p. 5.

Anker, R. 1998. *"Gender and Jobs : Sex Segregation of Occupations in the World"*. Geneva, International Labour Office. 444 p. [A comprehensive empirical review of gender inequality.]

Appiah, K.A.; Gates, H.L., Jr. 1997. *The Dictionary of Global Culture*. New York, Alfred A. Knopf. 717 p.

Banerjee, A.K., "Fisheries of Bengal", *Calcutta Geographical Review*, 1942.

Barker, R. *"And the Waters Turned to Blood"*, 1997. New York : Simon and Schuster.

Barry C. Field, Environmental Economics — An Introduction, 1997, The McGraw-Hill Companies, Inc.

Bartlett, A. A., "Colorado's Population Problem", Population Press, 1999, 5(6) : 8-9.

Basu, B., "Darjeeling Coals : Utilization Possibilities", *Calcutta Geographical Review*, 1943.

Belosouv, I.I., "Transportation and the Formation of Economic Regions", *Soviet Geography*, Review and Translation, Nov., 1964, p. 19.

Berlin, I., *Historical Inevitability*. New York, Oxford University Press. 1954, 79 p. [A classic.]

Berry, B.J.L., "Recent Studies Concerning the Role of Transportation in the Space Economy", Review Article, *Annals of the Association of American Geographers*, Vol. 49, 1959, p. 329.

Berry, B.J.L., and A.M. Baker, "Geographic Sampling" in B.J.L. Berry and D.F. Marble, (Eds.), *Spatial Analysis, A Reader in Statistical Geography*, Englewood Cliffs, N.J., 1968, pp. 91-100.

Berry, B.J.L., "Sampling, Coding and Storing Flood Plain Data", *United States Department of Agriculture Hand Book*, No. 237, Washington, DC., 1962.

Binmore, K. 1998. *Game Theory and the Social Contract : Just Playing* (Volume 2). Cambridge, Mass., MIT Press, p. 298.

Bor, N.L., "The Relief Vegetation of the Shillong Plateau, Assam", *Indian Forester*, 1938, 64.

Borenstein, S., Coastal Waters have big problems, Harvard study says. *The Buffalo News*, 1998, 25 August.

Bose, S.C., "Plant Geography of Damodar Valley", *Calcutta Geographical Review*, 1948.

Boulding, K., *Ecodynamics*. Beverly Hills, Calif., Sage Publications, 1981. 368 p. [An early classic on development and the environment.]

Brown, L., China's Water Crisis Linked to Global Security. *Population Press*, 1999, 5(5): 5.

Bryan, P.W., "A Technique for Recording Land Utilization in City and Rural Surveys", *Geography*, Vol. 16, 1930, pp. 211-15.

Buck, J.L., *Land Utilization in China*, London, Vol. 3, 1937.

Carr-Saunders, A.M. 1922. *The Population Problem : A Study in Human Evolution*. Oxford, Clarendon. 516 p. [An early attempt to examine population dynamics in the context of cultural evolution.]

Cassirer, E. 1944. *An Essay on Man: An Introduction to a Philosophy of Human Culture*. New Haven and London, Yale University Press. 287 p. [A comprehensive statement of symbolic cultural evolution.]

Center for New American Dream. 2000. [Online]. Available at http://www.newdream.org/monthly/aug00.html.

Chand Mahesh and R.K. Vishwakarma, "Urbanization as a Strategy of Rural Development", Urban System and Rural Development, The Director, Prasaranga, University of Mysore, Mysore-6, 1972, p. 141.

Chandrashekhar, C.S., G.D. Mathur and K.V. Sundaram, "Growth Foci in Regional Strategy", University of Mysore, 1972, p. 55.

Chatterjee, S.P., "Regional Pattern of Density and Distribution of Population in India", *Geographical Review of India*, 1962, p. 24.

Chatterjee, S.P., "Physical Features and Population Distribution in West Bengal", *Calcutta Geographical Review*, 1961, p. 23.

Chatterjee, S.P., "Minerals and Mineral Products", *Calcutta Geographical Review*, 1942, p. 43.

Chatterjee, S.P. and A.T. Ganguli, "Population Distribution in Two Districts of India", *Calcutta Geographical Review*, 1943, p. 5.

Chaudhari, M.R., "Power Resources of West Bengal", *Geographical Review of India*, 1965.

Choudhary, P.C.R., *Bihar District Gazetteers,* Munger, Patna, 1960, p. 74.

Christaller, W., "Central Places in Southern Germany", Englewood Cliff, 1965, Translated from the Original German Version Published in June, 1933.

Ciriacy-Wantrup, S.V., "Resource Conservation and Policies" (rec. ed., Berkeley: University of California Press), 1963 p. 42.

Clark, W.C. and Munn, R.E. (eds.) 1986. *Sustainable Development of the Biosphere,* Cambridge University Press, Cambridge.

Clark, Collin, "Urban Population Densities", *Journal of Statistical Society* (a), Vol. 1, 1951, pp. 492-596.

Coale, A.J.; Hoover E.M. 1958. *Population Growth and Economic Development in Low-income Countries : A Case Study of India's Prospects.* Princeton, N.J., Princeton University Press, 389 p. [A pioneer study in modeling population and development.]

Coleman, J.S. 1999. Social Capital in the Creation of Human Capital. In: P. Dasgupta and I. Serageldin, *Social Capital : A Multifaceted Perspective,* pp. 13-40. Washington, DC., The World Bank. [The classical statement about social capital. Recent analysis takes issue with Coleman's methodology and conclusions.]

Collier, C. 1999. Save Some for Tomorrow. Environment News Service. [Online]. Available at http://ens.lycos.com/ewire/July99/20july9903.html.

Collinson, D. 1987. *Fifty Major Philosophers : A Reference Guide.* London and New York, Croom Helm, 170 p.

Common, M. and Perrings, C. (1992). Towards an ecological economics and sustainability. *Ecological Economics* 6, 7-34.

Commoner, C. 1991. Rapid Population Growth and Environmental Stress. In : *Consequences of Rapid Population Growth in Developing Countries,* pp. 161–90. New York, Taylor and Francis (published on behalf of the United Nations).

Cornwell, J. (ed.) 1995. *Nature's Imagination : The Frontiers of Scientific Vision.* Oxford, Oxford University Press. 212 p. [Conference proceedings that include key studies on the boundaries of neuroscience, its potential development, philosophical base, and mathematical foundation.]

Coupland, H., Final Report on the Survey and Settlement Operations of Munger (North), 1920, p. 16.

Crick, F. 1989. *What Mad Pursuit*. London, Penguin Books. 182 p. [Skeptical about the potential of computers to mimic the human mind.]

Criteria and Indicators for Sustainable Forest Management : A Compendium. Paper compiled by Froylán Castañeda, Christel Palmberg-Lerche and Petteri Vuorinen, May 2001. Forest Management Working Papers, Working Paper 5. Forest Resources Development Service, Forest Resources Division. FAO, Rome (unpublished).

Dahrendorf, R. 1959. *Class and Class Conflict in Industrial Society*. Stanford, Calif., Stanford University.

Daly, H.E. and J. B. Cobb. 1994. *For the Common Good*. Boston : Beacon Press.

Daly, H. (1973). *Toward a Steady State Economy*, Freeman, San Francisco.

Daly, H.E. (1991). Towards an environmental macroeconomics. *Land Economics* 67, 255-59.

Daly, H.E. (1992). Allocation, distribution and scale : towards an economics that is efficient, just, and sustainable. *Ecological Economics* 6, 185-94.

Darling, F.F., "A Wider Environment of Ecology and Conservation", Deadalus, 96 (4), 1967, p. 1003.

Darwin, C. 1859/1964. *On the Origin of Species*. A facsimile of the 1st edn. Cambridge, Mass., Harvard University Press. [A classic.]

Das, H.P., *Agar Perfume Industry in Assam*, Gauhati University (Science) Journal, 1959.

Davis, K.; Barnstam, M. S.; Ricardo-Campbell, R. (eds.) 1986 *Below Replacement Fertility in Industrial Sector*.

Dawkins, R. 1976. *The Selfish Gene*. Oxford, Oxford University Press, 224 p.

Dayal, P. "Aspects of Industrial Location in India," Transactions of Indian Council of Geographers, Vol. I, Dec. 1964.

Dayal, P., "The Bihar Plain—Regional Study", Transactions of the Council of Indian Geographers Vol. 5, 1968, p. 20.

Dayal, P., "Industrial Location in India", Transactions of the Council of Indian Geographers, 1964, p.1.

Dayal, P. "The Energy Resources of India", Geology and Geography Section, 64th Session of Indian Science Congress, 1977.

Dewett, K.K. Singh, Gurucharan and J.D., Verma, "Irrigation and Power", *Indian Economics,* S. Chand and Company Limited, New Delhi, 1971, p. 157.

Draft Five Year Plan : 1978-83, Government of India, Planning Commission, 1978, p. 241.

Dubhashi, P.R., "Communication and Rural Development", *Yojana,* Vol. XXIII/10, 19, p. 13.

Duncan, C., Resource Utilization and Conservation Concept Economic Geography. Vol. XXXVIII, 1962, pp. 39, 119.

Durrani, P.K., "Industrial Development and Locational Factors in Rajasthan", Annals of Social Science ,1965, Vol.1, p.1.

Dutta, S.K., "Geohydrological Report on Thermal Spring", Rural Electrification Corporation, Government of India, 1973, p. 3.

Dyson, F. 1955. Introduction : The scientists as rebels. In : J. Cornwell (ed.), *Nature's Imagination : The Frontiers of Scientific Vision*, p. 4. Oxford, Oxford University Press.

Edie Summaries. 2000. Israel : Water Not Land is Key to Shepherdstown Talks. [Online]. Available at http://www.edie.net/news/Archive/2223.html.

ENN (Environmental News Network). 1998. Heavy Rains Drive Home Pollution Problem. [Online]. Available at http://www.enn.com/archives/1998/02/020698/yotostry.asp.

ENN (Environmental News Network). 1999. Beware, Don't Eat the Fish. [Online]. Available from http://www.enn.com/enn-news-archive /1999/03/032999/croaker_2349.asp.

ENN (Environmental News Network). 2003. Malaysian Leader Defends Government's Campaign Against Singapore Over Water Use. [Online]. Available at http://www.enn.com/news/2003-07-18?s_6707.asp.

ENS (Environment News Service). 1999a. Canada, U.S. Consider Great Lakes Water Export Ban. [Online]. Available at http://ens.lycos.com/ens/aug99/1999L-08-24-06.html.

ENS (Environment News Service). 1999b. Growing Population Faces Shrinking Water Supply. [Online]. Available at http:/ /ens.lycos.com/ens/jul99/1999L-07-20-01.html.

ENS (Environment News Service). 2003. Threats Rising for U.S. Public Water Supplies. [Online]. Available at http://ens-news.com/ens/jun2003/2003-06-11-10.asp.

Ellul, J., *The Technological Society*. New York, Vintage, 1964, 469 p. [An anthropologist's view of the direction scientific development and globalization are leading human development.]

Evans, K., De Jong, W., and Cronkleton, P. (2008) "Future Scenarios as a Tool for Collaboration in Forest Communities". *S.A.P.I.E.N.S.* 1 (2).

Fagg, C.C. and G.E., Hutchings, "An Introduction to Regional Surveying", Cambridge, 1930.

Fai, G.K. and Rains, G., "Innovation Capital Accumulation and Economic Review", Vol. LIII, p. 248, 1963.

Fananathan, V.S. and I.V. Bhanumati, Manufacturing Industries in and Around Poona, Indian Geographical Journal, 1963, Vol. 38, pp. 3-4.

Fieldman, G.A., "A Soviet Model of Growth", in Eassys in Economic Growth, Oxford Univ. Press, p. 223, 1979.

Flint, R.W, "The Sustainable Development of Water Resources : Water Security in the 21st Century", Five E's Unlimited, 28 Randolph Place, NW, Washington, DC 20001 (202)232-4591, rwflint@eeeee.nethttp://www.sustainabledevelopmentsolutions.com

Flint R.W., "The Sustainable Development of Water Resources" Universities Council on Water Resources Update, Issue 127, pages 41-51, February 2004, Five E's Unlimited, Washington, DC.

Flint, R.W. 2003. Sustainable Development : What does sustainability mean to the individual in the conduct of their life and business. In *Handbook of Development Policy*

Strategies, edited by G. M. Mudacumura. New York : Marcel Dekker, Inc.

Flint, R.W., S.B. Sterrett, W.G. Reay, G.F. Oertel, and W.M. Dunstan. 1996. Agricultural and environmental sustainability : A watershed study of Virginia's Eastern Shore, pp. 172-175. In: Tetra Tech (ed.), *Watershed '96 : A National Conference on Watershed Management*. Water Environment Federation, Washington, DC, 1163 p.

Flint, R.W. and M.J.E. Danner. 2001. The nexus of sustainability and social equity. *International Journal of Economic Development* 3(2). [Online]. Available at http://spaef.com/IJEC_PUB/v3n2.html.

Flint, R.W. and W.L. Houser. 2001. Living a Sustainable Lifestyle for Our Children's Children. IUniverse, Campbell, CA . 288 p.

Flint, R.W., F.C. Frick, A. Duffy, J. Brittingham, K. Stephens, P. Graham, and C. Borgmeyer. 2002. Characteristics of Sustainable Destination Resort Communities. Resort Municipality of Whistler, BC, Canada. [Online]. Available at http://www.whistlerfuture.com/thinkit/pdfs/Whistler_Sustainit3.pdf).

Fox, J.W., "Land Use Survey : General Principles and a Newzealand Example", Auckland, 1956.

Freeman. T.W. "Geography and Regional Planning", *National Geographer*, Vol. V (Planning Number), 1962.

Gadgil, D.R., "Economic Effects of Irrigation", 1945, p. 173.

Ganguli, B.N., "Iron and Steel Industry in Bengal Bihar Industrial Belt, *Calcutta Geographical Review*, 1949, Vol. II, pp. 3-4

Genes, W.D,."Ratios and Ispleth Maps in Regional Investigation of Agricultural Land Occupance", *Annals of Association Geographer*, Vol. 20, 1930.

Ghosh, A.B., "Some Aspects of Input Technology in Agricultural Production", Presidential Address at the 66th Session of Indian Science Congress, 1979, p. 1.

Ghosh, Shankar, "The Planning of Our Forest Resources" in *Yojana*, Planning Commission, 1976, Vol. XX/16, p. 10.

Gips, T. 1998. The Natural Step Four Conditions for Sustainability or System Conditions. Minneapolis, MN : Alliance for Sustainability and Sustainability Associates.

Gosal, G.S., "A Geographical Analysis of India's population", (Unpublidhed) Ph.D. Thesis, Wisconsin University, 1956.

Goudie, A. 1993. *The Human Impact on the Natural Environment*, Oxford, Blackwell. 454 pp. [An important contribution to the philosophical base of environmental analysis.]

Gould, S.J. 1978. Biological Potential vs. Biological Determinism. In : A.L. Kaplan, *The Sociobiology Debate : Readings on the Ethical and Scientific Issues Concerning Sociobiology*. New York, Harper and Row. pp. 343–51. [A classic article.]

Gowda, K.S. Rame : "Regional Planning for Integrated Rural Urban Development in Mysore State", Urban Systems and Rural Development, p. 154.

GreenBiz.com. 2003. Humanity Wages War with Nature for Water. Environmental News Network, New York. [Online]. Available at http://www.enn.com/news/2003-02-05/s_2417.asp).

Greenway, D., Bleaney, M. and Stewar, I. (1991), *Economics in Perspective*, Routledge, London.

Grossman, R. 1998. Can Corporations Be Held Accountable? Part I. *Rachel's Environment and Health Weekly* 609. [Online]. Available at http://www.rachel.org.

Guha, S.K. and P. Nag, "Studies on the Hot Springs of Munger District", Bihar, 1971.

Guidelines for Developing, Testing and Selecting Criteria and Indicators for Sustainable Forest Management, Ravi Prabhu, Carol J.P. Colfer and Richard G. Dudley. 1999. CIFOR. The Criteria and Indicators Toolbox Series.

Gupta, A.K., "Sustainable Development of Indian Agriculture : Green Revolution Revisited", Indian Institute of Management, Ahmedabad-380 015 (India), 1990, W.P. No. 896.

Gupta, K.L., "Tribeni-Bansberia, Geographical Review of India", 1960, p. 22.

Haffman, W.G., "The Growth of Operational Industrial Economics", Manchester Univ. Press, 1988.

Hagget, P. and C. Board, Rotational and Parallel Traverses in the Rapid Integratin of Geographic Areas, Annals of ths Association of American Geographers, 54, 1964, pp. 406-410.

Hall, K. 2003. Tackling World's Water Crisis Could Cost Up To US $100 Billion a Year. In Environmental News Network [Online]. Available at http://www.enn.com/news/2003-03-06/s_3213.asp).

Hameed, S.A., "Industrial pattern of Telangana : A Geograohical Analysis", *Indian Geographical Journal,* 1962, Vol. 37, p. 2.

Harris, M., "Steps in carrying out Formal Reseaech Project", *Land Economics Research,* Ackerman, J., *et al.,* (Eds.), Washington, 1962, p. 231.

Hart, W.E., "Forest of Madurai District", Journal of Madras Geographical Association, 1932.

Hasegawa, Sukehiro, "Development Cooperation : Global Issues and the United Nations", 2001.

Hayek, F.A. 1955. *The Political Ideal of the Rule of Law,* Fiftieth Anniversary Commemorating Lectures of the National Bank of Egypt, Cairo. Egypt, National Bank of Egypt. [One of the early analyses of governance that introduces the concept of isonomy and its importance to democratic government.]

Heintz, T., Personal Communication, May 12, 2003.

Hettena, S., California Desert Residents Use Water Like There's No Tomorrow—But Tomorrow Is Coming. In Environmental News Network (ENN). 2003, [Online]. Available at http://www.enn.com/news/2003-06-11/s_4917.asp.

Hirsch, F., *Social Limits to Growth,* Routledge, London, 1976.

Hora, S.L., "Presidential Address", Zoological Society of India, 1951.

Hotelling, H., "The economics of exhaustible resources", *Journal of Political Economy,* 1931, 39, 137-75.

Howarth, R.B. and Norgaard, R.B., Economics of sustainability or the sustainability of economics : Different Paradigms. *Ecological Economics,* 1992, 4, 93-116.

Hughes, D., Gaia : An ancient view of our planet. *The Ecologist,* 1983, 13, 2-3.

Husley-Tanslcy Report, "Conservation of Nature in England and Wales", Her Majesty's Stationery Office, 1947.

Instructions for Integrated Rural Development Programme in Bihar, 1979-80, Government of Bihar, Rural Development Department, 1979.

Isard, W., "Methods of Regional Analysis", An Introduction to Regional Science, New York, 1960.

Jacob, M.C., "Regeneration of the Bonsum Amare Forests of Assam", *Indian Forester,* 1952, 71.

Jacob, V.C., "A Note on the Hot Springs of Bhimbandh Area, District Munger", *Geological Survey of India,* 1962, p. 2.

Jain, J. K., "Water Resources of the Arid Region of Rajasthan", *Indian Journal of Geography,* 1967.

Jennings, I. 1932–3. The report on Ministers' Power. *Public Administration* 10 (1932) and 11 (1933) (quoted in Hayek, 1955, p. 54).

Jha, Premshankar, "The role of the Mass Media in Promoting Integrated Rural Development", Science and Integrated Rural Development (Eds.) M.S. Swaminathan, P.K. Bose and Archana Sharma, 1978, p. 65.

Jones, W.D. and V.C Finch, "Detailed Field Mapping in the Study of the Economic Geographers of an Agricultnral Area", Annals of the Association of American Geographers Vol. 15, 1926, pp. 148-57.

Karan, P.P., "Patna and Jamshedpur", *Geograbhecal Review of India,* 1952, 14(2)

Karnik, C.R., "Flora of Tapti Valley", *Bombay Geographical Magazine,* 1956.

Kayastha, S.L., "Conservation of Natural Resources in Himalaya, A Vital Need", Seminar on Himalayas, New Delhi, December, 1965.

Keynes, J.M. 1936. Some Economic Consequences of a Declining Population. *Eugenics Review* (Edinburgh), Vol. 29, No. 1, pp. 13-18. [A classic article.]

Kerry, R. Turner (ed.) (1993): "Sustainable Environmental Economics and Management", Belhaven Press.

Khaldun, I., 1967. *An Introduction to History (The Mugaddimah).* Trans. from the Arabic by F. Rothenthal; abridged and edited by N.J. Dawood. London and Henley, Routledge and Kegan Paul in association with Secker and Warburg. 465 p. [A classic on the sociology of history.]

Khan, M.A.W., "Ecology and Potentiality of the Bamboo Forests of Bilaspur Division, Madhya Pradesh", *Indian Forester,* 1960, p. 10.

Khasnabis, R., "On the Economic Problem of Sustainable Development".

Khusro, "Economic Development with no Population Transfer", Asia Publishing House, Bombay.

Knowles, "Economic Development of British Empire Overseas", Vol. I, pp. 367-68.

Knees, A.V.; Sweeny, J.L. (eds.) 1985. *Handbook of Natural Resource and Energy Economics,* Vols. 1 and 2. New York, North Holland. 755 p. [Important reference.]

Kolars, P., John, F. and Nystuen, D. John, "Ways to Weigh the World, A Preface to Man Environment Systems Geography", The Study of Location, Culture and Environment, 1964, p. 247.

Krishan, Gopal, "Distribution and Density of Population in Orissa", the *National Geograbhical Journal of India,* 1968, vol. XXIV, pp. 250-257.

Krishan, Gopal, "Regionalism in Growth of Population in Punjab's Border Districts of Amritsar and Gurdaspur, 1951-61", *Geographical Review of India,* 1968, 30, pp.1-17.

Krishna, M.S., "Geographical Control of Mineral Industries", *Bulletin of the National Geographical Society of India,* 1952, no.16.

Krishnamurthy, N., "Papanasam Hydro-electric Project", *Journal of the Madras Geographical Association,* 1940.

Kumar, A., and G.P. Sharma, "Changing Pattern of Land Use Categories, Land Use Efficiency in Haspura Anchal", in Dimensions in Geography, (eds.) V.N.P. Sinha and R.B. Mandal, Associated Book Agency, Patna, 1979, p. 217.

Kuriyan, G., "Population and its Distribution in Kerala", *Journal of Madras Geographical Association*, 1938, p. 13.

Lakshmi, B., "Pattern of Rail Traffic Flow in Madhya Pradesh", the *National Geographical Journal of India*, 1967, Vol. 13, p. 2.

Landes, D.S. 1999. *The Wealth and Poverty of Nations : Why Some are so Rich and Some so Poor*. New York, W.W. Norton. 658 p.

Latham, J.P., "Methodology for an Instrumental Geographic Analysis", Annals of the Association of American Geographers, 1963, pp. 194-211.

Lazaroff, C. 2000. Growing population faces diminishing resources. In : Environment News Service. [Online]. Available at http://ens.lycos.com/ens/jan2000/2000L-01-18-06.html.

Lewis, W.A., "Development with Unlimited Supplies of Labour", The Manchester School, p. 132, 1958.

Losch, August, Die Raumliche and Nug Der Wirfschaft, 2nd Edition, 1944, Translated into English as Economics of Location, Yale University Press, 1954.

Maddox, J. 1999. The Unexpected Science to Come. *Scientific American*, December, pp. 62–7. [Scientific future prospects by an outstanding scientist.]

Mahalanobis, P.C., "The Apporach of Operational Research to Planning in India", Asia Pub. House, Bombay, 1963.

Majumdar, S.C., "Rivers of Bengal Delta", 1943.

Mallick, U.C., "Development of Small Towns for Balanced Urbanisation and Economic Growth", Place of Small Towns in India, 1979, p. 55.

Malthus, T.R. 1926/1978. *First Essay on Population*. London, Macmillan, 396 p. [A classic.]

Malthus, T.R., A Summary, View of the principle of population in Encyclopedia Britannica Supplement, 132-3, reprinted G.S. Drmko *et al., Population Geography : A Reader*, New York, 1970, McGraw-Hill, pp. 44-70.

Mamoria, C.B., "Forest Resource and Forestry Development in Agricultural Problem of India", 17th Edition, Kitab Mahal, Allahabad, 1973, p. 123.

Mamoria, C.B., "Irrigation and Floods", Agricultural Problems of India, Kitab Mahal, Allahabad, 1973, p. 161.

Mamoria, C.B., "Agricultural Problems of India", Kitab Mahal, Allahabad, 1973.

Mandal, R.B., "Nature of Population Geography", Dimensions in Geography, (Eds. Sinha and Mandal), Associated Book Agency, Patna, 1979, p. 6.

Mathur, "Anatomy of Disguised Unemployment", Oxford Economic Paper, p. 187, 1964.

Mayer and Kohen, "Readings in Urban Geography", Central Book Depot, Allahabad, 1969.

Mayr, E. 1988. *Toward a New Philosophy of Biology : Observations of an Evolutionist*. Cambridge, Mass., Harvard University Press. 564 p. [An attempt to ground the philosophy of science into the living world of Biology. An important contribution to evolutionary Biology.]

Mill, J.S. 1848/1965. *Principles of Political Economy with Some of their Applications to Social Philosophy*, Vol. 2 (ed. V. W. Bladen and J. M. Robson). Toronto, University of Toronto Press. 388 p.

Mishra, S.D., A Note on the Socio-Historical Geography of Mathura, *The National Geograpical Journal of India*, 1958.

Mithal, R.S., "Geotectonic Evaluation for Development of Water Resources in India", Presidential Address at the 66th Session; Indian Science Congress Association, Calcutta, 1979, pp. 1-2

Montague, P. 1998. Landfills are Dangerous. *Rachel's Environment and Health Weekly*. 617: 2. [Online]. Available at http://www.rachel.org.

Morgan, J.; Sirageldin, I.; Baerwaldt, N. 1966. *Productive American*. Ann Arbor, Mich., Institute for Social Research, University of Michigan. 546 p. [One of the early empirical studies that integrate economic, sociological, and psychological analysis of human development.]

Morgan and R.J.C., Munton, "Agricultural Activity and Physical Environment", *Agricultural Geography,* 1974, p. 39.

Morlok, E.K., "The Effects of Reduced Fares for the Elderly on Transit System Routes", The Transportation Centre North Western University, 1971.

Mozgeris, G. (2008) "The continuous field view of representing forest geographically: from cartographic representation towards improved management planning". *S.A.P.I.E.N.S.* 1 (2).

Muggiati, A. 2003. Four Nations Guard Giant South American Aquifer. In Environment News Service. [Online]. Available at http://ens-news.com/ens/may2003/2003-05-29-03.asp.

Mukherjee, A.B., "Vegetation of the Upper Doab", *Indian Geographical Journal,* 1962, Vol. 37, p. 1.

Mukherjee, B.N., "A Rail-Road development Between Howrah and Bishnupur", *Observer,* 1958, p. 4.

Mukherjee, H.K., "River Fisheries of Bengal", *Calcutta Geographical Review,* 1946. National Technical Information Service, 1968, p. 28.

Munsi, S., "Railway Network Growth in Eastern India—1854-1919", Centre for Studies in Social Sciences, Calcutta, No. 3, 1975.

Murthy, Y.K., "Irrigation Research". Irrigation and Power in the Fourth Plan, Publication Division Ministry, Patiala House, New Delhi, 1970, p. 13.

Mustikhan, A. 1999. "Pakistan Provinces Feud Over Water". In Environment News Service (ENS), 7/28/99. [Online]. Available at http://ens.lycos.com/ens/jul99/1999L-07-28-02.html.

Myers S., "Personal Transportation Confluence on poverty and transportation", American Academy of Arts and Sciences, J.G., *et al.*

Nagel, E., "Teleology Revisited : Goal Directed Processes in Biology". *Journal of Philosophy,* 1977, Vol. 74, pp. 261-301. [A philosopher's critique of teleology.]

Nand, N., "Distribution and Spatial Arrangement of Rural Population in East Rajasthan", Annals of the Association of American Geographers, 1966, p. 6.

National Atlas of India Forest and Land Use Map, 1957, p.11.

Negi, B.S., "The Mineral Resources of Kumaon and Garhwal", *Geographical Review of India*, 1960, Part I, "The Forest Wealth of Uttar Pradesh", *The Geographer*, Aligarh, 1950, 3(1).

Nehru, J.N., "Speeches", New Delhi, March 1963.

Nietzsche, F.W. 1956. *The Genealogy of Morals*. Trans. F. Golffing. New York, Doubleday Anchor.

Nurkse, R., "Problem of Capital Formation in UDC", New York, 1953.

Page, T., *Conservation and Economics Efficiency*. Johns Hopkins University Press, 1977, Baltimore, MD.

Parrington, V.L. 1930. *Main Currents in American Thought : An Interpretation of American Literature from the Beginnings to 1920*. New York, Harcourt, Brace, 413 p.

Pasupatinathan, K., "Minerals of India", *Geographical Teacher*, pp. 1965-66.

Pathak, C.R., Spatial variation in Urban and Industrial Growth in India, Dimension in Geograpy, (Eds.) V.N.P. Sinha and R.B. Mandal 1979, pp. 433-450.

Pearce, D.W. and Turner, R.K. (1990). *Economics of Natural Resources and the Environment*. Harvest Wheatsheaf, Hemel Hempstead.

Pearce, D.W., Markandya, A. and Barbier, E.B. (1990). *Sustainable Development*. Earthscan, London.

Pearce, D., Economics and Environment; *Edward Elgar*, 1998.

Pezzy, J., Sustainability : An interdisciplinary guide. *Environmental values*, 1992, 321-62.

Pinker, S. 1997. *How the Mind Works*. New York, W. W. Norton. 660 p. [A psychologist's view of how the brain-mind functions.]

Popper, K., *The Poverty of Historicism*. London, Routledge and Kegan Paul. 1957, 166 p. [A critique of the directionality of history.]

Portes, A., Social Capital : Its Origins and Applications in Modern Sociology. *Annual Review of Sociology*, 1998, No. 24, pp. 1-25.

Prasad, Ishwari, "Location of Industries Dimensions in Geography (Eds.) V.N.P. Sinha and R.B. Mandal, 1979, pp. 407-431.

Prasad, I., Population Distribution and Dynamics of Supply in Bihar, *Geographical Outlook*, 1956, p. 2.

Prasad, S.D., "Census of India, 1961, Vol. IV, Part-I-A-(i)", 1967, pp. 5-8.

Prasad, S.D., District Census Hand Book, Munger, Vol. IV, Part-viii, 1961.

Project Monghyr : An Integrated Programme for the Development of the Rural Economy of Monghyr District Rural Development Department, Patna, 1977.

Punmia, B.A. and Bansilal Brij Pande, "Irrigation and Water Power Engineering", Standard Publishers and Distributors, Delhi-6, 1972, pp. 14, 59-60.

Putkayastha C.C., Working plan for Shillong Pine Forest, Shillong, 1948.

Quaid, L., Missouri River Ruling Could Hinder Water Quality. In : Environmental News Network [Online]. 2003, Available at http://www.enn.com/news/2003-07-15/s_6555.asp.

Quintal, O., "Agriculture and Sustainable Development".

Raja, M., Is India Over Populated, *Geographer*, 1950.

Rajkhowa, S., Forest Types of Assam, *Indian Forester*, 1961.

Ramchandran, R., 'Population Trend in the Malnad', *Deccan Geographer*, 1965, 3.

Rao A.S., The Vegetation of Khasi and Jaintia Hills, Proceedings of the Pre-Congress Symposium of Meghalaya and Eastern Himalayas, International Geographical Union, Department of Geography, Gauhati University, 1968.

Reily, R.C., "Industrial Geography", London, 1973, p. 10.

Renne, R.R., Land Economics, Harper and Brothers, New York, 1947, p. 26.

Resource Survey and Integrated Area Development Planning of Monghyr District, Government of Bihar, Bihar State Planning Board, 1977.

Richard, M., Technological Change in Plantation Transport System : A Caribbean Example, 1930-70, *The Journal of Tropical Geography*, Vol. 41, December, 1975, pp. 1-8.

Ridley, M. 2000. *Genome : The Autobiography of a Species in 23 Chapters*. New York, HarperCollins, 344 p.

Robert, K.H. 1991. Educating a Nation : The Natural Step. In *Context* 28, In Context Institute [Online]. Available at http://www.context.org/ICLIB/IC28/Robert.htm.

Ronald, A., Introductory Economics, Harper and Row Publishers, New York, 1971, p. 15.

Rousseau, J.J. 1973. *The Social Contract and Discourses*, New York, E. P. Dutton, 362 p.

Roy Choudhary, P.C., "Forest and Wild Life in Bihar", *Indian Geographer*, 1959.

Roy Chaudhary, P.C., ''Bihar District Gazetteers Munger", Secretariats Press, Patna, 1960.

Roy, M.P., "Yield Regulation in the Forest of West Bengal", *Indian Forester*, 1961, p. 2.

Russell, B. 1934. *Freedom versus Organization: 1814–1914*. New York, W.W. Norton, 471 p.

Russell, B. 1945/1972. *A History of Western Philosophy*. New York, Simon and Schuster, 895 p.

Russell, B. 1988. *The Problems of Philosophy*. Buffalo, N.Y., Prometheus, 161 p.

Russell, B. 1995. *Authority and the Individual*, New York, Routledge, 95 p.

Ryan, J.C. 1997. *Stuff—The Secret Lives of Everyday Things*. Seattle, WA: Northwest Environmental Watch. Stevenson, M. 2003. The "Other" Gulf War, Dispute Over Water Rights threatens Environment. In Environmental News Network (ENN). [Online]. Available at http://www.enn.com/news/2003-05-23/s_4585.asp.

Sahay, S.S., "Working Plan of the Divisional Forest Office", Munger, 1976, p. 12.

Saika, M.L., Grasslands of Assam, Symposium on Vegetation Types of India, 1955.

Sastry, S., "Location of Crops", *The Indian Geographical Journal*, Vol. XVIII, 1942, pp. 110-114.

Schumpeter, J. 1954. *History of Economic Analysis*, New York, Oxford University Press, 1260 p.

Sdasyuk, G., "Transport and Formation of Economic Regions in India", quoted from T.P. Patil, "Transport in Sholapur District : A Geographical Survey", *The Deccan Geographer*, Vol. X, No. 2, 1972, p. 1.

Sen, A., The Concept of Development. In : H. Chenery and T. N. Srinivasan, *Handbook of Development Economics*, 1996, Vol. 1, Amsterdam, North Holland, 848 p.

Sengupta, P., "Some Aspects of Industrial Growth of Hooghly Region", *The National Geographical Journal of India*, 1958, pp. 79-139.

Serageldin, I.; Martin-Brown, J. (eds.) 1999. *Culture in Sustainable Development : Investing in Cultural and Natural Endowments.* Proceedings of a Conference sponsored by the World Bank and the United Nations Educational, Scientific, and Cultural Organization. Washington, D.C., World Bank.

Sharma, R.P., "Forestry in Rajasthan", *Indian Forester*, 1955, p. 1.

Simons, C., Resource System : Aspect of Geography, 1974, p. 7.

Singh, D.N., "Road Planning in North Bihar", *Uttar Bharat Bhoogol Patrika*, Vol. V, No. I, 1969, pp. 33-35.

Singh, G.N., Problems and Trends of Industrial Growth in Bihar, Dimension in Geography, (Eds.) V.N.P. Sinha and R.B. Mandal, 1979, pp. 471-478.

Singh, J., "Research Methodology in Spatial Functional Interaction", *The Deccan Geographer*, Vol. XII, No. 2, 1974, p. 119. Mandal, R.B., "Role of Transportational Linkages in Rural Community of Development", *Geographical Outlook*, Vol. XIII, 1977-78, pp. 68-74.

Singh, R.C.P., Agricultural Land Use Planning in India, Dimensions in Geography, (Eds.) V.N.P. Sinha and R.B. Mandal, Associated Book Agency, Patna, 1979, p. 317.

Singh, V.R., The Research Methodology in Rural Land Use Analysis in India : A Systematic Approach, in Dimension in Geography (Eds.) V.N.P. Sinha and R.B. Mandal, Associated Book Agency, Patna, 1979, p. 216.

Singh, Harihar, "Evolution of the Townscape of Jaunpur", *The National Geographical Journal of India*, 1958, Vol. 4, pp. 35-46.

Singh, Madhusudan, "Evolution of Meerut", *The National Geographical Journal of India*, 1965, Vol. II, pp. 144-158.

Sinha, B.N., "Forest Economy of Orissa", *Indian Geographer*, 1958.

Sinha, B.N., Transport and Communication Problems of Orissa, *The National Geographical Journal of India,* 1957, Vol. 3, p. 2.

Sinha, R.P., Power Development in the State of Bihar, Abstracts of the International Geographical Union Congress, 1968.

Sinha, V.N.P., Changes in the Geographic Patterns of Population in Ghana, Dimensions in Geography, (Eds.) V.N.P. Sinha and R.B. Mandal, 1979, 118-139.

Sirageldin, Ismail, "Sustainable Human Development in the twenty-first century : an evolutionary perspective, in Sustainable Human Development", [ed. Ismail Sirageldin], in *Encyclopedia of Life Support Systems (EOLSS),* Developed under the Auspices of the UNESCO, Eolss Publishers, Oxford, UK, 2003 [http://www.eolss.net]

Sirageldin, I., 1975. The Demographic Aspects of Income Distribution. In : W. Robinson, *Population and Development Planning*. New York, Population Council, pp. 153–87.

Sirageldin, I. (ed.) 2001. *Population Challenges in the Middle East and North Africa : Towards the Twenty-First Century*. London, Routledge, jointly with the American University in Cairo Press, for the Economic Research Forum for the Arab Countries Iran and Turkey (ERF).

Societies : Causes, Consequences, Policies. A Supplement to Vol. 12, *Population and Development Review,* New York, Population Council, 360 p.

Sourirajan, V.K., "The Buckingham Canal", *Journal of Madras Geographical Association,* 1929, Vol. 3, p. 4.

Srinivasachari, C.S., "Growth of Madras City", *Journal of Madras Geographical Association,* 1934, p. 14.

Srivastava, B.P. and Pant, M.M., "Social Forestry in India", *Yojana,* Planning Commission, 1979, Vol. XXIII/ I2, pp. 18-23, Ministerial Conference on the Protection of Forests in Europe.

Stamp, L. Dudley, "Land Use and Resources : Applied Geograhpy", Penguin Books, 1960.

Stamp, L.D., *Man and the Land,* 1955.

Stebbins, G.L. 1982. *Darwin to DNA: Molecules to Humanity*. San Francisco, W. H. Freeman. 491 p. [A classic.]

Stracey, P.D. and H.P. Das, Assam's Economy and Forest, Shillong, 1949.

Subrahamanyam, N., "Some Aspects of Growth of Greater Madras", *Journal of the Madras Geographical Association*, 1938, 13, pp. 55-64.

Sundar Raj, B., "Pearl Fisheries of South India", *Journal of the Madras Geograophical Association*, 1930.

Sundaram, K.V., "Growth of the Urban Economy of small and Medium Size Towns", Urban and Regional Planning in India, Vikas Publishing House Pvt. Ltd., 1972, New Delhi, p. 213.

Sutherland, D. 1999. Tug-o-War: Cancer kids vs water pollution. Environment News Service. [Online]. Available at http:// ens.lycos.com/ens/sep99/1999L-09-27-01.html.

Swaminathan, M.S., P.K. Bose and Archana Sharma, "Science and Integrated Rural Development", 63rd Indian Science Congress, 1976.

Taori, Kamal, "People's participation in Sustainable Human Development" (A Unified Search), Concept Publishing Company Pvt. Ltd., New Delhi, 1998.

Tattersall, I. 1995. *The Last Neanderthalensis : The Rise, Success, and Mysterious Extinction of our Closest Human Relatives*. New York, Macmillan, West View, 208 p. [An empirical critique of the directionality of history.]

Teilhard de Chardin, P. 1969. *The Future of Man*, New York, Harper Torchbooks, 319 p. [A classic.]

Teitelbaum, M.S.; Winter, J. M. (eds.) 1988. *Population and Resources in Western Intellectual Traditions*. Supplement to Vol. 14, *Population and Development Review*. New York, Population Council. [An important reference.]

Thacker, M.S., "India's Urban Problm", University of Mysore, 1965, pp. 1-49.

The United States Army Corps of Engineers in Oregon, Water Resources Development, North Pacific Division, Custom House, Portland, 1975, p. 37.

Thomas, N., Presidential Address to the Calcutta Geographical Society, Calcutta Geographical Review, 1943, p.1.

Thomas, L. 1974. *The Lives of a Cell : Notes of a Biology Watcher.* New York, Viking. 153 p. UNDP (United Nations Development Programme). 1999. *Human Development Report, 1999.* New York, Oxford University Press. [Essential reading and reference.]

Thunen, J.H. Von, Wirtschaft and Nationale Isolated State : Peter Hall, Pergamon Press, 1966, pp. 72-73.

Tiwari, A.K., "Mathura : A Study of site and Evolution", *The National Geographical Journal of India,* 1963, Vol. 9, pp. 41-50.

Trefethan, J.T., "A Method for Geographical Surveying", *American Journal of Science,* 32, 1936, pp. 454-64.

Tripathi, R.N., "Public Finance and Underdeveloped Countries", World Press, 2nd revised edition, p. 99, 1968.

United Nations Forum on Forests (2004).

Uyanga, J.T., Measuring the perception of Farming problems by Small Scale Farmers in a Nigerian Locality, in Dimensions in Geography, (Eds.) V.N.P. Sinha and R.B. Mandal, Associated Book Agency, Patna, 1979, 216.

UNESCO, 2003. Water for People, Water for Life: The World Water Development Report. World Water Assessment Programme, United Nations, NY. [On-line]. Available at http://ens-news.com/ens/mar2003/2003-03-05-02.asp).

Verma, P., History of Resource Use, Applied Geography, 1968, pp. 135-36.

Verma, S.D., "Density and Pattern of Population in Punjab", *The National Geographical Journal of India,* 1956, p. 2.

Victor, P.A. (1991). "Indications of sustainable development : Some lessons from capital theory", *Ecological Economics 4,* 191-213.

Vohra, B.B., "Hand Book on Irrigation Water Management", Water Management Division, Ministry of Agriculture, New Delhi, 1971, pp. 5-8.

Vorosmarty, C.J., P. Green, J. Salisbury, and R. B. Lammers. 2000. "Global water resources : vulnerability from climate change and population growth". *Science* 289: 284-287.

Waal, F. de, 1997. *Good Natured : The Origins of Right and Wrong in Humans and Other Animals.* Cambridge, Mass., Harvard University Press. 296 p. [On the origin of the moral code.]

Wackernagel, M. and W. Rees. 1996. *Our Ecological Footprint.* Gabriola Island, Canada : New Society Publishers.

Waldrop, M.M. 1992. *Complexity : The Emerging Science at the Edge of Order and Chaos.* New York, Touchstone and Simon and Schuster. 380 p. [A good, accessible review of the emerging science of complexity.]

Wilson, A. and P. Yost. 2001. "Buildings and the environment : the numbers". *Environmental Building News,* 1 May.

Wilson, Art and Tyrchniewicz, Allen, "Agriculture and sustainable development : policy analysis on the Great Plains", International Institute for Sustainable Development, Winnipeg, Manitoba, Canada, 1995.

Wilson, E.O. 1975. *Sociobiology.* Cambridge, Mass., Harvard University Press. 697 p. [A classic.]

World Bank. 1992. *World Development Report 1992 : Development and the Environment,* Oxford, Oxford University Press (published for the World Bank). [The series of World Development Reports by the World Bank are essential reference material.]

World Bank. *Claiming the Future : Choosing Prosperity in the Middle East and North Africa,* 1995, Washington, D.C., World Bank.

World Commission on Environment and Development. 1987. *Our Common Future.* New York : Oxford University Press. [A comprehensive statement.]

Zimmerman, E.W., "World Resources and Industries", Harper and Row, New York, 1961, p. 402.

Zobler, L., "An Economic Historical view on Natural Resource Use and Conservation", *Economic Geography,* Vol. XXXVIII, 1962, p. 191.

Index